THEORY OF LINEAR OPERATORS IN HILBERT SPACE

N. I. Akhiezer and
I. M. Glazman

Translated from the Russian by
MERLYND NESTELL

TWO VOLUMES BOUND AS ONE

DOVER PUBLICATIONS, INC.
NEW YORK

Bibliographical Note

This Dover edition, first published in 1993, is an unabridged and unaltered republication of the edition published by Frederick Ungar Publishing Co., New York, in two separate volumes in 1961 and 1963.

Library of Congress Cataloging-in-Publication Data

Akhiezer, N. I. (Naum Il'ich), 1901–
[Teoriia lineĭnykh operatorov v gil'bertovom prostranstve. English]
Theory of linear operators in Hilbert space / N.I. Akhiezer and I.M. Glazman ; translated from the Russian by Merlynd Nestell.
p. cm.
Originally published: New York : F. Ungar Pub. Co., c1961–c1963.
"Two volumes bound as one."
Includes bibliographical references and indexes.
ISBN-13: 978-0-486-67748-4 (pbk.)
ISBN-10: 0-486-67748-6 (pbk.)
1. Hilbert space. 2. Linear operators. I. Glazman, I. M. (Izrail' Markovich) II. Title.
QA322.4.A3813 1993
515'.733—dc20 93-6143
CIP

Manufactured in the United States by Courier Corporation
67748605
www.doverpublications.com

TRANSLATOR'S PREFACE

The present volume comprises the first five chapters of the Russian text of seven chapters and two appendices. It is hoped that it will be a worthwhile addition to the few books on linear operators in English. Present plans call for bringing out the balance of the work in a second volume.

A few words should be said regarding the translation. It was the translator's desire to produce a faithful translation and to convey the intent of the authors in clear English. When conflicts arose, they were resolved in favor of the latter objective. Therefore, a fair degree of freedom was exercised in the translation. Substantial alterations are accompanied by translator's notes.

The translator owes much to Dr. Philip M. Anselone, of the United States Army Mathematics Research Center, University of Wisconsin, who critically read the text, contributed the translator's notes, and did much to polish both the mathematics and style of the text. Thanks go also to Dr. Hans Bueckner of the United States Army Mathematics Research Center, University of Wisconsin, and to Professor Phillip Hartman of The John Hopkins University, for reading critically the entire text. However, for any errors in translation the translator assumes full responsibility.

M.K.N.

AUTHOR'S PREFACE

The present book stems from lectures and papers which were presented by the authors at the Kharkov Mathematics Institute. It is not exhaustive and perhaps should be called "An Introduction to Linear Operators in Hilbert Space." However, the geometry of Hilbert space and the spectral theory of unitary and self-adjoint operators is presented in detail. The book contains some results obtained by Soviet mathematicians.

According to the original plan, we had hoped also to present as applications the theory of Jacobi matrices in connection with the exponential moment problem on the whole axis and the theory of integral equations with Carleman kernel. In any case, the size of the book without these forced us to abandon this intention. It was with special regret that we decided against the presentation of the moment problem since, on one hand, it plays a big role in the development of the theory of operators and, on the other, is a beautiful application for illustrating the theory.

We recommend to the reader that, after the study of this book, he become acquainted with both the moment problem and the theory of the Carleman integral operator.[1]

The book consists of seven chapters of basic text and two appendices. The first appendix is devoted to generalized extensions of symmetric operators. Here results obtained by M. A. Naimark (Moscow) and M. A. Krein (Odessa) have a classical character; these results were placed in an appendix only so that the basic text could be kept comparatively elementary. The second supplement is devoted to the theory of differential operators, which are here an application of the general theory. We do not give a complete list of literature on the theory of operators, but cite only basic papers and recent or little known books. We list a few titles from a general course.[2]

The attentive reader will notice that we occasionally repeat some formulas or proofs. We do this to keep the reader from being exhausted by too much reference to preceding material.

Further, we might have reduced the size of the book somewhat by omitting chapter five, which contains the spectral analysis of completely continuous self-adjoint operators; however, we decided to keep this material as a very natural introduction to the general spectral theory.

[1] Cf. for example, N. I. Akhiezer [2], [3], M. G. Krein and M. A. Krasnoselski [1].
[2] M. H. Stone [3], A. I. Plesner [1], A. I. Plesner and V. H. Rokhlin [1], V. I. Smirnov [1].

The first three chapters, chapter five and chapter six (with the exception of sections 69–73) were written by N. I. Akhiezer and the remainder by I. M. Glazman. However, the final editing of all the material was done jointly. Hence each of us bears responsibility for the whole book.

We consider it our duty to express deep thanks to M. G. Krein for a number of valuable remarks. We are also very grateful to G. I. Drinfeld, M. A. Krasnoselski, V. A. Martchenko and A. Ia. Povzner for remarks and corrections to improve individual parts of the book.

CONTENTS

Chapter II (continued)

Chapter IV (continued)

THEORY OF LINEAR OPERATORS IN HILBERT SPACE

VOLUME I

Chapter I

HILBERT SPACE

1. Linear Spaces

A set R of elements $f, g, h, \ldots,$ (also called points or vectors) forms a *linear space* if

(a) there is an operation, called addition and denoted by the symbol $+$, with respect to which R is an abelian group (the zero element of this group is denoted by 0);

(b) multiplication of elements of R by (real or complex) numbers $\alpha, \beta, \gamma, \ldots$ is defined such that

$$\alpha(f+g)=\alpha f+\alpha g,$$

$$(\alpha+\beta)f=\alpha f+\beta f,$$

$$\alpha(\beta f)=(\alpha\beta)f,$$

$$1\cdot f=f, \quad 0\cdot f=0.$$

Elements $f_1, f_2, \ldots, f_n$ in R are *linearly independent* if the relation

$$\alpha_1 f_1+\alpha_2 f_2+\ldots+\alpha_n f_n=0 \qquad (1)$$

holds only in the trivial case with $\alpha_1=\alpha_2=\ldots=\alpha_n=0$; otherwise $f_1, f_2, \ldots, f_n$ are *linearly dependent*. The left member of equation (1) is called a *linear combination* of the elements $f_1, f_2, \ldots, f_n$. Thus, linear independence of $f_1, f_2, \ldots, f_n$ means that every nontrivial linear combination of these elements is different from zero. If one of the elements $f_1, f_2, \ldots, f_n$ is equal to zero, then these elements are evidently linearly dependent. If, for example, $f_1=0$, then we obtain the nontrivial relation (1) by taking

$$\alpha_1=1, \alpha_2=\alpha_3=\ldots=\alpha_n=0.$$

A linear space R is *finite dimensional* and, moreover, *n-dimensional* if R contains n linearly independent elements and if any $n+1$ elements of R are linearly dependent. Finite dimensional linear spaces are studied in linear algebra. If a linear space has arbitrarily many linearly independent elements, then it is *infinite dimensional.*

2. The Scalar Product

A linear space R is *metrizable* if for each pair of elements $f, g \in R$ there is a (real or complex) number (f, g) which satisfies the conditions:[1]

(a) $(g,f) = \overline{(f,g)}$,
(b) $(\alpha_1 f_1 + \alpha_2 f_2, g) = \alpha_1 (f_1, g) + \alpha_2 (f_2, g)$,
(c) $(f,f) \geqq 0$, with equality only for $f = 0$.

The number (f,g) is called the *scalar product*[2] of f and g.

Property (b) expresses the linearity of the scalar product with respect to its first argument. The analogous property with respect to the second argument is

($\bar{\text{b}}$) $$(f, \beta_1 g_1 + \beta_2 g_2) = \bar{\beta}_1 (f, g_1) + \bar{\beta}_2 (f, g_2).$$

Property ($\bar{\text{b}}$) is derived as follows:

$$(f, \beta_1 g_1 + \beta_2 g_2) = \overline{(\beta_1 g_1 + \beta_2 g_2, f)} = \overline{\beta_1 (g_1, f) + \beta_2 (g_2, f)} = \\ = \bar{\beta}_1 (f, g_1) + \bar{\beta}_2 (f, g_2).$$

The positive square root $\sqrt{(f,f)}$ is called the *norm* of the element (vector) f and is denoted by the symbol $\|f\|$. The norm is analogous to the length of a line segment. As with line segments, the norm of a vector is zero if and only if it is the zero vector. In addition, it follows that

1° $$\|\alpha f\| = |\alpha| \cdot \|f\|.$$

This is shown by using properties (b) and ($\bar{\text{b}}$) of the scalar product:

$$(\alpha f, \alpha f) = \alpha(f, \alpha f) = \alpha \bar{\alpha} (f,f) = |\alpha|^2 (f,f),$$

from which 1° follows.

We shall prove that for any two vectors f and g,

2° $$|(f,g)| \leqq \|f\| \cdot \|g\|,$$

with equality if and only if f and g are linearly dependent. We call 2° the Cauchy-Bunyakovski inequality[3], because in the two most important particular cases, about which we shall speak below, it was first used by Cauchy and Bunyakovski.

In the proof of 2°, we may assume that $(f,g) \neq 0$. Letting

$$\theta = \frac{(f,g)}{|(f,g)|},$$

we find that for any real λ,

$$0 \leqq (\bar{\theta} f + \lambda g, \bar{\theta} f + \lambda g) = \lambda^2 (g,g) + 2\lambda |(f,g)| + (f,f).$$

---

[1] A bar over a complex number denotes complex conjugation.

[2] Translator's Note: The phrase *inner product* is also used. Henceforth, we shall call R an *inner product space*.

[3] Translator's Note: This is often called the Schwarz or the Cauchy-Schwarz inequality.

On the right we have a quadratic in λ. For real λ this polynomial is non-negative, which implies that

$$|(f,g)|^2 \leqq (f,f)\cdot(g,g),$$

and this proves 2°. The equality sign will hold only in case the polynomial under consideration has a double root, in other words, only if

$$\bar{\theta} f + \lambda g = 0$$

for some real λ. But this equation implies that the vectors f and g are linearly dependent.

We shall derive one more property of the norm, the inequality,

$$3^\circ \qquad \|f+g\| \leqq \|f\| + \|g\|.$$

There is equality if $f=0$ or $g=\lambda f$, $\lambda \geqq 0$. This property is called the *triangle inequality*, by analogy with the inequality for the sides of a triangle in elementary geometry.

In order to prove the triangle inequality, we use the relation

$$\|f+g\|^2 = (f+g, f+g) = (f,f) + (f,g) + (g,f) + (g,g).$$

Hence, by the Cauchy-Bunyakovski inequality

$$\|f+g\|^2 \leqq \|f\|^2 + 2\|f\|\cdot\|g\| + \|g\|^2 = \{\|f\| + \|g\|\}^2$$

which implies that

$$\|f+g\| \leqq \|f\| + \|g\|.$$

For equality, it is necessary that

$$(g,f) = \|f\|\cdot\|g\|.$$

If $f \neq 0$, then, by 2°, it is necessary that

$$g = \lambda f$$

for some λ. From this it is evident that

$$\lambda(f,f) = \|f\|\cdot\|\lambda f\|,$$

whence it also follows that $\lambda \geqq 0$.

An inner product space R becomes a *metric space*, if the distance between two points $f, g \in$ R is defined as

$$D[f,g] = \|f-g\|.$$

It follows from the properties of the norm that the distance function satisfies the usual conditions.[4]

The scalar product yields a definition for the angle between two vectors. However, for what follows, this concept will not be needed. We confine ourselves to the more limited concept of *orthogonality*. Two vectors f and g are *orthogonal* if

$$(f,g) = 0.$$

[4] These conditions are

(a) $D[f,g] = D[g,f] > 0$ (for $f \neq g$),

(b) $D[f,f] = 0$,

(c) $D[f,g] \leqq D[f,h] + D[h,g]$ (triangle inequality).

3. Some Topological Concepts

In the present section we consider some general concepts which are introduced in the study of point sets in an arbitrary metric space. We denote a metric space by E, and speak of the distance $D[f, g]$ between two elements of E. Let us bear in mind that in what will follow we shall consider only the case with E = R and $D[f, g] = \|f - g\|$, i.e., the case with the metric generated by a scalar product.

If f_0 is a fixed element of E, and ρ is a positive number, then the set of all points f for which

$$D[f, f_0] < \rho$$

is called the *sphere* in E with *center* f_0 and *radius* ρ. Such a sphere is a neighborhood, more precisely a ρ-neighborhood of the point f_0.

We say that a sequence of points $f_n \in \mathrm{E}$ $(n = 1, 2, 3, \ldots)$ has the *limit point* $f \in \mathrm{E}$, and we write

$$f_n \to f \text{ or } \lim_{n\to\infty} f_n = f \tag{1}$$

when

$$\lim_{n\to\infty} D[f_n, f] = 0. \tag{2}$$

It is not difficult to see that (1) implies

$$\lim_{m,\, n\to\infty} D[f_m, f_n] = 0 \tag{3}$$

where m and n tend to infinity independently. In fact, by the triangle inequality,

$$D[f_m, f_n] \leqq D[f_m, f] + D[f_n, f].$$

But the converse is not always correct, i.e., if for the sequence $f_n \in \mathrm{E}$ $(n = 1, 2, 3, \ldots)$ relation (3) holds, then there may not exist an element $f \in \mathrm{E}$ to which the sequence converges. If (3) is satisfied, then the sequence is called *fundamental*. Thus, a fundamental sequence may or may not converge to an element of the space.

A metric space E is called *complete* if every fundamental sequence in E converges to some element of the space. If a metric space is not complete, then it is possible to complete it by introducing certain new elements. This operation is similar to the introduction of irrational numbers by Cantor's method.

If each neighborhood of $f \in \mathrm{E}$ contains infinitely many points of a set M in E, then f is called a limit point of M. If a set contains all its limit points, then it is said to be *closed*. The set consisting of M and its limit points is called the *closure* of M and is denoted by $\overline{\mathrm{M}}$.

If the metric space E is the closure of some countable subset of E, then E is said to be *separable*. Thus, in a separable space there exists a countable set N such that, for each point $f \in E$ and each $\varepsilon > 0$, there exists a point $g \in N$ such that

$$D[f, g] < \varepsilon.$$

4. Hilbert Space

A *Hilbert space* H is an infinite dimensional inner product space which is a complete metric space with respect to the metric generated by the inner product. This definition, similar to those in preceding sections, has an axiomatic character. Various concrete linear spaces satisfy the conditions in the definition. Therefore, H is often called an *abstract* Hilbert space, and the concrete spaces mentioned are called *examples* of this abstract space.

One of the important examples of H is the space l^2. The construction of the general theory, to which the present book is devoted, was begun for this particular space by Hilbert in connection with his theory of linear integral equations. The elements of the space l^2 are sequences (of real or complex numbers)

$$f = \{x_n\}_1^\infty, \qquad g = \{y_n\}_1^\infty, \ldots,$$

such that

$$\sum_{n=1}^\infty |x_n|^2 < \infty, \qquad \sum_{n=1}^\infty |y_n|^2 < \infty, \ldots \quad .$$

The numbers $x_1, x_2, x_3, \ldots,$ are called *components* of the vector f or *coordinates* of the point f. The zero vector is the vector with all components zero. The addition of vectors is defined by the formula

$$f + g = \{x_n + y_n\}_1^\infty.$$

The relation

$$\sum_{n=1}^\infty |x_n + y_n|^2 < \infty$$

follows from the inequality

$$|x + y|^2 \leqq 2\,|x|^2 + 2\,|y|^2.$$

The multiplication of a vector f by a number λ is defined by

$$\lambda f = \{\lambda x_n\}_1^\infty.$$

The scalar product in the space l^2 has the form

$$(f, g) = \sum_{n=1}^\infty x_n \bar{y}_n.$$

The series on the right converges absolutely because

$$|xy| \leqq \tfrac{1}{2}|x|^2 + \tfrac{1}{2}|y|^2.$$

The inequality

$$|(f,g)| \leqq \|f\| \cdot \|g\|$$

now has the form

$$\left|\sum_{n=1}^{\infty} x_n \bar{y}_n\right| \leqq \sqrt{\sum_{n=1}^{\infty} |x_n|^2} \cdot \sqrt{\sum_{n=1}^{\infty} |y_n|^2}$$

and is due to Cauchy.

The space l^2 is separable. A particular countable dense subset of l^2 consists of all vectors with only finitely many nonzero components and with these components rational, i.e., the components are of the form $\xi + i\eta$ where ξ and η are rational numbers.

In addition, the space l^2 is complete. In fact, if the sequence of vectors

$$f^{(k)} = \{x_n^{(k)}\}_{n=1}^{\infty} \quad (k = 1, 2, 3, \ldots)$$

is fundamental, then each of the sequences of numbers

$$\{x_n^{(k)}\}_{k=1}^{\infty} \quad (n = 1, 2, 3, \ldots)$$

is fundamental and, hence, converges to some limit x_n $(n = 1, 2, 3, \ldots)$. Now, for each $\varepsilon > 0$ there exists an integer N such that for $r > N, s > N$

$$\sqrt{\sum_{n=1}^{\infty} |x_n^{(r)} - x_n^{(s)}|^2} < \varepsilon.$$

Consequently, for every m,

$$\sqrt{\sum_{n=1}^{m} |x_n^{(r)} - x_n^{(s)}|^2} < \varepsilon.$$

Let s tend to infinity to obtain

$$\sqrt{\sum_{n=1}^{m} |x_n^{(r)} - x_n|^2} \leqq \varepsilon.$$

But, because this is true for every m,

$$\sqrt{\sum_{n=1}^{\infty} |x_n^{(r)} - x_n|^2} \leqq \varepsilon.$$

Hence, it follows that

$$f = \{x_n\}_1^{\infty} \in l^2,$$

and, since $\varepsilon > 0$ is arbitrary,

$$f^{(k)} \to f.$$

Thus, the completeness of the space l^2 is established.

As we demonstrated, the space l^2 is separable. Originally, the requirement of separability was included in the definition of an abstract Hilbert space. However, as time passed it appeared that this requirement was not necessary for a great deal of the theory, and therefore, it is not included in our definition of the space H.

But the requirement of completeness is essential for almost all of our considerations. Therefore, it is included in the definition of H. The appropriate reservation is made in the cases for which this requirement is superfluous.

The space l^2 is infinite dimensional because the *unit vectors*

$$\begin{aligned} e_1 &= \{1, 0, 0, \ldots\}, \\ e_2 &= \{0, 1, 0, \ldots\}, \\ e_3 &= \{0, 0, 1, \ldots\}, \\ &\ldots\ldots\ldots, \end{aligned}$$

are linearly independent. The space l^2 is the infinite dimensional analogue of E_m, the (complex) m-dimensional *Euclidean space*, the elements of which are finite sequences

$$f = \{x_n\}_1^m,$$

and most of the theory which we present consists of generalizations to H of well-known facts concerning E_m.

5. Linear Manifolds and Subspaces

One often considers particular subsets of R (and, in particular, of H). Such a subset L is called a *linear manifold* if the hypothesis $f, g \in$ L implies that $\alpha f + \beta g \in$ L for arbitrary numbers α and β. One of the most common methods of obtaining a linear manifold is the construction of a *linear envelope*. The point of departure is a finite or infinite set M of elements of R. Consider the set L of all finite linear combinations

$$\alpha_1 f_1 + \alpha_2 f_2 + \ldots + \alpha_n f_n$$

of elements $f_1, f_2, \ldots, f_n$ of M. This set L is the smallest linear manifold which contains M. It is called the linear envelope of M or the linear manifold spanned by M. If R is a metric space, then the closure of the linear envelope of a set M is called the *closed linear envelope* of M.

In what follows, closed linear manifolds in H will have a particularly important significance. Each such manifold G is a linear space, metrizable with respect to the scalar product defined in H. Furthermore, G is complete. In fact, every fundamental sequence of elements of G has a limit in H because H is complete, and this limit must belong to G because G is closed. From what has been said, it follows that G itself is a Hilbert space if it contains an infinite number of linearly independent elements; otherwise G is a Euclidean space. Therefore, G is called a *subspace* of the space H.

6. The Distance from a Point to a Subspace

Consider a linear manifold L which is a *proper subset* of H. Choose a point $h \in \mathrm{H}$ and let

$$\delta = \inf_{f \in \mathrm{L}} \| h - f \|.$$

The question arises as to whether there exists a point $g \in \mathrm{L}$ for which

$$\| h - g \| = \delta.$$

In other words, is there a point in L nearest to the point h?[5]

We prove first that there exists at most one point $g \in \mathrm{L}$ such that $\delta = \| h - g \|$. Assume that there exist two such points, g' and g''. Since $\frac{1}{2}(g' + g'') \in \mathrm{L}$, we have

$$\left\| h - \frac{g' + g''}{2} \right\| \geqq \delta;$$

on the other hand

$$\left\| h - \frac{g' + g''}{2} \right\| \leqq \frac{1}{2} \| h - g' \| + \frac{1}{2} \| h - g'' \| = \delta.$$

Consequently,

$$\left\| h - \frac{g' + g''}{2} \right\| = \delta$$

and therefore

$$\left\| h - \frac{g' + g''}{2} \right\| = \frac{1}{2} \| h - g' \| + \frac{1}{2} \| h - g'' \|.$$

But this is the triangle inequality with the sign of equality. Since

$$h - g' \neq 0$$

we have

$$h - g'' = \lambda (h - g')$$

for some $\lambda \geqq 0$. If $\lambda = 1$, the proof is complete. If $\lambda \neq 1$, then

$$h = \frac{g'' - \lambda g'}{1 - \lambda}$$

so that $h \in \mathrm{L}$, which contradicts our assumption. Thus, our assertion is proved.

But, in general, does there exist a point $g \in \mathrm{L}$ nearest to the point h? In the most important case, the answer is yes, and the following theorem holds.

THEOREM: *If* G *is a subspace of the space* H *and if*

$$\delta = \inf_{f \in \mathrm{G}} \| h - f \|,$$

[5] Translator's Note. The case with $h \in \mathrm{L}$ is trivial. Henceforth, the author assumes without saying so that $h \notin \mathrm{L}$.

then there exists a vector $g \in G$ (its uniqueness was proved above) for which

$$\|h-g\| = \delta.$$

Proof: According to the definition of the greatest lower bound, there exists an infinite sequence of vectors $\{g_n\}_1^\infty$, in G, for which

$$\lim_{n\to\infty} \| h - g_n \| = \delta.$$

Now,

$$\left\| h - \frac{g_m + g_n}{2} \right\| \leqq \frac{1}{2} \| h - g_m \| + \frac{1}{2} \| h - g_n \|.$$

Therefore

$$\overline{\lim_{m,\, n\to\infty}} \left\| h - \frac{g_m + g_n}{2} \right\| \leqq \delta,$$

and since

$$\left\| h - \frac{g_m + g_n}{2} \right\| \geqq \delta$$

we have

$$\lim_{m,\, n\to\infty} \left\| h - \frac{g_m + g_n}{2} \right\| = \delta.$$

In the easily proved relation,

$$2\|f'\|^2 + 2\|f''\|^2 = \|f' + f''\|^2 + \|f' - f''\|^2,$$

let

$$f' = h - g_m, \qquad f'' = h - g_n$$

to obtain

$$\| g_n - g_m \|^2 = 2\|h - g_m\|^2 + 2\|h - g_n\|^2 - 4\left\| h - \frac{g_m + g_n}{2} \right\|^2.$$

Therefore,

$$\lim_{m,\, n\to\infty} \|g_n - g_m\| = 0.$$

So the sequence of vectors $\{g_n\}_1^\infty$ converges to some vector $g \in G$. It remains to prove that

$$\|h - g\| = \delta.$$

Now

$$\lim_{n\to\infty} \|g - g_n\| = 0, \qquad \lim_{n\to\infty} \|h - g_n\| = \delta$$

and

$$\|h - g\| \leqq \|h - g_n\| + \|g - g_n\|;$$

consequently

$$\|h - g\| \leqq \delta.$$

But, by the hypothesis of the theorem, $\| h - g \| \geqq \delta$. Thus, the theorem is proved.

7. Projection of a Vector on a Subspace

Let G be a subspace of H. By the preceding section, to each element $h \in \mathrm{H}$ there corresponds a unique element $g \in \mathrm{G}$ such that

$$\|h - g\| = \inf_{g' \in \mathrm{G}} \|h - g'\|. \tag{1}$$

Considering h and g as points, we say that g is the point of the subspace G nearest the point h. If the elements g and h are considered as vectors, then it is said that g is the particular vector of G which deviates least from h. Now, using (1), we show that the vector $h-g$ is orthogonal to the subspace G; i.e., orthogonal to every vector $g' \in \mathrm{G}$.

For the proof, we assume that the vector $h-g$ is not orthogonal to every vector $g' \in \mathrm{G}$. Let

$$(h - g, g_0) = \sigma \neq 0 \qquad (g_0 \in \mathrm{G}).$$

We define the vector

$$g^* = g + \frac{\sigma}{(g_0, g_0)} g_0 \in \mathrm{G}.$$

Then

$$\begin{aligned} \| h - g^*\|^2 &= \left(h - g - \frac{\sigma}{(g_0, g_0)} g_0, h - g - \frac{\sigma}{(g_0, g_0)} g_0\right) = \\ &= \|h - g\|^2 - \frac{\bar{\sigma}}{(g_0, g_0)} (h - g, g_0) - \frac{\sigma}{(g_0, g_0)} (g_0, h-g) + \frac{|\sigma|^2}{(g_0, g_0)} = \\ &= \|h - g\|^2 - \frac{|\sigma|^2}{(g_0, g_0)}, \end{aligned}$$

so that

$$\|h - g^*\| < \|h - g\|,$$

which contradicts (1).

From the proof it follows that h has a representation of the form[6]

$$h = g + f,$$

where $g \in \mathrm{G}$ and f is orthogonal to G (in symbols, $f \perp \mathrm{G}$). It follows easily that

$$\|h\|^2 = \|g\|^2 + \|f\|^2.$$

It is natural to call g the *component* of h in the subspace G or the *projection* of h on G.[7]

---

[6] Translator's Note: It is easy to prove that this representation is unique. The author uses the uniqueness later.

[7] This contrasts with the situation in analytic geometry, where a projection is a number and a component is a vector. Here projection and component are equivalent terms.

We denote by F the set of all vectors f orthogonal to the subspace G. We show that F is closed, so that F is a subspace. In fact, let $f_n \in$ F ($n = 1, 2, 3, \ldots$) and $f_n \to f$. Then $(f_n, g) = 0$ and

$$(f, g) = (f - f_n, g).$$

In absolute value, the right member does not exceed

$$\|f - f_n\| \cdot \|g\|,$$

which converges to zero as $n \to \infty$. Hence, $(f, g) = 0$, so that $f \in$ F and the manifold F is closed.

The subspace F is called the *orthogonal complement* of G and is expressed by

$$\text{F} = \text{H} \ominus \text{G}. \tag{2}$$

It is easy to see that

$$\text{G} = \text{H} \ominus \text{F}. \tag{2'}$$

Both relations (2) and (2′) are expressed by the equation

$$\text{H} = \text{G} \oplus \text{F},$$

because H is the so-called direct sum of the subspaces F and G (in the given case, the *orthogonal sum*).

In general, a set M $\subset$ H is called the *direct sum* of a finite number of linear manifolds $\text{M}_k \subset$ H ($k = 1, 2, 3, \ldots, n$) and is expressed by

$$\text{M} = \text{M}_1 \oplus \text{M}_2 \oplus \ldots \oplus \text{M}_n$$

if each element g $\in$ M is represented uniquely in the form

$$g = g_1 + g_2 + \ldots + g_n$$

where $g_k \in \text{M}_k$ ($k = 1, 2, 3 \ldots, n$). It is evident that M is also a linear manifold.

It will be necessary for us to consider direct sums of an infinite number of linear manifolds only in cases for which the manifolds are pairwise orthogonal subspaces of the given space. This is done as follows.

DEFINITION: *Let* $\{\text{H}_\alpha\}$ *be a countable or uncountable class of pairwise orthogonal subspaces of* H. *Their orthogonal sum*

$$\sum_\alpha \oplus \text{H}_\alpha$$

is defined as the closed linear envelope of the set of all finite sums of the form

$$\text{H}_{\alpha'} \oplus \text{H}_{\alpha''} \oplus \ldots.$$

Often it is necessary to determine the projection of a vector on a finite dimensional subspace. We consider this question in some detail. Let G be an n-dimensional subspace and let

$$(3) \qquad g_1, g_2, \ldots, g_n$$

be n linearly independent elements of G. Since any $n+1$ elements of G are linearly dependent, each vector $g' \in$ G can be represented (uniquely) in the form

$$g' = \lambda_1 g_1 + \lambda_2 g_2 + \ldots + \lambda_n g_n.$$

In other words, G is the linear envelope of the set of vectors (3).

We choose an arbitrary vector $h \in$ H and denote by g its projection on G. The vector g has a unique representation,

$$g = a_1 g_1 + a_2 g_2 + \ldots + a_n g_n.$$

According to the definition of a projection, the difference $h-g=f$ must be orthogonal to the subspace G, i.e., f is orthogonal to each of the vectors $g_1, g_2, \ldots, g_n$. Therefore,

$$(4) \quad (f, g_k) \equiv (h, g_k) - a_1 (g_1, g_k) - a_2 (g_2, g_k) - \ldots - a_n (g_n, g_k) = 0$$
$$(k = 1, 2, 3, \ldots, n).$$

This is a system of n linear equations in the unknowns $a_1, a_2, \ldots, a_n$. We have shown that it has a unique solution for each vector h. Therefore, the determinant of this system is different from zero[8]. This determinant

$$\Gamma(g_1, g_2, \ldots, g_n) = \begin{vmatrix} (g_1, g_1) & (g_2, g_1) & \ldots & (g_n, g_1) \\ (g_1, g_2) & (g_2, g_2) & \ldots & (g_n, g_2) \\ \ldots & \ldots & \ldots & \ldots \\ \ldots & \ldots & \ldots & \ldots \\ (g_1, g_n) & (g_2, g_n) & \ldots & (g_n, g_n) \end{vmatrix}$$

is called the *Gram determinant* of the vectors $g_1, g_2, \ldots, g_n$. It is easy to see that if the vectors $g_1, g_2, \ldots, g_n$ are linearly dependent, then the Gram determinant is equal to zero. Hence, for the linear independence of the vectors it is necessary and sufficient that their Gram determinant be different from zero.

We proceed to determine the number

$$\delta = \min_{g' \in G} \| h - g' \|.$$

We shall express δ by means of the Gram determinant. As above let g be the projection of h on G and let $f = h - g$. Then $\delta = \| f \| = \| h - g \|$ and

$$\delta^2 = (f, f) = (f, h),$$

[8] Translator's Note: In (4), suppose that $(h, g_k) = 0$ for $k = 1, 2, 3, \ldots, n$. In other words, let $h \perp$ G. Then $\| g \|^2 = \| a_1 g_1 + \ldots + a_n g_n \|^2 = \sum_{k=1}^{n} \bar{a}_k [(a_1 g_1, g_k) + \ldots + (a_n g_n, g_k)] = \sum_{k=1}^{n} \bar{a}_k (h, g_k) = 0$ so that $g = a_1 g_1 + a_2 g_2 + \ldots + a_n g_n = 0$. Since $g_1, \ldots, g_n$ are linearly independent, $a_1 = a_2 = \ldots a_n = 0$. This proves that the homogeneous system $a_1(g_1, g_k) + \ldots + a_n(g_n, g_k) = 0$ $(k = 1, 2, 3, \ldots, n)$ has only the trivial solution $a_1 = \ldots = a_n = 0$. Consequently, the determinant of the system is not zero.

since $(f,g)=0$. Let $g=\alpha_1 g_1+\alpha_2 g_2+\ldots+\alpha_n g_n$, where the g_k are as in equation (3), to obtain

$$\delta^2=(h,h)-\alpha_1(g_1,h)-\alpha_2(g_2,h)-\ldots-\alpha_n(g_n,h). \tag{5}$$

The determination of δ^2 is reduced to the elimination of the quantities α_i from equations (4) and (5). This elimination yields

$$\begin{vmatrix} (h,h)-\delta^2 & (g_1,h) & (g_2,h) & \ldots & (g_n,h) \\ (h,g_1) & (g_1,g_1) & (g_2,g_1) & \ldots & (g_n,g_1) \\ \ldots & \ldots & \ldots & \ldots & \ldots \\ (h,g_n) & (g_1,g_n) & (g_2,g_n) & \ldots & (g_n,g_n) \end{vmatrix}=0.$$

Hence,

$$\delta^2=\frac{\Gamma(h,g_1,g_2,\ldots,g_n)}{\Gamma(g_1,g_2,\ldots,g_n)}. \tag{6}$$

This is the formula we wished to obtain.

Since $\Gamma(g_1)=(g_1,g_1)>0$ (for $g_1\neq 0$), it follows from formula (6) that the Gram determinant of linearly independent vectors is always positive. This fact can be regarded as a generalization of the Cauchy-Bunyakovski inequality, which asserts that

$$\Gamma(g_1,g_2)>0$$

for linearly independent vectors g_1 and g_2.

8. Orthogonalization of a Sequence of Vectors

Two sets M and N of vectors in H are said to be *equivalent* if their linear envelopes coincide. Therefore, the sets M and N are equivalent if and only if each element of one of these sets is a linear combination of a finite number of vectors belonging to the other set.

If the elements of the set M are pairwise orthogonal vectors, and if each of the vectors is *normalized*, i.e., if each has norm equal to one, then the set M is called an *orthonormal system*. If, in addition, the set M is countable, then it is also called an *orthonormal sequence*.

Suppose given a finite or infinite sequence of independent vectors

$$g_1, g_2, \ldots, g_n, \ldots . \tag{1}$$

We show how to construct an equivalent orthonormal sequence of vectors

$$e_1, e_2, \ldots, e_n, \ldots . \tag{2}$$

For the first vector, we take

$$e_1=\frac{g_1}{\|g_1\|},$$

the norm of which is equal to one. The vectors e_1 and g_1 generate the same (one dimensional) subspace E_1. The vector e_2 is constructed in two steps. First, we subtract from the vector g_2 its projection on E_1 to get

$$h_2 = g_2 - (g_2, e_1)e_1,$$

which is orthogonal to the subspace E_1. Since the vectors (1) are linearly independent, g_2 does not belong to E_1, so that $h_2 \neq 0$. Now let

$$e_2 = \frac{h_2}{\|h_2\|}.$$

The vectors e_1 and e_2 generate the same (two dimensional) subspace E_2 as do the vectors g_1 and g_2. We now construct the vector e_3. First, we subtract from g_3 its projection on E_2 to get

$$h_3 = g_3 - (g_3, e_1)e_1 - (g_3, e_2)e_2,$$

which is different from zero and orthogonal to the subspace E_2, i.e., h_3 is orthogonal to each of the vectors e_1 and e_2. Next we let

$$e_3 = \frac{h_3}{\|h_3\|}.$$

We continue in the same way. If the vectors

$$e_1, e_2, \ldots, e_n$$

have been constructed, then we let

$$h_{n+1} = g_{n+1} - \sum_{k=1}^{n} (g_{n+1}, e_k)\, e_k,$$

and

$$e_{n+1} = \frac{h_{n+1}}{\|h_{n+1}\|}.$$

The method described is called *orthogonalization.*[9]

In the solution of many problems concerning manifolds generated by a given sequence of vectors, preliminary orthogonalization of the sequence turns out to be very useful. We illustrate this in the problem considered in the preceding paragraph. That problem concerned the determination of the distance from a point $h \in H$ to a linear manifold G, which was the closed linear envelope of the given sequence (1). We shall show how elegantly this problem is solved if the system (1) is orthogonalized beforehand.

[9] Often, in particular cases, one does not bother to normalize the system (2) of pairwise orthogonal elements which is equivalent to the system (1). The transition from (1) to such an orthogonal system likewise is called orthogonalization. (See Section 11 below.)

Assume given the orthogonal sequence (2) and a vector $h \in \mathrm{H}$. For each integer n, the vector h can be expressed in the form

$$h = \sum_{k=1}^{n} (h, e_k)\, e_k + f_n$$

where the vector f_n is orthogonal to each of the vectors $e_1, e_2, \ldots, e_n$. The vector

$$s_n = \sum_{k=1}^{n} (h, e_k)\, e_k \tag{3}$$

belongs to the set of vectors

$$\lambda_1 e_1 + \lambda_2 e_2 + \ldots + \lambda_n e_n \tag{4}$$

and, of these vectors, s_n is nearest to the vector h. The distance from s_n to h is

$$\delta_n = \min_{\lambda_k} \| h - \lambda_1 e_1 - \lambda_2 e_2 - \ldots - \lambda_n e_n \| = \| f_n \| = \tag{5}$$
$$= \sqrt{\|h\|^2 - \sum_{k=1}^{n} |(h, e_k)|^2}.$$

This is the distance from the point h to the linear envelope G_n of the set consisting of the first n vectors of the sequence (2). If instead of the linear combination of the nth order (4), we wish to find the linear combination of the $(n + 1)$th order,

$$\mu_1 e_1 + \mu_2 e_2 + \ldots + \mu_{n+1} e_{n+1}, \tag{4'}$$

which is nearest to the vector h, then we must take the vector

$$s_{n+1} = \sum_{k=1}^{n+1} (h, e_k)\, e_k.$$

Thus, we do not change the coefficients in the linear combination (3). Rather, we merely add one more term,

$$(h, e_{n+1}) e_{n+1}$$

to the right member of (3).

These considerations show that, being given the infinite orthonormal sequence (2), it is appropriate to associate with each vector $h \in \mathrm{H}$ the *infinite series*

$$\sum_{k=1}^{\infty} (h, e_k) e_k. \tag{6}$$

Equation (5) yields the important inequality

$$\sum_{k=1}^{\infty} |(h, e_k)|^2 \leqq \| h \|^2. \tag{7}$$

The convergence of the series

$$\sum_{k=1}^{\infty} |(h, e_k)|^2$$

implies that

$$\|s_n - s_m\|^2 = \sum_{k=m+1}^{n} |(h, e_k)|^2 \qquad (m < n)$$

converges to zero as $m, n \to \infty$, i.e., the series (6) converges in H.[10] We see that the square of the distance from the point h to the manifold G is

$$\|h\|^2 - \sum_{k=1}^{\infty} |(h, e_k)|^2$$

and that the vector h belongs to the manifold G if and only if there is equality in formula (7).

We shall say that a system of vectors is *closed* in H if its linear envelope is dense in H. From our considerations, the orthonormal system (2) is closed in H if and only if

$$\|h\|^2 = \sum_{k=1}^{\infty} |(h, e_k)|^2 \tag{8}$$

for each $h \in$ H. Following V. A. Steklov,[11] we call this equality the *closure relation.*[12] We show next that if the Parseval relation holds for each vector $h \in$ H, then for any pair of vectors $g, h \in$ H, the general Parseval relation

$$(g, h) = \sum_{k=1}^{\infty} (g, e_k)(e_k, h) \tag{9}$$

holds. In fact, we have the Parseval relation for each vector $g + \lambda h$:

$$\|g + \lambda h\|^2 = \sum_{k=1}^{\infty} |(g + \lambda h, e_k)|^2,$$

which yields

$$(g, g) + \lambda(h, g) + \bar{\lambda}(g, h) + |\lambda|^2 (h, h) =$$

$$= \sum_{k=1}^{\infty} \{|(g, e_k)|^2 + \lambda(h, e_k)(e_k, g) + \bar{\lambda}(g, e_k)(e_k, h) + |\lambda|^2 |(h, e_k)|^2\}$$

and

$$\lambda(h, g) + \bar{\lambda}(g, h) = \lambda \sum_{k=1}^{\infty} (h, e_k)(e_k, g) + \bar{\lambda} \sum_{k=1}^{\infty} (g, e_k)(e_k, h).$$

Since λ is arbitrary, equation (9) follows.

[10] We remark that convergence in H of any series $f_1 + f_2 + \dots$, where $(f_i, f_k) = 0$ $(i \neq k)$ is equivalent to the convergence of the numerical series $\|f_1\|^2 + \|f_2\|^2 + \|f_3\|^2 + \dots$.

[11] V. A. Steklov showed for the first time the important significance of the closure relation in various problems of analysis and mathematical physics. Before the work of V. A. Steklov, the relation under consideration was known only for systems of trigonometric functions (the so-called Parseval-Liapunov equation).

[12] Translator's Note: Following rather common English practice, we shall refer to (8) henceforth as the *Parseval relation.* Some authors refer to it as the completeness relation.

9. Complete Orthonormal Systems

The vectors of an orthonormal system cannot be linearly dependent. Therefore, in n-dimensional Euclidean space each orthonormal system of vectors contains at most n vectors.

We say that an orthonormal system M is *complete* in H if M is not contained in a larger orthonormal system in H, i.e., if there is no nonzero vector in H which is orthogonal to every vector of the system M. In Euclidean n-space any orthonormal system of n vectors is complete. In Hilbert space a complete orthonormal system contains an infinite number of elements, and there arises the problem of the cardinality of such systems. This problem is solved easily for separable spaces. We begin with them.

THEOREM 1: *If the space* H *is separable, then every orthonormal system of vectors in* H *consists of a finite or countable number of elements.*

Proof: Let

$$f_1, f_2, f_3, \ldots \tag{1}$$

be a sequence of vectors which is dense in H, and let M be an orthonormal system of vectors. We proceed to show that M can be enumerated. Let e and e' be distinct vectors in M. From (1) choose vectors f_k and $f_{k'}$ such that

$$\|e - f_k\| < \tfrac{1}{2}\sqrt{2} \tag{2}$$

and similarly for e' and k'. We show that $k' \neq k$. In fact,

$$\|e - e'\|^2 = \|e\|^2 + \|e'\|^2 = 2$$

so that

$$\sqrt{2} = \|e - e'\| \leqq \|e - f_k\| + \|e' - f_k\| < \tfrac{1}{2}\sqrt{2} + \|e' - f_k\|.$$

Therefore,

$$\|e' - f_k\| > \tfrac{1}{2}\sqrt{2}$$

so that $f_{k'} \neq f_k$ and $k \neq k'$. Thus, we can associate with each vector of M a different integer k. This proves that the set M is finite or countable.

The existence of a nondenumerable orthonormal system of vectors in H implies that the space is not separable. An important example of this kind will be considered later.

THEOREM 2: *An infinite orthonormal sequence*

$$e_1, e_2, e_3, \ldots \tag{3}$$

is complete in H *if and only if the sequence is closed in* H.

Proof: (A) Let the system (3) be closed in H. Then, for each vector of H, the Parseval relation holds. Assume that the system (3) is not complete, and denote by h a nonzero vector which is orthogonal to each of the vectors (3). Thus, $(h, e_k) = 0$ $(k = 1, 2, 3, \ldots)$ and the Parseval relation for h reduces to the contradiction

$$0 \neq \|h\|^2 = 0.$$

(B) We suppose now that the system (3) is complete. We choose an arbitrary vector $h \in H$ and consider the sequence of vectors

$$s_n = \sum_{k=1}^{n} (h, e_k) e_k \qquad (n = 1, 2, 3, \ldots).$$

By the preceding section, $\{s_n\}$ is fundamental, which implies that it converges to some vector g. Then

$$(g, e_k) = \lim_{n \to \infty} (s_n, e_k) = (h, e_k) \qquad (k = 1, 2, 3, \ldots) \tag{4}$$

and g belongs to the closed linear envelope of the sequence (3). Consequently, the Parseval relation is valid for g:

$$\|g\|^2 = \sum_{k=1}^{\infty} |(g, e_k)|^2 = \sum_{k=1}^{\infty} |(h, e_k)|^2. \tag{5}$$

It follows from (4) that the vector $g - h$ is orthogonal to each vector of the sequence (3). The assumption that this sequence is complete implies that $g - h = 0$, so that $g = h$, and (5) takes the form

$$\|h\|^2 = \sum_{k=1}^{\infty} |(h, e_k)|^2.$$

We have shown that for an arbitrary vector $h \in H$, the Parseval relation holds. Thus, it is proved that (3) is closed in H.

THEOREM 3: *The space* H *contains a complete orthonormal sequence if and only if it is separable.*

Proof: (A) We assume that the space H is separable, and let N denote a countable set of vectors which is dense in H. Deleting from the sequence N any vector which is a linear combination of the preceding ones, and orthogonalizing the resulting sequence, we obtain an orthonormal sequence M. This sequence is complete. For, let the vector $h \in H$ be orthogonal to each element of the sequence M. Then h is orthogonal to each vector of N. For each $\varepsilon > 0$ there exists a vector $f \in N$ such that

$$\|h - f\| < \varepsilon$$

which implies that

$$\|h\|^2 = (h, h) = (h - f, h) \leq \|h - f\| \cdot \|h\| < \varepsilon \|h\|$$

and

$$\|h\| < \varepsilon.$$

Since $\varepsilon > 0$ is arbitrary, $h = 0$ and the orthonormal sequence M is complete in H.

(B) We assume that (3) is a complete orthonormal sequence in H. Let N be the set of all linear combinations of the form

$$\gamma_1^{(n)} e_1 + \gamma_2^{(n)} e_2 + \ldots + \gamma_n^{(n)} e_n \qquad (n = 1, 2, 3, \ldots),$$

where $\gamma_k^{(r)} = \alpha_k^{(r)} + i\beta_k^{(r)}$, and $\alpha_k^{(r)}, \beta_k^{(r)}$ are rational numbers. The set N is countable. For each $h \in$ H and each $\varepsilon > 0$ there exists an integer n such that

$$\left\| h - \sum_{k=1}^{n} (h, e_k) e_k \right\| < \frac{\varepsilon}{2}.$$

It is possible to approximate the complex numbers (h, e_k), $(k = 1, 2, 3, \ldots, n)$, by numbers of the form $\gamma_k^{(n)}$ such that

$$\left\| \sum_{k=1}^{n} \{(h, e_k) - \gamma_k^{(n)}\} e_k \right\| < \frac{\varepsilon}{2}.$$

Thus, there exists a vector,

$$f = \sum_{k=1}^{n} \gamma_k^{(n)} e_k,$$

in N, for which

$$\| h - f \| < \varepsilon,$$

and this implies that H is separable.

The question of the cardinality of a complete orthonormal system in a separable space now can be answered completely: every complete orthonormal system in a separable space is necessarily an infinite sequence --- a so-called *orthonormal basis* of the space.

Now we consider arbitrary Hilbert spaces. First, we remark that whatever the cardinality of an orthonormal system M, each vector f has no more than a countable set of nonzero projections on the elements of the system M. This follows from the fact that for any sequence of elements $e', e'', e''', \ldots$ of M the inequality

$$|(f, e')|^2 + |(f, e'')|^2 + |(f, e''')|^2 + \ldots \leqq \|f\|^2$$

holds, which shows that it is possible to enumerate the set of all nonzero numbers (f, e) with $e \in$ M. Further, we have

THEOREM 4: *Any two complete orthonormal systems in a Hilbert space have the same cardinal number.*

Proof: Let M and N be two orthonormal systems, each complete in H, with cardinalities $\mathfrak{m}$ and $\mathfrak{n}$, respectively. Choose $e \in$ M. At least one of the scalar products (e, f), $f \in$ N, is different from zero because otherwise it would be possible to extend the orthonormal system N by appending to it the vector e. On the other hand, by the remark of the previous paragraph, there exists no more than a countable set of elements $f \in$ N for which $(e, f) \neq 0$. We denote these elements by

$$f_1, f_2, \ldots, f_n \qquad (1 \leqq n \leqq \infty). \tag{6}$$

We define a function φ with domain M such that $\varphi(e)$ is the set of vectors, (6), for which $(e, f) \neq 0$. The function φ is at least single valued and at most countably valued. This function maps M onto a set of countable subsets of N. Each $f^* \in$ N satisfies $f^* \in \varphi(e^*)$ for some $e^* \in$ M, since for each $f^* \in$ N there exists an element $e^* \in$ M which is not orthogonal to the element f^*. From what has been said it follows that

$$\mathfrak{m} \geqq \mathfrak{n}.$$

Reversing the roles of the systems M and N, we get

$$\mathfrak{n} \geqq \mathfrak{m}.$$

And so

$$\mathfrak{m} = \mathfrak{n},$$

and the theorem is proved.[13] The following definition is based on this theorem.

DEFINITION: *The dimension of a Hilbert space* H *is the cardinality of a complete orthonormal system in* H.

It is not necessary to make a separate definition of the dimension of a subspace G $\subset$ H. The dimension of an arbitrary linear manifold L $\subset$ H is defined as the dimension of the corresponding subspace $\bar{\text{L}}$.

If two Hilbert spaces H and H$'$ have the same dimension, then they are *isomorphic* in the sense that there exists a one-to-one correspondence between H and H$'$ having the following property. If the elements $f, g \in$ H correspond to the elements $f', g' \in$ H$'$, respectively, then (1) $\alpha f + \beta g$ corresponds to $\alpha f' + \beta g'$ and (2) $(f, g)_{\text{H}} = (f', g')_{\text{H}'}$. In fact, since the spaces H and H$'$ have the same dimension they possess complete orthonormal systems of identical cardinalities. We choose any one-to-one correspondence between the elements of these two orthonormal systems and extend this correspondence to the linear envelopes of the orthogonal systems under consideration in such a way that condition (1) is satisfied. Then condition (2) is automatically satisfied, which permits one to get, by passage to the limit, the required correspondence for all the elements of the spaces H and H$'$.

From the proof it follows that any separable space is isomorphic to the space l^2. It is evident that two Hilbert spaces of different dimensions are not isomorphic. Therefore, two abstract Hilbert spaces (similarly, two abstract Euclidean spaces) differ from each other only in their dimensions.

[13] The actual construction of a complete orthonormal system in a non-separable space requires transfinite induction.

10. The Space L^2

Let (a, b) denote a finite or infinite interval[14] on the real axis. We denote by $L^2(a, b)$ (or simply by L^2) the set of all complex valued Lebesgue measurable functions f defined on (a, b) such that $|f|^2$ is Lebesgue integrable on (a, b). We do not regard as distinct elements of L^2 a pair of functions which differ only on a set of measure zero.

It follows by means of the inequality

$$|\alpha + \beta|^2 \leqq 2|\alpha|^2 + 2|\beta|^2$$

that $f + g \in L^2$ whenever $f, g \in L^2$. Furthermore, for each complex number λ and each $f \in L^2$, it follows that $\lambda f \in L^2$. Thus, L^2 is a linear space and the zero element is a function which is equal to zero almost everywhere in (a, b). In this linear space the scalar product is defined by the formula

$$(f, g) = \int_a^b f(t)\overline{g(t)}\,dt.$$

The existence of the integral on the right side is a consequence of the inequality

$$|\alpha\beta| \leqq \tfrac{1}{2}|\alpha|^2 + \tfrac{1}{2}|\beta|^2.$$

In the present case, the inequality

$$|(f, g)| \leqq \|f\| \cdot \|g\|$$

has the form

$$\left| \int_a^b f(t)\overline{g(t)}\,dt \right| \leqq \sqrt{\int_a^b |f(t)|^2\,dt}\,\sqrt{\int_a^b |g(t)|^2\,dt}.$$

This inequality was obtained by Bunyakovski for Riemann integrals.

Now we show that L^2 is complete, from which it will follow that L^2 is a Hilbert space. Let the sequence of functions $f_n \in L^2$ $(n = 1, 2, 3, \ldots)$ be fundamental, i.e., let

$$\lim_{m, n \to \infty} \int_a^b |f_n(t) - f_m(t)|^2\,dt = 0.$$

Then there exists an infinite sequence of integers

$$k_1 < k_2 < k_3 < \ldots < k_r < \ldots,$$

[14] Instead of an interval, the domain of definition of the function could be any measurable set (of finite or infinite measure) on the real axis, on the plane, or in Euclidean n-space.

for which

$$\int_a^b |f_{k_{r+1}}(t) - f_{k_r}(t)|^2 dt < \frac{1}{8^r} \qquad (r = 1, 2, 3, \ldots).$$

From this inequality it follows that the set of points of the interval (a, b) for which

$$|f_{k_{r+1}}(t) - f_{k_r}(t)| \geqq \frac{1}{2^r}$$

has measure less than $\frac{1}{2^r}$. For $s = 1, 2, 3, \ldots$, let I_s denote the set of points $t \in (a, b)$ such that

$$|f_{k_{s+1}}(t) - f_{k_s}(t)| < \frac{1}{2^s},$$

$$|f_{k_{s+2}}(t) - f_{k_{s+1}}(t)| < \frac{1}{2^{s+1}},$$

$$\cdots\cdots\cdots\cdots\cdots\cdots$$

The complement of I_s with respect to the interval $I = (a, b)$ has measure

$$m(I - I_s) < \sum_{n=s}^{\infty} \frac{1}{2^n} = \frac{1}{2^{s-1}}.$$

Since $I_n \subset I_{n+1} \subset \ldots$, $\lim_{n\to\infty} I_n = I^*$ exists and

$$m(I - I^*) = 0.$$

The sequence $\{f_{k_r}(t)\}_{r=1}^{\infty}$ converges uniformly on each set I_s. This follows from the inequality

$$|f_{k_n}(t) - f_{k_m}(t)| \leqq \sum_{r=m}^{n-1} |f_{k_{r+1}}(t) - f_{k_r}(t)| < \sum_{r=m}^{n-1} \frac{1}{2^r} < \frac{1}{2^{m-1}} \quad (n > m > s)$$

which is valid for $t \in I_s$. Consequently, $\{f_{k_r}(t)\}_{r=1}^{\infty}$ converges on I^* (i.e., almost everywhere in I). Let

$$f(t) = \begin{cases} \lim_{r\to\infty} f_{k_r}(t) & (t \in I^*), \\ 0 & (t \in I - I^*). \end{cases}$$

Since $\{f_n\}$ was assumed to be a fundamental sequence, for each $\varepsilon > 0$ there exists an integer $N(\varepsilon)$ such that

$$\int_{I_s(\alpha)} |f_m(t) - f_{k_r}(t)|^2 dt \leqq \|f_m - f_{k_r}\|^2 < \varepsilon$$

for $m, k_r > N(\varepsilon)$, where $I_s(\alpha) = I_s$ if the interval (a, b) is finite and $I_s(\alpha) = I_s \cap (-\alpha, \alpha)$ otherwise. In $I_s(\alpha)$, convergence of the sequence $\{f_{k_r}(t)\}_{r=1}^{\infty}$ is uniform. Therefore, in the integral, passage to the limit is legitimate and we obtain

$$\int_{I_s(\alpha)} |f_m(t) - f(t)|^2 dt \leqq \varepsilon \qquad (m > N(\varepsilon)).$$

It follows that

$$\int_{I_s} |f_m(t) - f(t)|^2 dt \leqq \varepsilon,$$

where s is arbitrary. Hence

$$\int_a^b |f_m(t) - f(t)|^2 dt \leqq \varepsilon.$$

This implies that $f_m - f \in L^2$, so that $f \in L^2$. Since $\varepsilon > 0$ is arbitrary, we have proved also that

$$\lim_{m \to \infty} \int_a^b |f_m(t) - f(t)|^2 dt = 0 .$$

In the process of the proof, we have obtained the following fact: if the sequence $f_n \in L^2$ $(n = 1, 2, 3, \ldots)$ converges to f in norm, i.e., if $\|f_n - f\| \to 0$ $(n \to \infty)$, then there exists a subsequence $\{f_{k_r}(t)\}_{r=1}^{\infty}$ which converges to $f(t)$ almost everywhere. Furthermore, if a proper set of arbitrarily small measure is removed from the interval (a, b), then in the remaining set the subsequence $\{f_{k_r}(t)\}_{r=1}^{\infty}$ converges to $f(t)$ uniformly.

We remark that it is possible to consider the space $L^2(a, b)$ as a subspace of $L^2(a_1, b_1)$ if $a_1 \leqq a < b \leqq b_1$ and, in particular, as a subspace of $L^2(-\infty, \infty)$. For this, it is necessary to extend each function $f \in L^2(a, b)$ beyond the limits of the interval (a, b) by defining $f(t)$ to be zero for t outside of (a, b).

Convergence in the metric space L^2 is called *convergence in the mean* and is denoted by

$$f(t) = \underset{n \to \infty}{\text{l.i.m.}} f_n(t),$$

if

$$\lim_{n \to \infty} \int_a^b |f(t) - f_n(t)|^2 dt = 0$$

(l.i.m.) is an abbreviation for limes in medio, i.e. *limit in the mean*).

11. Complete Orthonormal Systems in L^2

In the present paragraph, we show that there exist complete orthonormal sequences in $L^2(a, b)$, where a and b are finite or infinite. Hence, by Theorem 2, Section 9, it will follow that the space L^2 is separable. It would be possible to prove this latter fact immediately. In fact, using the definition of the Lebesgue integral, it is not difficult to prove that the linear envelope of the set of functions f such that $f \equiv 1$ on some finite interval and $f \equiv 0$ outside, is complete in L^2. Hence, the separability of L^2 follows.

A. We begin with the space $L^2(0, 2\pi)$. In this space, the functions

$$\frac{1}{\sqrt{2\pi}} e^{ikt} \qquad (\pm k = 0, 1, 2, \ldots)$$

form an orthonormal system. We wish to show that this *trigonometric system* is complete. Assume there exists $f \in L^2(0, 2\pi)$ such that

$$\int_0^{2\pi} |f(t)|\, dt \neq 0$$

and

$$\int_0^{2\pi} f(t) e^{-ikt} dt = 0 \qquad (\pm k = 1, 2, 3, \ldots). \tag{1}$$

It follows by means of integration by parts that the function

$$F(t) = \int_0^t f(t)\, dt$$

satisfies the equations

$$\int_0^{2\pi} \{F(t) - C\} e^{-ikt}\, dt = 0 \qquad (\pm k = 1, 2, 3, \ldots) \tag{2}$$

for any constant C. We specify this constant so that equation (2) holds also for $k = 0$. Since the function

$$\Phi(t) = F(t) - C$$

is continuous, the well-known theorem of Weierstrass applies: for each $\varepsilon > 0$ there exists a trigonometric sum

$$\sigma(t) = \sum_{k=-n}^{n} A_k e^{ikt}$$

such that

$$|\Phi(t) - \sigma(t)| < \varepsilon.$$

Therefore, using relation (2), we obtain

$$\int_0^{2\pi} |\Phi(t)|^2 dt = \int_0^{2\pi} \overline{\Phi(t)}\{\Phi(t) - \sigma(t)\}\, dt \leqq$$

$$\leqq \varepsilon \int_0^{2\pi} |\Phi(t)|\, dt \leqq \varepsilon\sqrt{2\pi}\sqrt{\int_0^{2\pi} |\Phi(t)|^2 dt}$$

whence

$$\int_0^{2\pi} |\Phi(t)|^2 dt \leqq 2\pi\varepsilon^2.$$

Since $\varepsilon > 0$ is arbitrary, this implies that $\Phi(t) = 0$, so that $F(t) = C$ and $f(t) = 0$ almost everywhere. Thus, the completeness of the trigonometric system is proved.

B. We consider now the space $L^2(a, b)$, where (a, b) is an arbitrary finite interval. The orthogonalization of the sequence of functions

$$1, t, t^2, \ldots$$

yields the sequence of polynomials

$$C_k \frac{d^k\{(t-a)(t-b)\}^k}{dt^k} \qquad (k = 0, 1, 2, \ldots),$$

where C_k are certain positive constants. These are the well-known *Legendre polynomials.* They are usually considered for $a = -1$, $b = 1$. The completeness of this orthonormal system may be proved in the same way as the completeness of the trigonometric system.

C. We consider the space $L^2(-\infty, \infty)$. The orthogonalization of the system

$$e^{-\frac{t^2}{2}}, te^{-\frac{t^2}{2}}, t^2e^{-\frac{t^2}{2}}, \ldots$$

yields the sequence of *Tchebysheff-Hermite functions,*

$$\varphi_k(t) = (-1)^k e^{\frac{t^2}{2}} \frac{d^k e^{-t^2}}{dt^k} = H_k(t) e^{-\frac{t^2}{2}} \qquad (k = 0, 1, 2, \ldots),$$

where $H_k(t)$ is the so-called *Tchebysheff-Hermite polynomial* of degree k. The Tchebysheff-Hermite functions satisfy the relations

$$\int_{-\infty}^{\infty} \varphi_k(t)\,\varphi_m(t)\, dt = \begin{cases} 0 & (k \neq m), \\ 2^m m! \sqrt{\pi} & (k = m), \end{cases}$$

so that they are pairwise orthogonal but not normalized. We prove next that the sequence of Tchebysheff-Hermite functions is complete. Assume there exists a nonzero function $f \in L^2(-\infty, \infty)$ such that

$$\int_{-\infty}^{\infty} f(t)\,\varphi_k(t)\,dt = 0 \qquad (k = 0, 1, 2, \ldots)$$

or, equivalently, such that

$$(3) \qquad \int_{-\infty}^{\infty} f(t)\,e^{-\frac{t^2}{2}} t^k\,dt = 0 \qquad (k = 0, 1, 2, \ldots).$$

We introduce the function

$$F(z) = \int_{-\infty}^{\infty} f(t)\,e^{-\frac{t^2}{2}} e^{itz}\,dt$$

which, it is evident, exists for every complex z. The function $F(z)$ has the finite derivative

$$F'(z) = \int_{-\infty}^{\infty} f(t)\,e^{-\frac{t^2}{2}} e^{itz}\,it\,dt.$$

Since this equation holds everywhere in the complex plane, $F(z)$ is an entire function. But, by (3)

$$F^{(k)}(0) \equiv \int_{-\infty}^{\infty} f(t)\,e^{-\frac{t^2}{2}} (it)^k\,dt = 0 \qquad (k = 0, 1, 2, \ldots)$$

so that $F(z)$ is identically zero. Therefore,

$$\int_{-\infty}^{\infty} f(t)\,e^{-\frac{t^2}{2}} e^{itx}\,dt = 0 \qquad (-\infty < x < \infty).$$

Multiplying this equality by e^{-ixy}, where y is real, and integrating with respect to x from $-\omega$ to ω, we get

$$\int_{-\infty}^{\infty} f(t)\,e^{-\frac{t^2}{2}} \frac{\sin \omega (t - y)}{t - y}\,dt = 0,$$

which is valid for every real y and ω. Hence, as is proved in analysis courses, it follows that $f(t) = 0$ almost everywhere, and this contradicts our original assumption.

D. In the space $L^2(0, \infty)$, we have the orthonormal system of *Tchebysheff-Laguerre functions*

$$\psi_k(t) = \frac{e^{-\frac{t}{2}} L_k(t)}{k!} \qquad (k = 0, 1, 2, \ldots),$$

where the $L_k(t)$ are the *Tchebysheff-Laguerre polynomials* which are defined by the formulas

$$L_k(t) = e^t \frac{d^k}{dt^k}(t^k e^{-t}) \qquad (k = 0, 1, 2, \ldots).$$

The completeness of this system can be proved by an argument based on the completeness of the Tchebysheff-Hermite system. We leave this to the reader.

12. The Space L^2_σ

Let a non-decreasing function of bounded variation $\sigma(t)(-\infty < t < \infty)$ be given. We assume that it is left-continuous:

$$\sigma(t-0) = \sigma(t).$$

Such a function is often called a distribution function. With the aid of the function $\sigma(t)$ it is possible to construct a measure analogous to the Lebesgue measure but differing from it in that the length $b-a$ of the interval $[a, b]$, $(a \leqq b)$[15] is replaced by the *σ-length* $\sigma(b+0)-\sigma(a)$. Some points may have σ-length different from zero (points of jumps of $\sigma(t)$) and some proper intervals may have σ-length equal to zero (intervals of constancy of $\sigma(t)$). The measure determined by the σ-length is called the *σ-measure*; the *σ-measurable functions* and the corresponding Lebesgue-Stieltjes integral are constructed from it.

We consider the linear space of all σ-measurable functions f for which the Lebesgue-Stieltjes integral

$$\int_{-\infty}^{\infty} |f(t)|^2 d\sigma(t)$$

exists, and metrize it by means of the metric generated by the scalar product

$$(f, g) = \int_{-\infty}^{\infty} f(t)\overline{g(t)}\, d\sigma(t).$$

This linear space is complete, so that it is a Hilbert space. It is denoted by L^2_σ. Special significance is possessed by *characteristic functions*. A characteristic function is equal to one in a certain finite interval of positive σ-length and is equal to zero outside of that interval. We do not exclude improper intervals here. The linear envelope of the set of all characteristic functions is dense in L^2_σ. Using this fact, it is easy to prove that L^2_σ is separable.

The case of a distribution function $\sigma(t)$ for which each integral

$$s_k = \int_{-\infty}^{\infty} t^k d\sigma(t) \qquad (k = 0, 1, 2, \ldots)$$

---

[15] For $a = b$, we get an *improper* interval, i.e., a point.

exists is of great interest. Because of mechanical considerations these integrals are called *moments* of the distribution function $\sigma(t)$. If $\sigma(t)$ has only a finite number of points of increase, then the Stieltjes integral becomes a finite sum. We do not consider this uninteresting case.

Orthogonalizing the sequence,

$$1, t, t^2, \ldots,$$

in L^2_σ, we get a sequence of polynomials $\{P_k(t)\}_{k=1}^{\infty}$ (where $P_k(t)$ is a polynomial of degree k), which satisfy the relations

$$\int_{\infty}^{\infty} P_k(t) P_m(t)\, d\sigma(t) = \begin{cases} 0 & (k \neq m), \\ 1 & (k = m). \end{cases}$$

This polynomial sequence is said to be orthonormal with respect to the distribution function $\sigma(t)$. If the function $\sigma(t)$ is absolutely continuous and if

$$\sigma'(t) = w(t)$$

then the orthonormality relations can be written in the form

$$\int_{-\infty}^{\infty} P_k(t) P_m(t)\, w(t)\, dt = \begin{cases} 0 & (k \neq m), \\ 1 & (k = m). \end{cases}$$

In this case, it is said that the polynomials $P_k(t)$ are orthonormal with respect to the *weight* $w(t)$. For instance, the Tchebysheff-Hermite polynomials are orthogonal (but not normalized) with respect to the weight e^{-t^2} $(-\infty < t < \infty)$.

If the interval of orthogonality is finite (i.e., if $\sigma(t)$ is constant for $t<a$ and for $t>b$), then the orthogonal polynomials $P_k(t)$ form a complete system. This is proved with the aid of the theorem of Weierstrass in exactly the same way as in the proof of the completeness in $L^2(a, b)$ of the sequence of Legendre polynomials.

If the interval of orthogonality is infinite, then the system of orthogonal polynomials $P_k(t)$ may fail to be complete. We restrict ourselves to one example, which is due to Hamburger.[16] Consider the interval $(0, \infty)$. The orthogonal polynomials on this interval with respect to the weight

$$w(t) = e^{-\frac{\pi\sqrt{t}}{\ln^2 t + \pi^2}}$$

is an incomplete system because the function

$$g(t) = e^{\frac{\ln t}{\ln^2 t + \pi^2}} \sin \frac{\sqrt{t} \ln t + \pi}{\ln^2 t + \pi^2}$$

[16] H. Hamburger [1].

satisfies the relations

$$\int_0^\infty t^k g(t) w(t)\, dt = 0 \qquad (k = 0, 1, 2, \ldots).$$

13. The Space of Almost Periodic Functions

We consider the set of all functions of the form $e^{i\lambda t}$ $(-\infty < t < \infty)$, where the parameter λ is real. We denote by L the linear envelope of this set, i.e., the collection of all "polynomials" of the form

$$\sum_k A_k e^{i\lambda_k t}.$$

Adding to L the limits of sequences of functions of L which are uniformly convergent on the entire real axis, we get a certain set B of continuous functions. As H. Bohr proved, a continuous function $f(t)$ defined on the real axis belongs to the collection B if and only if it is *almost periodic*, i.e., if for each $\varepsilon > 0$ there exists a real number $l = l(\varepsilon)$ such that in every interval of length l there is at least one number τ for which

$$|f(t+\tau) - f(t)| < \varepsilon \qquad (-\infty < t < \infty).$$

We can metrize the linear system L, by defining the scalar product of two polynomials

$$f(t) = \sum_{r=1}^{m} A_r e^{i\lambda_r t},$$

$$g(t) = \sum_{s=1}^{n} B_s e^{i\mu_s t}$$

as

$$(f, g) = \lim_{T\to\infty} \frac{1}{2T} \int_{-T}^{T} f(t)\overline{g(t)}\, dt =$$

$$= \lim_{T\to\infty} \sum_{r,s=1}^{m,n} A_r \bar{B}_s \frac{1}{2T} \int_{-T}^{T} e^{it(\lambda_r - \mu_s)}\, dt = \sum_{r,s=1}^{m,n} \delta(\lambda_r, \mu_s) A_r \bar{B}_s$$

where

$$\delta(\lambda, \mu) = \begin{cases} 0 & (\lambda \neq \mu), \\ 1 & (\lambda = \mu). \end{cases}$$

When L is closed by means of the metric generated by this scalar product, we get a certain complete Hilbert space B^2 which contains B as a linear manifold. The space B^2 is not separable. This follows from the fact that in B^2 there exists a continuum of orthonormal vectors $e^{i\lambda t}$ $(-\infty < \lambda < \infty)$, whereas (see paragraph 9) every orthonormal system in a separable space contains a finite or countable number of vectors.

Chapter II

LINEAR FUNCTIONALS AND BOUNDED LINEAR OPERATORS

14. Point Functions

Two kinds of point functions are considered in elementary treatments of 3- and n-dimensional spaces: *scalar functions*, the values of which are (real or complex) numbers, and *vector functions*, which relate the points of a space to other points of the same or another space. In the present book we shall study point functions in Hilbert space. In correspondence with the indicated division of the functions of elementary analysis into scalar functions and vector functions, we introduce in H so-called *functionals* and *operators*. The appropriate definitions follow.

Let D denote a subset of the space H. A function Φ which relates to each point $f \in \mathrm{D}$ a definite complex number $\Phi(f)$ is called a *functional* in the space H with *domain* D. A function T which relates to each element $f \in \mathrm{D}$ a particular element $Tf = g \in \mathrm{H}$ is called an operator[1] in the space H with *domain* D. The set Δ, consisting of all $g = Tf$, where f runs through D, is called the *range* of T. Sometimes we shall denote the domain of a functional Φ by D_Φ and, correspondingly, the domain and range of an operator T by D_T and Δ_T respectively.

The identity operator, i.e., the operator which maps each vector into itself, we shall denote by E. The operator which maps every vector into zero, we shall denote by O.

If the operator T maps each pair of different elements of D into a pair of different elements of Δ, then T has an inverse operator, which maps the elements of Δ into elements of D. The inverse operator is denoted by the symbol T^{-1}, and $T^{-1}g = f$ if and only if $f = Tg$. Moreover,

$$\mathrm{D}_{T^{-1}} = \Delta_T, \quad \Delta_{T^{-1}} = \mathrm{D}_T.$$

We shall consider two functionals (or operators) to be equivalent if their domains coincide and if for each element of their common domain, the values of these functionals (or operators) coincide.

---

[1] Sometimes it is necessary to consider also functions which map elements of the space H into elements of some other Hilbert space. These functions are also called operators. They will not be encountered often in this book.

If the domain D_T of the operator T contains the domain D_S of the operator S, i.e., if $D_S \subset D_T$, and if

$$Tf = Sf$$

for each $f \in D_S$, then T is called an *extension* of S and we write

$$S \subset T.$$

The concept of an extension of a functional is defined analogously.

Motivated by the notion of a continuous function, we make the following definition of the *continuity of an operator:* T is continuous at a point $f_0 \in D_T$ if

$$\lim_{f \to f_0} Tf = Tf_0 \qquad (f \in D_T).$$

An equivalent condition is that for each $\varepsilon > 0$, there exists $\delta = \delta(\varepsilon) > 0$ such that if f satisfies the inequality,

$$\|f - f_0\| < \delta, \quad f \in D_T,$$

then

$$\|Tf - Tf_0\| < \varepsilon.$$

The *continuity of a functional* is defined analogously.

If the element f_0 does not belong to D_T but $\lim Tf = g_0$ exists as $f \to f_0$ with $f \in D_T$, then the operator T can be defined for f_0, by letting $Tf_0 = g_0$. Proceeding in exactly the same way with all such elements f_0, we arrive at the so-called *extension by continuity of the operator T*. This extension is uniquely defined for each operator T. The *extension by continuity of a functional* is defined analogously.

We shall introduce below notation relating to so-called linear functionals and linear operators, which are the basic objects of our study.

Here we consider only the operator analogue of a "function of a function". Let S and T be two operators such that the range of T intersects the domain of S (i.e., let $\Delta_T \cap D_S \neq 0$). In this case we define the *product* ST of the operators S and T as the operator such that

$$STf = S(Tf)$$

for each element f in its domain, which is defined as the set of all $f \in D_T$ for which $Tf \in D_S$. The product TS is defined analogously whenever the set $\Delta_S \cap D_T$ is non-empty. It is clear that ST and TS are not generally equivalent because their domains may be different, and, moreover, even if g is an element belonging to both domains, it is possible that

$$STg \neq TSg.$$

Since it is difficult to give a reasonable completely general definition of the commutativity of two operators, we restrict ourselves here to the case in which at least one of the two operators S and T is defined everywhere in H. Let $D_S = H$. Then S and T are commutative if

$$ST \subset TS$$

i.e., if $f \in D_T$ implies both $Sf \in D_T$ and

$$STf = TSf.$$

In particular, if both operators are defined everywhere in H, then S and T are commutative if and only if $ST = TS$.

15. Linear Functionals

A functional Φ is said to be *linear* if:

(a) its domain D is a linear manifold and

$$\Phi(\alpha f + \beta g) = \alpha \Phi(f) + \beta \Phi(g)$$

for $f, g \in D$ and any complex numbers α and β;

(b) the inequality

$$\sup_{f \in D, \|f\| \leqq 1} |\Phi(f)| < \infty$$

is satisfied.

The left member of this inequality is called the *norm* of the functional Φ and is denoted by the symbol $\|\Phi\|_D$ or, if $D = H$, simply by $\|\Phi\|$.

If $f \in D$ and $f \neq 0$ then, by the definition of the norm of a functional,

$$\left| \Phi\left(\frac{f}{\|f\|}\right) \right| \leqq \|\Phi\|_D.$$

Hence, for $f \in D$,

$$|\Phi(f)| \leqq \|\Phi\|_D \cdot \|f\|. \tag{1}$$

Relation (1) shows that the linear functional Φ is continuous. In fact, by (1),

$$|\Phi(f) - \Phi(f_0)| = |\Phi(f - f_0)| \leqq \|\Phi\|_D \cdot \|f - f_0\|$$

for $f, f_0 \in D$.

From (1) it also follows that, if $f \in D$ and $\|f\| \leqq 1$, then

$$|\Phi(f)| \leqq \|\Phi\|_D,$$

with strict inequality if $\|f\| < 1$. Therefore, the norm $\|\Phi\|_D$ can be defined by

$$\sup_{f \in D, \|f\| = 1} |\Phi(f)| = \|\Phi\|_D, \tag{2}$$

or, equivalently, by[2]

$$\sup_{f \in D} \frac{|\Phi(f)|}{\|f\|} = \|\Phi\|_D. \tag{2'}$$

[2] Translator's Note: The condition $f \neq 0$ should be included in (2′). Similar conditions should be added in many other relations which occur below. This task is left to the reader.

If Φ and Ψ are two linear functionals with domains of definition D_Φ and D_Ψ, then $\alpha\,\Phi + \beta\,\Psi$, where α and β are constants, is also a linear functional, with the intersection $D_\Phi \cap D_\Psi$ of the domains D_Φ and D_Ψ as domain (of course, only the case when $D_\Phi \cap D_\Psi$ contains elements different from $f = 0$ offers much interest).

If the functional Φ satisfies the condition (a) listed above, which states that Φ is *homogeneous and additive*, and if Φ is continuous at any one point $f_0 \in D$, then Φ also satisfies condition (b) above, i.e., Φ is bounded and, therefore, Φ is a linear functional. In fact, if Φ is continuous at f_0 then for each $\delta > 0$ there exists $\varepsilon > 0$ such that

$$|\Phi(h) - \Phi(f_0)| < \delta$$

for $\|h - f_0\| \leqq \varepsilon$ and $h \in D$. For each $f \in D$ such that $f \neq 0$,

$$\Phi(f) = \frac{\|f\|}{\varepsilon}\,\Phi\left(\frac{\varepsilon f}{\|f\|}\right) = \frac{\|f\|}{\varepsilon}\left\{\Phi\left(\frac{\varepsilon f}{\|f\|} + f_0\right) - \Phi(f_0)\right\}.$$

Since the vector $\dfrac{\varepsilon f}{\|f\|} + f_0 = h$ satisfies the relation $\|h - f_0\| = \varepsilon$ we have, for $f \in D$,

$$|\Phi(f)| < \frac{\delta}{\varepsilon}\|f\|;$$

in other words, for $f \in D$ and $f \neq 0$,

$$\frac{|\Phi(f)|}{\|f\|} < \frac{\delta}{\varepsilon}.$$

This proves that Φ is bounded.

If the linear manifold D on which the linear functional Φ is defined is not closed, then it is possible to extend Φ by continuity to the closure of D. This extension, as is easy to see, leads to a unique linear functional with the same norm as the initial functional.

16. The Theorem of F. Riesz

The following theorem of F. Riesz provides a representation for each linear functional in H.

THEOREM: *Each linear functional* Φ *in the Hilbert space* H *can be expressed in the form*

$$\Phi(h) = (h, f),$$

where f *is an element of* H *which is uniquely determined by the functional* Φ; *furthermore,*

$$\|\Phi\| = \|f\|.$$

Proof: We denote by G the set of all elements $g \in H$ for which

$$\Phi(g) = 0.$$

By the linearity of the functional Φ, the set G is a linear manifold. Furthermore, G is closed, so that G is a subspace. In fact, if $g_n \in G$, $n \geqq 1$, and $g_n \to g$ then, by the continuity of Φ,

$$\Phi(g) = \lim_{n\to\infty} \Phi(g_n),$$

so that $\Phi(g)=0$ and $g \in G$. If $G=H$, then the functional Φ is equal to zero everywhere, and the theorem of Riesz is proved by taking $f=0$. We now suppose that $G \neq H$. Then there exists a nonzero element $f_0 \in H \ominus G$. We consider elements of the form

$$\Phi(h) f_0 - \Phi(f_0)\, h,$$

where h runs through H. These elements belong to G because

$$\Phi\,[\Phi(h) f_0 - \Phi(f_0)\, h] = \Phi(h)\,\Phi(f_0) - \Phi(f_0)\,\Phi(h) = 0.$$

Since $f_0 \in H \ominus G$,

$$(\Phi(h) f_0 - \Phi(f_0)\, h, f_0) = 0$$

and

$$\Phi(h) = \left(h, \frac{\overline{\Phi(f_0)}}{(f_0, f_0)} f_0\right).$$

If we set

$$f = \frac{\overline{\Phi(f_0)}}{(f_0, f_0)} f_0,$$

then it follows from the equality just obtained that

$$\Phi(h) = (h, f).$$

This is the required representation of the functional Φ.

We now prove that f is unique. Assuming the contrary, we have the equation

$$(h, f') = (h, f''),$$

for $h \in H$, where $f' \neq f''$. But this is impossible since the substitution of $h = f' - f''$ yields the contradiction,

$$\|f' - f''\|^2 = 0.$$

It remains to be proved that

$$\|\Phi\| = \|f\|.$$

It follows from the equation

$$\Phi(h) = (h, f)$$

that

$$\|\Phi(h)\| \leqq \|f\| \cdot \|h\|$$

which yields

$$\|\Phi\| \leqq \|f\|.$$

On the other hand, taking $h = f$, we get

$$\Phi(f) = \|f\|^2,$$

whence it follows that

$$\| \Phi \| \geqq \| f \|.$$

Thus the theorem of F. Riesz is proved.

We consider now a linear functional Ψ with domain D_Ψ closed in H. Then D_Ψ is a subspace of H and the theorem of F. Riesz asserts the existence of a unique element $g \in D_\Psi$ such that

$$\Psi(h) = (h, g) \qquad (h \in D_\Psi) \tag{3}$$

and

$$\| \Psi \|_{D_\Psi} = \| g \|.$$

By means of (3), the linear functional Ψ may be extended to the whole space H without increasing the norm.[3] Any other extension of the linear functional Ψ to the whole space H increases the norm of the functional. In fact, if Φ is any extension of Ψ to the whole space, then

$$\Phi(h) = (h, f)$$

and

$$\| \Phi \| = \| f \|.$$

For $h \in D_\Psi$,

$$(h, g) = (h, f)$$

so that $f - g \perp D_\Psi$. Because $g \in D_\Psi$,

$$\| f \|^2 = \| g \|^2 + \| f - g \|^2$$

which implies that

$$\| \Phi \| \geqq \| \Psi \|_{D_\Psi},$$

where there is strict inequality if $f \neq g$.

17. A Criterion for the Closure in H of a Given System of Vectors

According to the definition in Section 8, a system M of vectors is closed in H if it is possible to approximate each $h \in H$ to any degree of accuracy by means of a linear combination of vectors belonging to M.

THEOREM: *In order that the system* M *be closed in* H, *it is necessary and sufficient that a linear functional* Φ *in* H *which vanishes for all* $g \in M$, *be identically equal to zero.*

Proof: The necessity is an immediate consequence of the continuity of the linear functional. In order to prove the sufficiency, let us assume that the system is not closed. Then there exists $\delta > 0$ and a vector $h_0 \in H$ for which

$$\inf_{n,\, \alpha_i} \| h_0 - \alpha_1 g_1 - \alpha_2 g_2 - \dots - \alpha_n g_n \| = \delta > 0 \qquad (g_i \in M).$$

[3] Since any linear functional can be extended to the whole space without increasing the norm, one usually considers a linear functional as being defined on the whole space when the domain is not specified.

We denote by G the closed linear envelope of the system M. On the basis of Section 6, there exists $g \in G$ such that

$$\| h_0 - g \| = \delta.$$

Let

$$f = h_0 - g.$$

Then $f \perp G$. Consider the functional Φ defined by

$$\Phi(h) = (h, f),$$

the norm of which is equal to $\| f \| = \delta > 0$. This nonzero functional vanishes for each vector of G and, in particular, for each vector of M. Thus, the sufficiency is also proved.

18. A Lemma Concerning Convex Functionals[4]

Definition: *A real functional $p(h)$ in H is said to be convex if*

$$p(f+g) \leqq p(f) + p(g) \tag{1}$$

and

$$p(\alpha f) = |\alpha| p(f), \tag{2}$$

for $f, g \in H$ and any complex number α.

From this definition it follows that $p(0) = 0$ and $p(h) \geqq 0$.

LEMMA: *If a convex functional $p(h)$ is lower semicontinuous*, i.e., *if for each $h_0 \in H$ and each $\varepsilon > 0$ there exists $\delta > 0$ such that*

$$p(h) - p(h_0) > -\varepsilon$$

for $\|h - h_0\| < \delta$, then the convex functional is bounded, i.e., *there exists $M > 0$ such that*

$$p(h) \leqq M \| h \|$$

for $h \in H$.

Proof:[5] First, we prove that if the functional is not bounded in the unit sphere ($\| h \| < 1$), then it will not be bounded in the sphere $S(\rho, g)$ with center $g \in H$ and radius $\rho > 0$, where g and ρ are arbitrary. For, assuming that $p(h) < C$ for $\|h - g\| < \rho$, we find that

$$p(h-g) \leqq p(h) + p(-g) = p(h) + p(g) < 2C$$

for $\|h - g\| < \rho$. Consequently, if

$$f = \frac{h-g}{\rho},$$

[4] In this section we follow I. M. Gelfand [1].

[5] This proof does not require that H be a Hilbert space; it goes through if H is any Banach space.

then

$$p(f) = \frac{2C}{\rho}$$

for $f \in S(1, 0)$, so that the functional $p(h)$ is bounded in the unit sphere. In view of property (2) it is sufficient to prove that the functional $p(h)$ is bounded in the sphere $S(1, 0)$. We assume the contrary. Then $p(h)$ is unbounded in every sphere. We choose a point $f_1 \in S(1, 0)$ such that $p(f_1) > 1$. The lower semicontinuity of the functional $p(h)$ implies that there is a sphere $S(\rho_1, f_1) \subset S(1, 0)$ with radius $\rho_1 < \frac{1}{2}$ at all points of which $p(h) > 1$. Since $p(h)$ is unbounded in every sphere, there exists a point $f_2 \in S(\rho_1, f_1)$ and also a sphere $S(\rho_2, f_2) \subset S(\rho_1, f_1)$ with radius $\rho_2 < \frac{1}{2}\rho_1$ in which $p(h) > 2$. Continuing this process, we get an infinite sequence of spheres,

$$S(1, 0) \supset S(\rho_1, f_1) \supset S(\rho_2, f_2) \supset \ldots$$

for which $\rho_n < \frac{1}{2}\rho_{n-1}$, $(n = 1, 2, 3, \ldots; \rho_0 = 1)$, and also $p(h) > n$ if $h \in S(\rho_n, f_n)$. But the sequence of centers $\{f_n\}_1^\infty$ is fundamental and, therefore, converges to some element f. Then $p(f) > n$ for each n, which is impossible. Thus, the lemma is proved.

We remark that this lemma can also be formulated as follows: if a convex functional is lower semicontinuous, then it is continuous.

COROLLARY: *Let $p_k(h)$, $(k = 1, 2, 3, \ldots)$ be a sequence of convex continuous functionals in* H. *If this sequence is bounded at each point $h \in$ H, then the functional*

$$p(h) = \sup_n p_n(h)$$

is also convex and continuous.

Proof: That $p(h)$ is a convex functional is evident. On the other hand, for each $h_0 \in$ H and each $\varepsilon > 0$, there exists N such that

$$p(h_0) - p_N(h_0) < \frac{\varepsilon}{2}.$$

Then there exists $\delta > 0$ such that

$$|p_N(h) - p_N(h_0)| < \frac{\varepsilon}{2}$$

for $\|h - h_0\| < \delta$. But if $\|h - h_0\| < \delta$ then

$$p(h) - p(h_0) > \sup_n p_n(h) - p_N(h_0) - \frac{\varepsilon}{2} \geqq p_N(h) - p_N(h_0) - \frac{\varepsilon}{2} > -\varepsilon.$$

This implies that the functional $p(h)$ is lower semicontinuous. It remains only to apply the lemma, and the corollary is proved.

We give two simple applications of the propositions just proved. We know that each linear functional in $L^2(a, b)$ can be expressed in the form

$$\Phi(h) = \int_a^b h(t)\,\varphi(t)\,dt \tag{1}$$

where $\varphi(t)$ is the function in $L^2(a, b)$ which "represents" the functional $\Phi(h)$. We shall prove that if a functional $\Phi(h)$ is defined everywhere in $L^2(a, b)$ by means of formula (1), where $\varphi(t)$ is some fixed function, then this functional is necessarily linear, so that $\varphi(t)$ belongs to $L^2(a, b)$. In other words we shall prove that if the integral (1) exists for each function $h(t) \in L^2(a, b)$, then $\varphi(t) \in L^2(a, b)$. This fact is a special case of a more general theorem of F. Riesz.[6]

For the proof we denote by e_n the set of all points t which belong to the intersection of the intervals $[a, b]$, $[-n, n]$ and for which

$$|\varphi(t)| \leqq n.$$

Further let

$$p_n(h) = \int_{e_n} |h(t)\varphi(t)|\,dt.$$

This is a convex continuous functional in $L^2(a, b)$. The quantity

$$p(h) = \sup_n p_n(h) = \lim_{n\to\infty} p_n(h) = \int_a^b |h(t)\varphi(t)|\,dt$$

is finite for any $h(t) \in L^2(a, b)$. So, by the corollary of the lemma, the functional $p(h)$ is continuous, i.e., $p(h) \leqq M \| h \|$ for $h \in \mathrm{H}$. But $|\Phi(h)| \leqq p(h)$ so that, since the homogeneity and additivity of the functional $\Phi(h)$ are evident, $\Phi(h)$ is a linear functional.

An analogous proposition is valid for the space l^2. We restrict ourselves to its formulation. Let a functional $\Phi(f)$ be defined everywhere in l^2 by means of the formula

$$\Phi(f) = \sum_{k=1}^{\infty} a_k x_k \qquad (f = \{x_k\}_1^\infty)$$

where $\{a_k\}_1^\infty$ is some fixed sequence. Then

$$\sum_{k=1}^{\infty} |a_k|^2 < \infty, \tag{2}$$

[6] Riesz's theorem pertains to the space $L^p(a, b)$ for any $p > 1$. (The space $L^p(a, b)$ is defined as the space of functions measurable in (a, b) for which $\int_a^b |f(x)|^p dx$ exists). See F. Riesz [1].

which implies that $\Phi(f)$ is a linear functional. In other words, if the series

$$\sum_{k=1}^{\infty} a_k x_k$$

converges for each sequence $\{x_k\}_1^\infty$ such that

$$\sum_{k=1}^{\infty} |x_k|^2 < \infty$$

then the inequality (2) must hold. This fact is a special case of a more general theorem of E. Landau.[7]

19. Bounded Linear Operators

An operator T is *linear* if its domain of definition D is a linear manifold and if

$$T(\alpha f + \beta g) = \alpha T f + \beta T g$$

for any $f, g \in$ D and any complex numbers α and β.

We emphasize the fact that, in contrast with the definition of a linear functional, this definition does not require that the operator be bounded. This is related to the fact that many important operations of analysis such as, for instance, the operation of differentiation, generate unbounded but homogeneous and additive operators, i.e., operators which are linear in the sense of the definition given here.

A linear operator T is *bounded* if

$$\sup_{f \in \mathrm{D}, \|f\| \leq 1} \|Tf\| < \infty.$$

The left member of this inequality is called the *norm* of the operator T in D and is denoted by the symbol $\|T\|$ or, sometimes, by $\|T\|_{\mathrm{D}}$.

It is easy to see that the properties of Section 15 relating to linear functionals are also valid for bounded linear operators:

1. The norm of a bounded linear operator T can be defined equivalently by

$$\|T\| = \sup_{f \in \mathrm{D}, \|f\| = 1} \|Tf\| = \sup_{f \in \mathrm{D}} \frac{\|Tf\|}{\|f\|}.$$

2. A bounded linear operator is continuous.
3. If a linear operator is continuous at one point, then it is bounded.
4. The extension by continuity of a bounded linear operator T leads to a unique linear operator with the same norm as the original operator.

[7] Landau's theorem pertains to the space l^p for any $p > 1$. (The space l^p is the space of numerical sequences $x_1, x_2, \ldots$ for which the series $\sum_{i=1}^{\infty} |x_i|^p$ converges.) See E. Landau [1].

5. If S and T are linear operators, then $\alpha S + \beta T$, where α and β are complex numbers, is a linear operator with the intersection $D_S \cap D_T$ of the domains D_S and D_T as the domain of definition. Each of the products ST and TS (cf. Section 14) is also a linear operator. If S and T are bounded linear operators defined everywhere in H, then the operators ST and TS are also bounded linear operators defined everywhere in H, and

$$\|ST\| \leqq \|S\| \cdot \|T\|, \qquad \|TS\| \leqq \|T\| \cdot \|S\|.$$

20. Bilinear Functionals

We shall say that Ω is a bilinear functional defined in H,[8] if to each pair of elements $f, g \in H$ there corresponds a definite complex number $\Omega(f,g)$, and

(a) $\Omega(\alpha_1 f_1 + \alpha_2 f_2, g) = \alpha_1 \Omega(f_1, g) + \alpha_2 \Omega(f_2, g)$,

(b) $\Omega(f, \beta_1 g_1 + \beta_2 g_2) = \bar{\beta}_1 \Omega(f, g_1) + \bar{\beta}_2 \Omega(f, g_2)$,

(c) $\sup\limits_{\|f\| \leqq 1, \|g\| \leqq 1} |\Omega(f,g)| < \infty$.

An example of a bilinear functional is the scalar product (f, g). The number

$$\sup_{\|f\| \leqq 1, \|g\| \leqq 1} |\Omega(f,g)|$$

is called the norm of the bilinear functional Ω, and is denoted by the symbol $\|\Omega\|$. It is not difficult to prove that

$$\|\Omega\| = \sup_{\|f\| = 1, \|g\| = 1} |\Omega(f,g)| = \sup \frac{|\Omega(f,g)|}{\|f\| \cdot \|g\|}.$$

Therefore, for any $f, g \in H$,

$$|\Omega(f,g)| \leqq \|\Omega\| \cdot \|f\| \cdot \|g\|.$$

A bilinear functional is a continuous function of each of its arguments, since

$$|\Omega(f,g) - \Omega(f_0, g_0)| = |\Omega(f - f_0, g - g_0) + \Omega(f - f_0, g_0) + \Omega(f_0, g - g_0)| \leqq$$
$$\leqq \|\Omega\| \{\|f - f_0\| \cdot \|g - g_0\| + \|f - f_0\| \cdot \|g_0\| + \|f_0\| \cdot \|g - g_0\|\}.$$

The following simple proposition is often useful.

THEOREM: *If a complex scalar function $\omega(f, g)$ satisfies the conditions*

(a) $\omega(\alpha_1 f_1 + \alpha_2 f_2, g) = \alpha_1 \omega(f_1, g) + \alpha_2 \omega(f_2, g)$,

(b) $\omega(f, \beta_1 g_1 + \beta_2 g_2) = \bar{\beta}_1 \omega(f, g_1) + \bar{\beta}_2 \omega(f, g_2)$,

(c) $|\omega(f,f)| \leqq C\|f\|^2$,

(d) $|\omega(f,g)| = |\omega(g,f)|$,

[8] It is possible to introduce bilinear functionals which are not defined everywhere in H, but in what follows they will not be considered.

where C is a constant, f, f_1, f_2, g, g_1, g_2 are arbitrary elements of H *and $\alpha_1, \alpha_2, \beta_1, \beta_2$ are arbitrary complex numbers, then ω is a bilinear functional with norm $\|\omega\| \leqq C$.*

Proof: It is immediately proved by means of (a) and (b) that[9]

$$\omega(f,h) + \omega(h,f) = \tfrac{1}{2}\{\omega(f+h,f+h) - \omega(f-h,f-h)\}.$$

This implies that

$$(1) \quad |\omega(f,h)+\omega(h,f)| \leqq \tfrac{1}{2}C\{\|f+h\|^2+\|f-h\|^2\} = C\{\|f\|^2+\|h\|^2\}.$$

Let $\|f\| \leqq 1$, $\|h\| \leqq 1$ and $h = \lambda g$ where λ is a complex number such that $|\lambda| = 1$ (λ will be specified later). Then (1) yields

$$(2) \qquad |\bar{\lambda}\omega(f,g) + \lambda\omega(g,f)| \leqq 2C.$$

We suppose that $\omega(f,g) \neq 0$ and, in accordance with (d), let

$$\omega(f,g) = |\omega(f,g)|e^{i\alpha}, \qquad \omega(g,f) = |\omega(f,g)|e^{i\beta}.$$

Then, by (2),

$$|\omega(f,g)| \cdot |\bar{\lambda}e^{i\alpha} + \lambda e^{i\beta}| \leqq 2C.$$

Letting

$$\lambda = e^{i\frac{\alpha-\beta}{2}},$$

we find that

$$\bar{\lambda}e^{i\alpha} + \lambda e^{i\beta} = e^{i\frac{\alpha+\beta}{2}} + e^{i\frac{\alpha+\beta}{2}} = 2e^{i\frac{\alpha+\beta}{2}}$$

which yields

$$|\omega(f,g)| \leqq C \qquad (\|f\| \leqq 1, \|g\| \leqq 1).$$

This proves the theorem, since this relation is also correct for $\omega(f,g) = 0$.

COROLLARY: *If the bilinear functional Ω satisfies the condition*

$$|\Omega(f,g)| = |\Omega(g,f)|,$$

for $f, g \in$ H, then

$$\|\Omega\| = \sup_{f \in H} \frac{|\Omega(f,f)|}{(f,f)}.$$

Proof: By the theorem,

$$\|\Omega\| \leqq \sup_{f \in H} \frac{|\Omega(f,f)|}{(f,f)},$$

but on the other hand

$$\sup_{f \in H} \frac{|\Omega(f,f)|}{(f,f)} \leqq \sup_{f,g \in H} \frac{|\Omega(f,g)|}{\|f\| \cdot \|g\|} = \|\Omega\|.$$

[9] From this equation and the analogous equation

$$\omega(f,h) - \omega(h,f) = \frac{i}{2}\{\omega(f+ih,f+ih) - \omega(f-ih,f-ih)\}$$

it follows by means of (c) that ω is a bilinear functional with norm $\leqq 2C$. But, by means of condition (d), it is established that the norm of ω does not exceed C.

21. The General Form of a Bilinear Functional

THEOREM: *Each bilinear functional $\Omega(f, g)$ has a representation of the form*

$$\Omega(f,g) = (Af,g).$$

In this equation A is a bounded linear operator with domain H *which is uniquely determined by Ω. Furthermore,*

$$\|A\| = \|\Omega\|.$$

Proof: For fixed f, the expression $\overline{\Omega(f, g)}$ defines a linear functional in g with domain H. Consequently, according to the theorem of F. Riesz (cf. Section 16), there exists an element h_f, uniquely determined by the element f, for which

$$\overline{\Omega(f,g)} = (g, h_f)$$

or

$$\Omega(f,g) = (h_f, g)$$

for each $g \in$ H. Define the mapping A from H into H by the equation $Af = h_f$ for $f \in$ H. Then

$$\Omega(f,g) = (Af,g).$$

Since

$$\Omega(a_1 f_1 + a_2 f_2, g) = a_1 \Omega(f_1, g) + a_2 \Omega(f_2, g)$$

we have

$$(A\{a_1 f_1 + a_2 f_2\} - a_1 Af_1 - a_2 Af_2, g) = 0$$

for $g \in$ H. Since g is arbitrary,

$$A(a_1 f_1 + a_2 f_2) = a_1 Af_1 + a_2 Af_2,$$

so that A is a linear operator. The domain of the operator A is the whole space H. Furthermore, since

$$|(Af,g)| \leqq \|Af\| \cdot \|g\|$$

we have

$$\|\Omega\| = \sup \frac{|\Omega(f,g)|}{\|f\| \cdot \|g\|} = \sup \frac{|(Af,g)|}{\|f\| \cdot \|g\|} \leqq \sup \frac{\|Af\|}{\|f\|}.$$

On the other hand

$$\|\Omega\| = \sup \frac{|(Af,g)|}{\|f\| \cdot \|g\|} \geqq \sup \frac{(Af, Af)}{\|f\| \cdot \|Af\|} = \sup \frac{\|Af\|}{\|f\|}.$$

These relations show that the operator A is bounded and that

$$\|\Omega\| = \|A\|.$$

The operator A is uniquely determined by the linear functional Ω. In fact, if

$$\Omega(f,g) = (A'f,g) = (A''f,g),$$

for $f, g \in H$, then

$$(A'f - A''f, g) = 0.$$

But this is possible only for $A' = A''$.

22. Adjoint Operators

Let A be an arbitrary bounded linear operator defined on H. The expression

$$(f, Ag)$$

defines a bilinear functional on H with norm $\| A \|$. According to the theorem proved in the preceding section, there exists a unique bounded linear operator A^* defined on H with norm $\| A^* \| = \| A \|$ such that

$$(f, Ag) = (A^*f, g) \tag{1}$$

for $f, g \in H$. This operator A^* is called the *adjoint* of A. It is easy to see that the operator $(A^*)^* = A^{**}$ is equivalent to the original operator A.

If A is bounded and $A^* = A$, then A is said to be *self-adjoint*. A bounded linear operator A, defined on H, is said to be *normal* if it commutes with its adjoint, i.e., if

$$A^*A = AA^*.$$

Let A and B be two bounded linear operators defined on H. Then

$$(ABf, g) = (Bf, A^*g) = (f, B^*A^*g),$$

which implies that

$$(AB)^* = B^*A^*.$$

Therefore, the product of two self-adjoint operators is self-adjoint if and only if the operators commute.

THEOREM: *If A is a bounded self-adjoint operator, then*

$$\sup_{\|f\| = \|g\| = 1} |(Af, g)| = \sup_{\|f\| = 1} |(Af, f)|.$$

In other words,[10]

$$\|A\| = \max\{|\Lambda|, |\lambda|\}$$

where

$$\Lambda = \sup_{\|f\|=1} (Af, f), \quad \lambda = \inf_{\|f\|=1} (Af, f).$$

Proof: The bilinear functional

$$\Omega(f, g) = (Af, g)$$

satisfies the condition

$$|\Omega(f, g)| = |\Omega(g, f)|.$$

Therefore, the corollary of the theorem of Section 20 applies and the theorem is proved.

[10] Translator's Note: It follows from (1) that (Af, f) is real for $f \in H$ if A is self-adjoint.

23. Weak Convergence in H

We say that the sequence of vectors $f_k \in \mathrm{H}$, $(k = 1, 2, 3 \ldots)$ *converges weakly* to the vector f and we write $f_k \overset{w}{\to} f$ if

$$\lim_{k\to\infty} (f_k, h) = (f, h)$$

for $h \in \mathrm{H}$. The concepts of a *weakly fundamental sequence* and of *weak completeness* are defined analogously.

If the sequence $\{f_k\}_1^\infty$ converges to f in the sense of Section 3, i.e., if

$$\lim_{k\to\infty} \|f_k - f\| = 0$$

then we shall continue to write $f_k \to f$, but we shall say, to avoid confusion, that the sequence converges *strongly* to f. Strong convergence implies weak convergence, but not conversely. Indeed, let $\{e_k\}_1^\infty$ be any infinite orthonormal sequence of vectors in H. Since, for any $h \in \mathrm{H}$,

$$\sum_{k=1}^{\infty} |(h, e_k)|^2 \leqq (h, h)$$

(see Section 8), then for any $h \in \mathrm{H}$,

$$\lim_{k\to\infty} (e_k, h) = 0.$$

Thus, the sequence $\{e_k\}_1^\infty$ converges weakly to the vector 0, but this sequence does not converge strongly since

$$\| e_k - e_i \|^2 = 2 \qquad (i \neq k)$$

so that $\|e_k - e_i\|$ does not converge to zero as $i, k \to \infty$. However, the following theorem is valid.

THEOREM 1: *If the sequence of vectors $\{f_k\}_1^\infty$ converges weakly to the vector f and if*

$$\lim_{k\to\infty} \|f_k\| = \|f\|,$$

then

$$\lim_{k\to\infty} \|f_k - f\| = 0,$$

i.e., the sequence $\{f_k\}_1^\infty$ converges strongly to the vector f.

Proof: The proof follows from the equation

$$\|f_k - f\|^2 = \|f_k\|^2 - (f_k, f) - (f, f_k) + \|f\|^2.$$

Indeed, by the hypothesis of the theorem,

$$\lim_{k\to\infty} \{\|f_k\|^2 - (f_k, f) - (f, f_k) + \|f\|^2\} = 0.$$

An important property of every weakly convergent sequence of vectors is boundedness. The proof of this property does not present any difficulty if the following general proposition is proved first.

THEOREM 2: *If the linear functionals $\Phi_1, \Phi_2, \Phi_3, \ldots$, defined on the space* H, *have the property that the numerical sequence $\{\Phi_k(h)\}_1^\infty$ is bounded for each $h \in$* H, *then the sequence $\{\|\Phi_k\|\}_1^\infty$ of the norms of the functionals is bounded.*

Proof: The proof follows almost immediately from the lemma concerning convex functionals proved in Section 18. For $h \in$ H, let

$$p_n(h) = |\Phi_n(h)| \qquad (n = 1, 2, 3, \ldots).$$

The p_n are convex continuous functionals in H. By the lemma just mentioned, the convex functional

$$p(h) = \sup_n p_n(h)$$

is continuous, i.e.,

$$\mathrm{M} = \sup_{\|h\| \leqq 1} p(h) < \infty.$$

Consequently,

$$\|\Phi_n\| \leqq \mathrm{M} \qquad (n = 1, 2, 3, \ldots)$$

and the theorem is proved.

COROLLARY 1: *Every weakly convergent sequence $\{f_k\}_1^\infty$ is bounded.*

Proof: Each vector f_k determines a functional $\Phi_k(h) = (h, f_k)$. Since the sequence $\{f_k\}_1^\infty$ converges weakly, the numerical sequence $\{\Phi_k(h)\}_1^\infty$ converges for each $h \in$ H and, hence, is bounded. It remains only to apply Theorem 2 and to use the fact that

$$\|\Phi_k\| = \|f_k\|.$$

COROLLARY 2: *Every Hilbert space is weakly complete.*

Proof: Let the sequence of vectors $\{f_k\}_1^\infty$ be fundamental in the sense of weak convergence, i.e., for each $h \in$ H, let

$$\lim_{m, n \to \infty} (f_n - f_m, h) = 0.$$

It follows that the sequence of numbers (f_k, h) $(k = 1, 2, 3, \ldots)$ converges for each fixed $h \in$ H. According to Theorem 2 the sequence $\{f_k\}_1^\infty$ is bounded:

$$\|f_k\| \leqq \mathrm{M} \qquad (k = 1, 2, 3, \ldots).$$

Therefore, the limit

$$\lim_{k \to \infty} (h, f_k)$$

defines a linear functional $\Phi(h)$ with norm $\leqq$ M. According to the representation theorem of F. Riesz, $\Phi(h) = (h, f)$, where f is a unique element of the space H. This element is the weak limit of the sequence $\{f_k\}_1^\infty$.

24. Weak Compactness

A point set is said to be *compact*[11] if every sequence belonging to it contains a convergent subsequence. Corresponding to the two types of convergence (strong and weak) are *strong* (or *ordinary*) *compactness* and *weak compactness*. The concept of compactness is associated with the most important theorem of elementary analysis — the Bolzano-Weierstrass theorem. The conclusion of this theorem is false for a Hilbert space if the theorem is stated in terms of strong convergence. To prove this, it is sufficient to take the infinite orthonormal sequence of vectors $e_1, e_2, e_3, \ldots$. This sequence is bounded, but none of its subsequences is strongly convergent. In connection with what has been said, it may be surprising that the following theorem holds.

THEOREM 1: *Every bounded point set in* H *is weakly compact.*

Proof: Let us take any sequence $\{g_k\}_1^\infty$ of points such that, for some C,

$$\|g_k\| \leqq C < \infty \qquad (k = 1, 2, 3, \ldots).$$

Let L denote the linear envelope of the set $\{g_k\}_1^\infty$, and let $G = \bar{L}$ be its closure. Define F by

$$F = H \ominus G.$$

Consider the numerical sequence

$$(1) \qquad (g_1, g_k), \qquad (k = 1, 2, 3, \ldots).$$

It is bounded because

$$|(g_1, g_k)| \leqq \|g_1\| \cdot \|g_k\| \leqq C^2 \qquad (k = 1, 2, 3, \ldots).$$

Therefore, the sequence (1) contains a convergent subsequence. In other words, $\{g_k\}_1^\infty$ contains a subsequence $\{g_{1k}\}_{k=1}^\infty$ for which

$$\lim_{k\to\infty} (g_1, g_{1k})$$

exists. Similarily, from the boundedness of the numerical sequence

$$(2) \qquad (g_2, g_{1k})$$

we conclude that $\{g_{1k}\}_{k=1}^\infty$ contains a subsequence $\{g_{2k}\}_{k=1}^\infty$ for which

$$\lim_{k\to\infty} (g_2, g_{2k})$$

exists. Repeating this argument, we get an infinite sequence of sequences

$$\begin{array}{l} g_{11}, g_{12}, g_{13}, \ldots, \\ g_{21}, g_{22}, g_{23}, \ldots, \\ g_{31}, g_{32}, g_{33}, \ldots, \\ \ldots \;\; \ldots \;\; \ldots \;\; \ldots \end{array}$$

each of which is a subsequence of the preceding. It is evident that the diagonal sequence

[11] Translator's Note: This general concept is often called *sequential compactness.*

$$g_{11}, g_{22}, g_{33}, \ldots$$

has the property that, for each integer r

$$\lim_{k\to\infty} (g_r, g_{kk})$$

exists. Hence, it follows that

$$\lim_{k\to\infty} (g, g_{kk})$$

exists for each $g \in \mathrm{L}$ and, therefore, for each $g \in \mathrm{G}$. If $f \in \mathrm{F}$, then

$$(f, g_{kk}) = 0, \qquad (k = 1, 2, 3, \ldots).$$

Consequently,

$$\lim_{k\to\infty} (f, g_{kk})$$

exists for each $f \in \mathrm{F}$. Since $\mathrm{H} = \mathrm{F} \oplus \mathrm{G}$, the results we have obtained imply the existence, for each $h \in \mathrm{H}$, of

$$\lim_{k\to\infty} (h, g_{kk}).$$

Therefore, the sequence $\{g_{kk}\}_{k=1}^{\infty}$ is fundamental in the sense of weak convergence. By the weak completeness of the space, this sequence converges weakly to some element of H, and this proves our theorem.

THEOREM 2: *For the weak convergence of the sequence of vectors* $\{g_k\}_1^{\infty}$ *it is necessary and sufficient that:*

1. *the numerical sequence*

$$(g_k, f) \qquad (k = 1, 2, 3, \ldots)$$

converge for each f *of some set* M *which is dense in* H; *and*

2. *the sequence* $\{g_k\}_{k=1}^{\infty}$ *be bounded, i.e, the inequality*

$$\|g_k\| \leqq C < \infty \qquad (k = 1, 2, 3, \ldots)$$

hold for some C.

Proof: The necessity of condition 1 is evident. The necessity of condition 2 is indicated by Corollary 1 of Theorem 2 of Section 23. We turn to the proof of the sufficiency of the conditions mentioned. By Theorem 1 of the preceding paragraph, $\{g_k\}_{k=1}^{\infty}$ has a weakly convergent subsequence $\{g_{k_i}\}_{i=1}^{\infty}$. Let g be the weak limit of this subsequence.
Then

$$\lim_{i\to\infty} (h, g_{ki}) = (h, g).$$

According to condition 1 of the theorem,

$$\lim_{k\to\infty} (f, g_k)$$

exists for each $f \in \mathrm{M}$. Therefore,

$$\lim_{k\to\infty} (f, g_k) = (f, g)$$

for $f \in \mathrm{M}$, and it remains to prove (we leave this to the reader) that this equation holds if f is any element of H.

25. A Criterion for the Boundedness of an Operator

THEOREM: *Let A and A^* be linear operators defined on* H *and assume that*

$$(Af, g) = (f, A^*g)$$

for $f, g \in$ H. Then A is bounded, and A^* is the adjoint of A.

Proof: We assume the contrary and suppose that there exists a sequence of vectors $\{f_k\}_1^\infty$ such that

$$\|f_k\| = 1, \quad \|Af_k\| > k \qquad (k = 1, 2, 3, \ldots).$$

The expressions

$$(g, Af_k) = \Phi_k(g) \qquad (k = 1, 2, 3, \ldots)$$

define linear functionals Φ_k in H. Since

$$\Phi_k(g) = (A^*g, f_k) \qquad (k = 1, 2, 3, \ldots),$$

the numerical sequence $\{\Phi_k(g)\}_{k=1}^\infty$ is bounded. By Theorem 2 of Section 23, the sequence of norms $\| \Phi_k \|$ $(k = 1, 2, 3, \ldots)$, i.e., the sequence of numbers $\|Af_k\|$, is also bounded, which is a contradiction. Thus, the theorem is proved.

An important special case of this theorem is due to Hellinger and Toeplitz. We mention it in the following section.

26. Linear Operators in a Separable Space

In this section we shall consider linear operators defined everywhere on a separable Hilbert space H. We show that bounded operators admit matrix representations completely analogous to the well known matrix representations of operators on finite dimensional spaces.

We choose any orthonormal basis $\{e_k\}_1^\infty$ in H and let[12]

$$Ae_k = c_k, \qquad (k = 1, 2, 3, \ldots)$$

and

$$(1) \qquad (Ae_k, e_i) = a_{ik} \qquad (i, k = 1, 2, 3, \ldots).$$

Thus

$$c_k = \sum_{i=1}^{\infty} a_{ik} e_i \qquad (k = 1, 2, 3, \ldots).$$

Moreover,

$$\sum_{i=1}^{\infty} | a_{ik} |^2 < \infty \qquad (k = 1, 2, 3, \ldots).$$

[12] We remark that if the operator A is not defined everywhere in H, but only on a set D which is dense in H, then there exists in H an orthonormal basis $\{e_k\}_1^\infty$, the elements of which belong to D.

We introduce the infinite matrix

$$\begin{pmatrix} a_{11} & a_{12} & a_{13} & \dots \\ a_{21} & a_{22} & a_{23} & \dots \\ a_{31} & a_{32} & a_{33} & \dots \\ \dots & \dots & \dots & \dots \end{pmatrix}$$

of which the elements of the kth column are the components of the vector into which the operator A maps the kth coordinate vector. If the operator A is bounded, then it is uniquely determined by the infinite matrix (a_{ik}). For the proof of this assertion it is necessary to show how to represent the operator in terms of the matrix and the orthonormal basis $\{e_k\}_1^\infty$. First, we have

$$Ae_k = \sum_{i=1}^{\infty} a_{ik} e_i \qquad (k = 1, 2, 3, \dots).$$

Since the operator A is linear, it is well defined on the linear envelope of the given basis, i.e., for all vectors each of which has only a finite number of nonzero components relative to the basis. Since A is continuous, the value of Af for an arbitrary vector $f \in \mathrm{H}$ may be found by means of a passage to a limit.

It is not difficult to write a simple formula for the components of the vector f; indeed, if

$$f = \sum_{k=1}^{\infty} x_k e_k \tag{2}$$

then

$$Af = \sum_{k=1}^{\infty} y_k e_k, \tag{3}$$

where

$$y_k = \sum_{i=1}^{\infty} a_{ki} x_i. \tag{4}$$

In fact, if

$$f_n = \sum_{k=1}^{n} x_k e_k,$$

then

$$Af_n = \sum_{k=1}^{\infty} y_k^{(n)} e_k$$

where

$$y_k^{(n)} = \sum_{i=1}^{n} a_{ki} x_i.$$

By the boundedness of the operator A,

$$y_k = (Af, e_k) = \lim_{n\to\infty} (Af_n, e_k) = \lim_{n\to\infty} y_k^{(n)} = \lim_{n\to\infty} \sum_{i=1}^{n} a_{ki} x_i = \sum_{i=1}^{\infty} a_{ki} x_i.$$

DEFINITION: *If the operator A is defined everywhere in* H *and if its value for any vector* (2) *is given by the formulas* (3) *and* (4), *then we say that the operator A admits a matrix representation relative to the orthogonal basis* $\{e_k\}_1^\infty$.

Thus, we have proved that every bounded linear operator defined on the entire space admits a matrix representation with respect to each orthogonal basis. This is the analogue, mentioned at the very beginning of the present section, between a separable Hilbert space and a finite-dimensional space, with respect to bounded linear operators.

THEOREM: *If an operator A, defined everywhere in a separable space* H, *admits a matrix representation with respect to some orthogonal basis, then it is bounded.*

(This proposition is a frequently used special case of the theorem of the preceding section, mentioned above, which is due to Hellinger and Toeplitz).[13]

Proof: By hypothesis, the series

$$(Af, e_k) = \sum_{i=1}^{\infty} a_{ki} x_i \qquad (k = 1, 2, 3, \ldots)$$

converges for each vector

$$f = \sum_{k=1}^{\infty} x_k e_k,$$

where $\{e_k\}_1^\infty$ is the orthonormal basis, mentioned in the theorem, with respect to which the operator A admits a matrix representation. Therefore, by the theorem of Landau (see Section 18),

$$\sum_{i=1}^{\infty} |a_{ki}|^2 < \infty \qquad (k = 1, 2, 3, \ldots). \tag{5}$$

We introduce the sequence of vectors

$$c_k^* = \sum_{i=1}^{\infty} \bar{a}_{ki} e_i \qquad (k = 1, 2, 3, \ldots)$$

and by means of them, define the linear operator A^*. First, let

$$A^* e_k = c_k^* \qquad (k = 1, 2, 3, \ldots)$$

and then use linearity to define A^* on the linear envelope of the set of vectors e_k. Finally, extend A^* by continuity to all of H. It is easy to prove that for any $f, g \in$ H,

$$(Af, g) = (f, A^* g)$$

after which, to complete the proof, it remains to apply the theorem of the preceding section.

[13] E. Hellinger and O. Toeplitz [1].

We shall not present all the details of the proof just outlined, but we mention another proof of the theorem, which is based directly on the lemma concerning convex functionals in Section 18. In view of inequality (5), the expression

$$\Phi_k(f) = \sum_{i=1}^{\infty} a_{ki} x_i \qquad (k = 1, 2, 3, \ldots)$$

defines a linear functional of

$$f = \sum_{k=1}^{\infty} x_k e_k.$$

Therefore,

$$p_n(f) = \sqrt{\sum_{k=1}^{n} |\Phi_k(f)|^2} \qquad (n = 1, 2, 3, \ldots)$$

defines a convex continuous functional of f. Since

$$\sum_{k=1}^{n} |\Phi_k(f)|^2 = \sum_{k=1}^{n} |(Af, e_k)|^2 \leqq \|Af\|^2$$

the sequence $\{p_n(f)\}_{n=1}^{\infty}$ is bounded for each $f \in \mathrm{H}$. On the basis of the corollary of the lemma concerning convex functionals, the functional

$$p(f) = \sup_n p_n(f) = \lim_{n\to\infty} p_n(f) = \sqrt{\sum_{k=1}^{\infty} |\Phi_k(f)|^2} = \|Af\|$$

is continuous, i.e., there exists a constant M such that

$$p(f) \leqq M \|f\|.$$

But this implies that the operator A is bounded.

The proof of the theorem can be formulated also in the following form: if for arbitrary numbers $x_k (k=1, 2, 3, \ldots)$ such that

$$\sum_{k=1}^{\infty} |x_k|^2 < \infty$$

the inequality

$$\sum_{k=1}^{\infty} \left| \sum_{i=1}^{\infty} a_{ki} x_i \right|^2 < \infty$$

holds, then there exists a constant M such that

$$\sum_{k=1}^{\infty} \left| \sum_{i=1}^{\infty} a_{ki} x_i \right|^2 \leqq M^2 \sum_{k=1}^{\infty} |x_k|^2.$$

This reduces to the theorem of E. Landau (see Section 18) if $a_{ki} = 0$ for $k > 1$.

Let us agree to write

$$A \sim (a_{ik})$$

if the bounded linear operator A, defined everywhere in H, corresponds to the matrix (a_{ik}) according to (1). Here the orthogonal basis $\{e_k\}_1^\infty$ is arbitrary but fixed.

If

$$A \sim (a_{ik}), \qquad B \sim (b_{ik})$$

then, as is easily verified,

$$AB \sim (c_{ik})$$

where

$$c_{ik} = \sum_{r=1}^{\infty} a_{ir} b_{rk} \qquad (i, k = 1, 2, 3, \ldots).$$

If we define matrix multiplication by means of this equation, then

$$AB \sim (a_{ir}) \cdot (b_{rk}).$$

Furthermore, if

$$A \sim (a_{ik})$$

and

$$A^* \sim (a^*_{ik})$$

then

$$a^*_{ik} = \bar{a}_{ki} \qquad (i, k = 1, 2, 3, \ldots).$$

Therefore, the condition that the bounded operator A be self-adjoint may be expressed in the form

$$a_{ik} = \bar{a}_{ki}. \tag{6}$$

Matrices for which equation (6) holds are called *symmetric* or *Hermitean.*

A bilinear functional is generated by the operator A by means of the equation

$$(Af, g) = \sum_{k=1}^{\infty} \left(\sum_{i=1}^{\infty} a_{ki} x_i \right) \bar{y}_k.$$

In this equation

$$f = \sum_{k=1}^{\infty} x_k e_k, \qquad g = \sum_{k=1}^{\infty} y_k e_k.$$

In the double sum appearing above it is possible to reverse the order of summation, since the equation

$$(Af, g) = (f, A^* g)$$

implies that

$$\sum_{k=1}^{\infty} \left(\sum_{i=1}^{\infty} a_{ki} x_i \right) \bar{y}_k = \sum_{i=1}^{\infty} \left(\sum_{k=1}^{\infty} a_{ki} \bar{y}_k \right) x_i .$$

From the inequality

$$|(Af, g)| \leqq M \|f\| \cdot \|g\| \tag{7}$$

it follows that

$$\left| \sum_{i=1}^{\infty} \sum_{k=1}^{\infty} a_{ki} x_i \bar{y}_k \right| \leqq M \sqrt{\sum_{i=1}^{\infty} |x_i|^2} \sqrt{\sum_{k=1}^{\infty} |y_k|^2}.$$

If each of the vectors f and g has only a finite number of nonzero components, then the last inequality may be written in the form

$$\left|\sum_{i=1}^{p}\sum_{k=1}^{q}a_{ki}x_i\bar{y}_k\right| \leqq M\sqrt{\sum_{i=1}^{p}|x_i|^2}\sqrt{\sum_{k=1}^{q}|y_k|^2}. \tag{8}$$

THEOREM: *In order that the matrix* (a_{ik}) *represent a bounded linear operator defined everywhere in* H, *it is necessary and sufficient that, for some constant* M, *the inequality* (8) *hold for any numbers* $x_1, x_2, \ldots, x_p$ *and* $y_1, y_2, \ldots, y_q$.

Proof: If the operator A is bounded and

$$a_{ik} = (Ae_k, e_i) \qquad (i, k = 1, 2, 3, \ldots),$$

then (7) implies (8). Now let (a_{ik}) be a matrix which satisfies condition (8). We shall show that this matrix determines a bounded linear operator A. First, from (8) with

$$x_1 = x_2 = \ldots = x_{k-1} = x_{k+1} = \ldots = 0, x_k \neq 0,$$
$$y_1 = y_2 = \ldots = y_{m-1} = y_{n+1} = y_{n+2} = \ldots = 0$$

we get

$$\left|\sum_{i=m}^{n}a_{ik}\bar{y}_i\right| \leqq M\sqrt{\sum_{i=m}^{n}|y_i|^2}.$$

This implies the convergence of the series

$$\sum_{i=1}^{\infty}a_{ik}\bar{y}_i$$

for any sequence $\{y_i\}_1^\infty$ in l^2. Hence, the theorem of Landau (see Section 18) implies the convergence of the series

$$\sum_{i=1}^{\infty}|a_{ik}|^2 \qquad (k = 1, 2, 3, \ldots).$$

We define the operator A_0, first for the basis elements by the formula

$$A_0e_k = \sum_{i=1}^{\infty}a_{ik}e_i \qquad (k = 1, 2, 3, \ldots),$$

and then, by means of linearity, for all vectors with only a finite number of components different from zero. Now we prove that the operator A_0 is bounded. We have, by (8) for f and g with only finite number of nonzero components,

$$|(A_0f, g)| \leqq M\|f\|\cdot\|g\|. \tag{9}$$

By the continuity of the scalar product, the inequality (9) is satisfied for all $g \in$ H. Let

$$g = A_0f$$

in (9) to get

$$\|A_0f\|^2 \leqq M\|f\|\cdot\|A_0f\|$$

so that

$$\|A_0 f\| \leqq M \|f\|.$$

Thus, A_0 is bounded. Extending A_0 by continuity to the whole space H, we get the bounded operator A and the correspondence

$$A \sim (a_{ik}).$$

The theorem is proved.

We note that if the matrix (a_{ik}) is symmetric (Hermitean), i.e., if

$$a_{ik} = \bar{a}_{ki}$$

then it is possible to replace the condition (8) by (see Section 22)

$$\left| \sum_{i,k=1}^{p} a_{ik} x_i \bar{x}_k \right| \leqq M \sum_{i=1}^{p} |x_i|^2.$$

We now give an example of the matrix representation of a bounded linear operator. Consider the *integral operator* in $L^2(-\infty, \infty)$ defined by the formula

$$g(s) = Af(s) = \int_{-\infty}^{\infty} K(s,t) f(t)\, dt$$

where the function $K(s, t)$ is called the *kernel* of the operator. If the kernel satisfies the condition

$$\text{(10)} \qquad \iint_{-\infty}^{\infty} |K(s,t)|^2 ds\, dt < \infty,$$

it is called a *Hilbert-Schmidt kernel*, and the operator determined by it is called a *Hilbert-Schmidt operator*. We suppose that the condition (10) is satisfied. Then, for almost all t and u

$$\int_{-\infty}^{\infty} |K(s,t) \cdot \overline{K(s,u)}|\, ds \leqq \sqrt{\int_{-\infty}^{\infty} |K(s,t)|^2 ds} \sqrt{\int_{-\infty}^{\infty} |K(s,u)|^2 ds}\,.$$

But since

$$\int_{-\infty}^{\infty} \int_{-\infty}^{\infty} |f(t)| \cdot |f(u)| \cdot \sqrt{\int_{-\infty}^{\infty} |K(s,t)|^2 ds} \sqrt{\int_{-\infty}^{\infty} |K(s,u)|^2 ds}\, dt\, du =$$

$$= \left\{ \int_{-\infty}^{\infty} |f(t)| \sqrt{\int_{-\infty}^{\infty} |K(s,t)|^2\, ds}\, dt \right\}^2 \leqq$$

$$\leqq \int_{-\infty}^{\infty} |f(t)|^2 dt \int_{-\infty}^{\infty} \int_{-\infty}^{\infty} |K(s,t)|^2 ds\, dt < \infty,$$

we have, by Fubini's theorem,

$$\|g\| = \left\{ \int_{-\infty}^{\infty} ds \left| \int_{-\infty}^{\infty} K(s,t) f(t)\, dt \right|^2 \right\}^{\frac{1}{2}} =$$

$$= \left\{ \int_{-\infty}^{\infty} ds \int_{-\infty}^{\infty} K(s,t) f(t)\, dt \int_{-\infty}^{\infty} \overline{K(s,u)}\, \overline{f(u)}\, du \right\}^{\frac{1}{2}} \leqq$$

$$\leqq \sqrt{\int_{-\infty}^{\infty} |f(t)|^2\, dt} \sqrt{\int_{-\infty}^{\infty} \int_{-\infty}^{\infty} |K(s,t)|^2 ds\, dt}\, .$$

We see that the Hilbert-Schmidt operator is bounded and that its norm does not exceed the quantity

$$\sqrt{\int_{-\infty}^{\infty} \int_{-\infty}^{\infty} |K(s,t)|^2 ds\, dt.}$$

Let us take in $L^2(-\infty, \infty)$ any complete orthonormal system of functions $\{\varphi_k(t)\}_1^\infty$ and define

$$a_{ik} = \iint_{-\infty}^{\infty} K(s,t)\, \overline{\varphi_i(s)}\, \varphi_k(t)\, ds\, dt \qquad (i, k = 1, 2, 3, \ldots).$$

Choose any $f(t) \in L^2(-\infty, \infty)$ and let

$$\int_{-\infty}^{\infty} f(t)\, \overline{\varphi_k(t)}\, dt = x_k \qquad (k = 1, 2, 3, \ldots).$$

Then the Fourier coefficients

$$y_i = \int_{-\infty}^{\infty} g(s)\, \overline{\varphi_i(s)}\, ds \qquad (i = 1, 2, 3, \ldots)$$

of the function

$$g(s) = \int_{-\infty}^{\infty} K(s,t) f(t)\, dt$$

are given by

$$y_i = \iint_{-\infty}^{\infty} K(s,t) f(t)\, \overline{\varphi_i(s)}\, ds\, dt =$$

$$= \int_{-\infty}^{\infty} f(t) \left\{ \int_{-\infty}^{\infty} K(s,t)\, \overline{\varphi_i(s)}\, ds \right\} dt \qquad (i = 1, 2, 3, \ldots).$$

But since

$$(11) \qquad \int_{-\infty}^{\infty} K(s,t)\,\overline{\varphi_i(s)}\,ds \sim \sum_{k=1}^{\infty} a_{ik}\,\overline{\varphi_k(t)} \qquad (i = 1, 2, 3, \ldots)$$

and

$$f(t) \sim \sum_{k=1}^{\infty} x_k\,\varphi_k(t),$$

we have, by the Parseval relation,

$$y_i = \sum_{k=1}^{\infty} a_{ik}\,x_k \qquad (i = 1, 2, 3, \ldots).$$

In a similar way, it follows from relation (11) that

$$\int_{-\infty}^{\infty} dt \left| \int_{-\infty}^{\infty} K(s,t)\,\overline{\varphi_i(s)}\,ds \right|^2 = \sum_{k=1}^{\infty} |a_{ik}|^2 \qquad (i = 1, 2, 3, \ldots).$$

But on the other hand, by the Parseval relation,

$$\int_{-\infty}^{\infty} |K(s,t)|^2\,ds = \sum_{i=1}^{\infty} \left| \int_{-\infty}^{\infty} K(s,t)\,\overline{\varphi_i(s)}\,ds \right|^2$$

and, therefore,

$$\iint_{-\infty}^{\infty} |K(s,t)|^2\,ds\,dt = \sum_{i,k=1}^{\infty} |a_{ik}|^2.$$

We see that each Hilbert-Schmidt operator is represented by a matrix operator for which

$$\sum_{i,k=1}^{\infty} |a_{ik}|^2 < \infty.$$

27. Completely Continuous Operators

Hilbert considered first the important class of completely continuous operators. A linear operator A defined everywhere in H is said to be *completely continuous* if it maps each bounded point set into a set which is compact in the sense of strong convergence.

A completely continuous operator A is bounded. In fact, otherwise there would exist a sequence of points f_k $(k=1, 2, 3, \ldots)$ for which

$$\|f_k\| = 1, \quad \|Af_k\| > k \qquad (k = 1, 2, 3, \ldots)$$

but this is impossible, since the point set $\{Af_k\}_1^{\infty}$ is compact.

Completely continuous operators have another definition: a linear operator A defined everywhere in H is completely continuous if it maps every weakly convergent sequence into a strongly convergent sequence. The proof of the equivalence of these definitions we leave to the reader.

We also leave to the reader the proofs of the following simple facts:

1. If A is a completely continuous operator and if B is a bounded operator defined everywhere in H, then the operators AB and BA are completely continuous.

2. If A_1 and A_2 are completely continuous operators, then $\alpha_1 A_1 + \alpha_2 A_2$ is a completely continuous operator.

THEOREM: *If A is a bounded linear operator defined everywhere in* H, *and if the operator A^*A is completely continuous, then the operator A is completely continuous.*

Proof: Let M be any bounded infinite set of points f ($\|f\| \leqq C$). Let $\{f_k\}_1^\infty$ be any sequence of elements of this set. This sequence is mapped by the operator A^*A into a strongly convergent sequence. Since

$$\|Af_n - Af_m\|^2 = (A(f_n - f_m), A(f_n - f_m)) =$$
$$= (A^*A(f_n - f_m), f_n - f_m) \leqq \|A^*Af_n - A^*Af_m\| \cdot \|f_n - f_m\|,$$

we have

$$\lim_{m, n \to \infty} \|A^*Af_n - A^*Af_m\| = 0$$

where

$$\|f_n - f_m\| \leqq 2C.$$

Therefore,

$$\lim_{m, n \to \infty} \|Af_n - Af_m\| = 0,$$

so that the sequence $\{Af_n\}_1^\infty$ converges and the theorem is proved.

COROLLARY: *If the operator A is completely continuous, then the operator A^* has the same property.*

Proof: In fact, if the operator A is completely continuous, then the operator $AA^* = (A^*)^*A^*$ is completely continuous, and it remains only to apply the theorem just proved.

28. A Criterion for Complete Continuity of an Operator

The following theorem is often used to prove that a given operator is completely continuous.

THEOREM: *If for each $\varepsilon > 0$ there exists a completely continuous operator A_ε such that*

$$\|(A - A_\varepsilon)f\| \leqq \varepsilon \|f\|$$

for $f \in$ H then the operator A is completely continuous.

Proof: We choose a sequence of positive numbers $\varepsilon_1 > \varepsilon_2 > \dots$ ($\lim_{n \to \infty} \varepsilon_n = 0$) and consider a sequence of completely continuous operators $A_{\varepsilon_1}, A_{\varepsilon_2}, \dots$ corresponding to it by the condition of the theorem. Let M be an arbitrary bounded set of points f($\|f\| \leqq C$) in the space H. Let us take an arbitrary sequence $\{f_k\}_1^\infty$ of points belonging to M. According to the complete continuity of A_{ε_1} there exists a subsequence

$$f_{11}, f_{12}, f_{13}, \ldots \tag{1}$$

which is mapped by the operator A_{ε_1} into a convergent sequence. From the sequence (1) we select a subsequence

$$f_{21}, f_{22}, f_{23}, \ldots \tag{2}$$

which is mapped into a convergent sequence by the operator A_{ε_2}. Continuing this process, we get the infinite sequence of sequences

$$\begin{array}{l} f_{11}, f_{12}, f_{13}, \ldots, \\ f_{21}, f_{22}, f_{23}, \ldots, \\ f_{31}, f_{32}, f_{33}, \ldots, \\ \ldots\ldots\ldots\ldots\ldots, \end{array}$$

such that each is a subsequence of the preceding. The diagonal sequence

$$f_{11}, f_{22}, f_{33}, \ldots$$

is mapped into a strongly convergent sequence by each of the operators A_{ε_i}. We prove next that the diagonal sequence $\{f_{kk}\}_{k=1}^\infty$ is mapped into a convergent sequence also by the operator A. For this it suffices to prove that

$$\lim_{m,\, n\to\infty} \|Af_{nn} - Af_{mm}\| = 0. \tag{3}$$

We have the inequality

$$\begin{aligned} \|Af_{nn} - Af_{mm}\| &\leqq \|(A - A_{\varepsilon_k})f_{nn}\| + \|A_{\varepsilon_k}f_{nn} - A_{\varepsilon_k}f_{mm}\| + \\ &+ \|(A - A_{\varepsilon_k})f_{mm}\| \leqq 2\varepsilon_k C + \|A_{\varepsilon_k}f_{nn} - A_{\varepsilon_k}f_{mm}\|. \end{aligned}$$

By taking k sufficiently large, we can make the term $2\varepsilon_k C$ as small as desired. After this, we can take N so large that the second term of the last member is made as small as desired for $m, n > N$. Thus, the relation (3) is proved.

We make use of the theorem just proved in order to establish the complete continuity of the matrix operator defined by

$$y_i = \sum_{k=1}^{\infty} a_{ik} x_k \qquad (i = 1, 2, 3 \ldots),$$

where

$$\sum_{i,\, k=1}^{\infty} |a_{ik}|^2 < \infty, \tag{4}$$

which implies the complete continuity of every integral operator with a Hilbert-Schmidt kernel. From (4) it follows that

$$\lim_{p\to\infty} \sum_{i=1}^{p} \sum_{k=1}^{\infty} |a_{ik}|^2 < \infty.$$

Therefore, for each $\varepsilon > 0$ there exists an integer $p = p(\varepsilon)$ such that

$$\sum_{i=p+1}^{\infty} \sum_{k=1}^{\infty} |a_{ik}|^2 \leqq \varepsilon^2.$$

Now we construct the operator A_ε with the aid of the relation

$$A_\varepsilon f = y_1 e_1 + y_2 e_2 + \ldots + y_p e_p$$

where

$$y_i = \sum_{k=1}^{\infty} a_{ik} x_k \qquad (i = 1, 2, 3, \ldots)$$

if

$$f = \sum_{k=1}^{\infty} x_k e_k.$$

Let

$$Af = \sum_{i=1}^{\infty} y_i e_i.$$

We have

$$\| Af - A_\varepsilon f \|^2 = \sum_{i=p+1}^{\infty} |y_i|^2 =$$

$$= \sum_{i=p+1}^{\infty} \left| \sum_{k=1}^{\infty} a_{ik} x_k \right|^2 \leqq \sum_{i=p+1}^{\infty} \sum_{k=1}^{\infty} |a_{ik}|^2 \|f\|^2 \leqq \varepsilon^2 \|f\|^2.$$

It remains only to verify that the operator A_ε is completely continuous. Choose any bounded set of vectors in H. The operator A_ε maps this set into a bounded set in a finite-dimensional subspace of H, and this set is compact by the classical Bolzano-Weierstrass theorem.

We emphasize the fact that the convergence of the series

$$\sum_{i,k=1}^{\infty} |a_{ik}|^2$$

is only a sufficient, but not a necessary condition for the complete continuity of the matrix operator. In the special case when the numbers a_{ik} satisfy the relation

$$a_{ik} = 0 \text{ for } |i - k| > r \qquad (i, k = 1, 2, 3, \ldots)$$

for some fixed r, it is possible to specify a necessary and sufficient condition for complete continuity. It is expressed by the relation

$$\lim_{i,k\to\infty} a_{ik} = 0.$$

For simplicity we sketch the proof only in the case with $r = 1$. In this case, the matrix defining the operator has the form

$$\begin{pmatrix} \alpha_1 & \beta_1 & 0 & 0 & 0 & \dots \\ \gamma_1 & \alpha_2 & \beta_2 & 0 & 0 & \dots \\ 0 & \gamma_2 & \alpha_3 & \beta_3 & 0 & \dots \\ 0 & 0 & \gamma_3 & \alpha_4 & \beta_4 & \dots \\ \dots & \dots & \dots & \dots & \dots & \dots \end{pmatrix} \tag{5}$$

and is called a *Jacobi matrix*. For $r > 1$ the matrix is called a *generalized Jacobi matrix*. Let the operator A, determined by the matrix (5), be completely continuous. Then the sequence of vectors

$$Ae_i = \beta_{i-1}e_{i-1} + \alpha_i e_i + \gamma_i e_{i+1}$$
$$(\beta_0 = 0, \ i = 1, 2, 3, \dots)$$

must converge strongly. Supposing that the matrix A does not satisfy the condition in question, we select a sequence $i_1, i_2, i_3, \dots$ such that

$$i_k \geqq i_{k-1} + 3$$

and

$$|\beta_{i_k-1}|^2 + |\alpha_{i_k}|^2 + |\gamma_{i_k}|^2 \to \delta > 0$$

where $\delta \leqq \infty$. A simple computation yields

$$\|Ae_{i_n} - Ae_{i_m}\|^2 = |\beta_{i_n-1}|^2 + |\alpha_{i_n}|^2 + |\gamma_{i_n}|^2 +$$
$$+ |\beta_{i_m-1}|^2 + |\alpha_{i_m}|^2 + |\gamma_{i_m}|^2 \to 2\delta \neq 0.$$

This contradicts the strong convergence of the sequence $\{Ae_k\}_1^\infty$.

We now prove the sufficiency of the assertion. Let

$$\alpha_k \to 0, \ \beta_k \to 0, \ \gamma_k \to 0 \qquad (k \to \infty)$$

and let the sequence $\{f^{(k)}\}_1^\infty$ converge weakly to f. Because

$$Af^{(n)} = \sum_{k=1}^\infty x_k^{(n)} Ae_k = \sum_{k=1}^\infty x_k^{(n)}(\beta_{k-1}e_{k-1} + \alpha_k e_k + \gamma_k e_{k+1}) =$$

$$= \sum_{k=1}^\infty (\beta_k x_{k+1}^{(n)} + \alpha_k x_k^{(n)} + \gamma_{k-1} x_{k-1}^{(n)}) e_k \qquad (\gamma_0 = 0)$$

we have

$$\|Af^{(n)} - Af^{(m)}\|^2 =$$

$$= \sum_{k=1}^\infty |\beta_k\{x_{k+1}^{(n)} - x_{k+1}^{(m)}\} + \alpha_k\{x_k^{(n)} - x_k^{(m)}\} + \gamma_{k-1}\{x_{k-1}^{(n)} - x_{k-1}^{(m)}\}|^2$$

$$= \sum_{k=1}^q + \sum_{k=q+1}^\infty.$$

The first term on the right side tends toward zero for fixed q as $m, n \to \infty$. Therefore, it is sufficient to show that it is possible to make the second term on the right side as small as desired for all m and n by taking q sufficiently large. But, if q is sufficiently large and $k > q$, then

$$|\beta_k| < \varepsilon, |\alpha_k| < \varepsilon, |\gamma_{k-1}| < \varepsilon.$$

Therefore,

$$\sum_{k=q+1}^{\infty} |\beta_k\{x_{k+1}^{(n)} - x_{k+1}^{(m)}\} + \alpha_k\{x_k^{(n)} - x_k^{(m)}\} + \gamma_{k-1}\{x_{k-1}^{(n)} - x_{k-1}^{(m)}\}|^2 \leqq$$
$$\leqq 9\varepsilon^2 \|f^{(n)} - f^{(m)}\|^2.$$

Thus, our assertion is proved.

29. Sequences of Bounded Linear Operators

We distinguish three modes of convergence of a sequence $\{A_n\}_1^\infty$ of bounded linear operators defined everywhere in H: *weak* convergence, *strong* convergence (or, simply, convergence), and *uniform* convergence. A sequence $\{A_n\}_1^\infty$

converges weakly to the operator A $A_n \xrightarrow{w} A$	if for each $f \in$ H $A_n f \xrightarrow{w} Af \quad (n \to \infty)$
converges strongly to the operator A $A_n \to A$	if for each $f \in$ H $A_n f \to Af \quad (n \to \infty)$
converges uniformly to the operator A $A_n \Rightarrow A$	if $\|A_n - A\| \to 0 \quad (n \to \infty)$

If a sequence of operators converges uniformly, then it also converges strongly; if it converges strongly, then it also converges weakly.

Using the results of Section 23 and the lemma about convex functionals in Section 18, it is possible to prove the following proposition: if the sequence $\{A_n\}_1^\infty$ of bounded linear operators defined everywhere in H converges weakly, then the sequence $\{\|A_n\|\}_1^\infty$ of the norms of these operators is bounded.

We mention one more proposition, which is analogous to Corollary 2 in Section 23: if the sequence of bilinear functionals $\{\Omega_n(f, g)\}_1^\infty$ has the property that for arbitrary f and g the limit

$$\lim_{n\to\infty} \Omega_n(f, g) = \omega(f, g)$$

exists and is finite, then this limit defines a bilinear functional. As is easily seen, it is sufficient to prove that

$$|\omega(f,g)| \leqq C < \infty$$

for $\|f\| \leqq 1$, $\|g\| \leqq 1$, and some number C. Each of the bilinear functionals $\Omega_n(f, g)$ is determined by a particular bounded linear operator:

$$\Omega_n(f, g) = (A_n f, g).$$

The hypothesis implies that the sequence of operators $\{A_n\}_1^\infty$ converges weakly. Consequently, for $f, g \in \mathrm{H}$,

$$|(A_n f, g)| \leqq C \|f\| \cdot \|g\|$$

or equivalently,

$$|\Omega_n(f,g)| \leqq C \|f\| \cdot \|g\|.$$

Hence, it follows that

$$|\omega(f,g)| \leqq C \|f\| \cdot \|g\|.$$

Chapter III

PROJECTION OPERATORS AND UNITARY OPERATORS

30. Definition of a Projection Operator

Let G be a subspace of the space H and let

$$F = H \ominus G,$$

so that

$$H = G \oplus F.$$

Then each vector $h \in H$ is uniquely representable in the form

$$h = g + f,$$

where $g \in G$ and $f \in F$. In Section 7 the vector g was called the projection of h on G. The operator which maps each $h \in H$ into its projection g on G is called the operator of projection on G or, simply, a projection operator. It is denoted by P_G or sometimes, when the subspace G is specified in advance, by P. Thus, if g and h are related as above,

$$g = Ph = P_G h.$$

A projection operator is evidently linear. In addition, it is bounded and its norm is equal to one. Indeed, since the equation

$$\| h \|^2 = \| g \|^2 + \| f \|^2$$

implies that

$$\| g \| \leqq \| h \|, \tag{1}$$

we have

$$\| P \| \leqq 1.$$

But if $h \in G$, then $g = h$, so that there can be equality in (1). Therefore,

$$\| P \| = 1.$$

31. Properties of Projection Operators

From the definition of a projection operator it follows easily that

1) $P^2 = P$,
2) $P^* = P$.

Indeed, if $P = P_G$ then for an arbitrary $h \in H$ the vector $g = Ph$ belongs to G, so that $Pg = g$ and $P^2h = Ph$, and this implies that $P^2 = P$.

In order to prove that P is self-adjoint, we choose two arbitrary vectors $h_1, h_2 \in \mathrm{H}$ and let

$$h_1 = g_1 + f_1, \ h_2 = g_2 + f_2,$$

where $g_1 = Ph_1$ and $g_2 = Ph_2$. Then

$$(g_1, h_2) = (g_1, g_2) = (h_1, g_2),$$

so that

$$(Ph_1, h_2) = (h_1, Ph_2)$$

for $h_1, h_2 \in \mathrm{H}$. This implies that $P^* = P$.

From the properties just established it follows that

$$(Ph, h) \geqq 0.$$

In fact,

$$(Ph, h) = (P^2h, h) = (Ph, P^*h) = (Ph, Ph) \geqq 0.$$

Now we prove that the properties 1), 2) characterize a projection operator.

THEOREM: *If P is any operator defined on* H *such that, for arbitrary* $h_1, h_2 \in \mathrm{H}$,

$$(1) \qquad (P^2h_1, h_2) = (Ph_1, h_2),$$

$$(2) \qquad (Ph_1, h_2) = (h_1, Ph_2)$$

then there exists a subspace $\mathrm{G} \subset \mathrm{H}$ *such that P is the operator of projection on* G.

Proof: The operator P is bounded. This follows from (2) and a theorem in Section 25. However, it can be proved also by the following simple argument. We have

$$\|Ph\|^2 = (Ph, Ph) = (P^2h, h) = (Ph, h)$$

and

$$\|Ph\|^2 \leqq \|Ph\| \cdot \|h\|,$$

so that

$$\|Ph\| \leqq \|h\|.$$

Thus, the operator P is bounded and its norm is not greater than 1. We denote by G the set of all vectors $g \in \mathrm{H}$ for which

$$Pg = g.$$

Clearly, G is a linear manifold. We shall prove that G is closed so that it is also a subspace. Let $g_n \in \mathrm{G}$ ($n = 1, 2, 3, \ldots$) and $g_n \to g$. Then

$$g_n = Pg_n$$

and

$$Pg - g_n = Pg - Pg_n = P(g - g_n),$$

so that

$$\|Pg - g_n\| \leqq \|g - g_n\|.$$

Let $n \to \infty$ to get

$$\|Pg - g\| \leqq 0,$$

so that

$$Pg = g.$$

Hence, $g \in G$, which implies that G is closed. We must prove that $P = P_G$, where P_G is the operator of projection on G. For each $h \in H$, the vector Ph belongs to G because $P(Ph) = Ph$. The subspace G also contains $P_G h$. Therefore, it is sufficient to prove that

$$(Ph - P_G h, g') = 0$$

or

$$(Ph, g') = (P_G h, g')$$

for each $g' \in G$. But this follows from the equations

$$(Ph, g') = (h, Pg') = (h, g'),$$
$$(P_G h, g') = (h, P_G g') = (h, g').$$

To conclude the present section, we remark that if G is a subspace and E is the identity operator, then $E - P$ is the operator of projection on $H \ominus G$.

32. Operations Involving Projection Operators

In the present section we shall prove a few simple propositions concerning the multiplication, addition and subtraction of projection operators.

THEOREM 1: *The product of two projection operators P_{G_1} and P_{G_2} is also a projection operator if and only if P_{G_1} and P_{G_2} commute, i.e., if*

$$P_{G_1}P_{G_2} = P_{G_2}P_{G_1}.$$

In this case,

$$P_{G_1}P_{G_2} = P_G,$$

where $G = G_1 \cap G_2$.[1]

Proof: First, let the product be a projection operator. Then

$$P_{G_1}P_{G_2} = (P_{G_1}P_{G_2})^* = P_{G_2}^* P_{G_1}^* = P_{G_2}P_{G_1}.$$

Fix $h \in H$ arbitrarily and let

$$g = P_{G_1}P_{G_2}h = P_{G_2}P_{G_1}h.$$

By the first representation $g \in G_1$ and, by the second, $g \in G_2$. Hence $g \in G_1 \cap G_2$. If $h \in G_1 \cap G_2$, then $P_{G_1}P_{G_2}h = h$. Thus, one half of the theorem is proved. Now assume that P_{G_1} and P_{G_2} commute. Let

$$P_{G_1}P_{G_2} = P_{G_2}P_{G_1} = P.$$

[1] A geometrical implication of the commutativity of the operators P_{G_1} and P_{G_2} is that the subspaces $G_1 \ominus (G_1 \cap G_2)$ and $G_2 \ominus (G_1 \cap G_2)$ are orthogonal.

It follows that

$$P^2 = (P_{G_1}P_{G_2})^2 = P_{G_1}P_{G_2}P_{G_1}P_{G_2} = P_{G_1}P_{G_1}P_{G_2}P_{G_2} = P_{G_1}P_{G_2} = P$$

and

$$(Ph_1, h_2) = (P_{G_1}P_{G_2}h_1, h_2) = (P_{G_2}h_1, P_{G_1}h_2)$$
$$= (h_1, P_{G_2}P_{G_1}h_2) = (h_1, P_{G_1}P_{G_2}h_2) = (h_1, Ph_2).$$

These equations show that the operator $P = P_{G_1}P_{G_2}$ satisfies the conditions of the theorem of the preceding section. Therefore, it is a projection operator.

COROLLARY: *Two subspaces* G_1 *and* G_2 *are orthogonal if and only if*

$$P_{G_1}P_{G_2} = 0.$$

THEOREM 2: *A finite sum of projection operators*

$$P_{G_1} + P_{G_2} + \dots + P_{G_n} = Q \qquad (n < \infty)$$

is a projection operator if and only if

$$P_{G_i}P_{G_k} = 0 \qquad (i \neq k)$$

i.e., if and only if the spaces G_j $(j = 1, 2, 3, \dots, n)$ *are pairwise orthogonal. In this case*

$$Q = P_G,$$

where

$$G = G_1 \oplus G_2 \oplus \dots \oplus G_n.$$

Proof: If the spaces G_j are pairwise orthogonal, then $Q^2 = Q$, and, therefore, the sufficiency of the condition is evident. The last part of the assertion of the theorem is also evident. It remains only to prove the necessity of the condition. Let Q be a projection operator. Then

$$\|f\|^2 \geqq (Qf, f) = \sum_{j=1}^{n}(P_{G_j}f, f) \geqq (P_{G_i}f, f) + (P_{G_k}f, f)$$

for any pair of distinct indices i and k. From this relation it follows that

$$\|P_{G_i}f\|^2 + \|P_{G_k}f\|^2 \leqq \|f\|^2.$$

In this inequality let

$$f = P_{G_k}h.$$

Then

$$\|P_{G_i}P_{G_k}h\|^2 + \|P_{G_k}h\|^2 \leqq \|P_{G_k}h\|^2$$

which yields

$$\|P_{G_i}P_{G_k}h\| = 0$$

for $h \in \mathrm{H}$. Thus,

$$P_{\mathrm{G}_i} P_{\mathrm{G}_k} = 0$$

so that the spaces G_i and G_k are orthogonal.

THEOREM 3: *The difference of two projection operators,*

(1) $$P_{\mathrm{G}_1} - P_{\mathrm{G}_2},$$

is a projection operator if and only if $\mathrm{G}_2 \subset \mathrm{G}_1$. *In this case* $P_{\mathrm{G}_1} - P_{\mathrm{G}_2}$ *is the operator of projection on* $\mathrm{G}_1 \ominus \mathrm{G}_2$.

Proof: In view of the remark at the end of the preceding section, we attempt to find conditions for which the difference

$$Q = E - (P_{\mathrm{G}_1} - P_{\mathrm{G}_2})$$

is a projection operator. Since the equation

$$Q = (E - P_{\mathrm{G}_1}) + P_{\mathrm{G}_2}$$

represents Q as the sum of two projection operators, it follows from Theorem 2 that

$$(E - P_{\mathrm{G}_1}) P_{\mathrm{G}_2} = 0$$

or, equivalently,

(2) $$P_{\mathrm{G}_2} = P_{\mathrm{G}_1} P_{\mathrm{G}_2}.$$

If $g \in \mathrm{G}_2$ then

$$g = P_{\mathrm{G}_2} g = P_{\mathrm{G}_1} P_{\mathrm{G}_2} g = P_{\mathrm{G}_1} g,$$

so that $g \in \mathrm{G}_1$. Since every element $g \in \mathrm{G}_2$ belongs to G_1, we have $\mathrm{G}_2 \subset \mathrm{G}_1$. This condition, which can be expressed in the form of (2), is necessary and sufficient in order that the difference (1) be a projection operator. It remains only to characterize the space G on which the operator (1) projects. The operator Q projects on

$$[\mathrm{H} \ominus \mathrm{G}_1] \oplus \mathrm{G}_2.$$

Hence, the operator (1) projects on

(3) $$\mathrm{H} \ominus \{[\mathrm{H} \ominus \mathrm{G}_1] \oplus \mathrm{G}_2\},$$

i.e., on the subspace of vectors orthogonal both to G_2 and to $\mathrm{H} \ominus \mathrm{G}_1$. Since this subspace consists of all the vectors of G_1 which are orthogonal to G_2, it is the subspace

(4) $$\mathrm{G}_1 \ominus \mathrm{G}_2.$$

We notice that the difference (4) can be obtained directly from (3) by formally removing the brackets.

33. Monotone Sequences of Projection Operators

We shall prove that the relation $G_2 \subset G_1$ is equivalent to the inequality

$$\| P_{G_2} f \| \leqq \| P_{G_1} f \| \tag{1}$$

for all $f \in H$. The inequality (1) is evidently equivalent to

$$(P_{G_2} f, f) \leqq (P_{G_1} f, f)$$

or

$$(\{P_{G_2} - P_{G_1}\} f, f) \leqq 0$$

for $f \in H$. The last two inequalities are generally expressed by

$$P_{G_2} \leqq P_{G_1}.$$

Thus, we wish to prove that the relation $G_2 \subset G_1$ is equivalent to the relation $P_{G_2} \leqq P_{G_1}$. This will permit us to introduce for consideration monotone sequences of projection operators.

First, let $G_2 \subset G_1$. Then it follows that

$$P_{G_2} = P_{G_2} P_{G_1}.$$

Therefore, for each $f \in H$,

$$P_{G_2} f = P_{G_2} P_{G_1} f$$

and

$$\| P_{G_2} f \| \leqq \| P_{G_1} f \|. \tag{2}$$

Conversely, assume (2) for each $f \in H$. Consider

$$f = (E - P_{G_1}) h,$$

where h is an arbitrary element of H. From (2) and

$$P_{G_1}(E - P_{G_1}) h = 0,$$

we obtain

$$P_{G_2}(E - P_{G_1}) h = 0$$

or

$$P_{G_2} h = P_{G_2} P_{G_1} h.$$

Since this equality holds for each $h \in H$,

$$P_{G_2} = P_{G_2} P_{G_1},$$

so that $G_2 \subset G_1$. This completes the proof.

THEOREM: *If P_{G_k} $(k = 1, 2, 3, \ldots)$ is an infinite sequence of projection operators and if $P_{G_k} \leqq P_{G_{k+1}}$ $(k = 1, 2, 3, \ldots)$, then, as $k \to \infty$, P_{G_k} converges strongly to some projection operator P.*

Proof: For $m < n$ the difference $P_{G_n} - P_{G_m}$ is a projection operator. Therefore, for each $f \in H$,

$$\| P_{G_n} f - P_{G_m} f \|^2 = \| (P_{G_n} - P_{G_m}) f \|^2 = \tag{3}$$
$$(\{P_{G_n} - P_{G_m}\} f, f) = \| P_{G_n} f \|^2 - \| P_{G_m} f \|^2.$$

Since, for fixed f, $\| P_{G_k} f \|^2$ increases with k but is bounded above by $\| f \|^2$, it has a finite limit. Hence, the right member of (3) tends to zero and the sequence $\{P_{G_n} f\}_{n=1}^{\infty}$ is fundamental in the sense of strong convergence. By the completeness of the space there exists the strong limit

$$f^* = \lim_{n \to \infty} P_{G_n} f.$$

We define the operator P by

$$f^* = Pf,$$

$f \in H$. The operator P is obviously linear. Since

$$(P_{G_k} f, P_{G_k} g) = (P_{G_k} f, g) = (f, P_{G_k} g)$$

a passage to the limit yields

$$(Pf, Pg) = (Pf, g) = (f, Pg).$$

Therefore,

$$P = P^* = P^2,$$

so that P is a projection operator.

34. The Aperture of Two Linear Manifolds[2]

The present section is devoted to a concept which was introduced by B. Nagy and, independently of him, by M. G. Krein and M. A. Krasnoselski.[3]

DEFINITION: *The aperture of two linear manifolds in* H *is defined as the norm of the difference of the operators which project* H *on the closures of these two linear manifolds.*

The aperture of the linear manifolds M_1 and M_2 is denoted by the symbol $\Theta(M_1, M_2)$. Thus,

$$\Theta(M_1, M_2) = \| P_1 - P_2 \| = \| P_2 - P_1 \|,$$

where P_1, P_2 are the operators of projection on the closed linear manifolds (subspaces) $\overline{M}_1$, $\overline{M}_2$, respectively. From the definition of aperture it follows that

$$\Theta(M_1, M_2) = \Theta(\overline{M}_1, \overline{M}_2) = \Theta(H \ominus M_1, H \ominus M_2).$$

Consider the identity

$$P_2 - P_1 = P_2(E - P_1) - (E - P_2) P_1.$$

[2] The results of this paragraph are necessary only for the construction of the theory of symmetric extensions (Chapter 7).

[3] M. G. Krein and M. A. Krasnoselski [1], B. Sz. Nagy [2].

It follows that, for $h \in \mathrm{H}$,

$$(P_2 - P_1)h = P_2(E - P_1)h - (E - P_2)P_1 h.$$

Since the vectors $P_2(E - P_1)h$ and $(E - P_2)P_1 h$ are orthogonal,

$$(1) \quad \|(P_2 - P_1)h\|^2 = \|P_2(E - P_1)h\|^2 + \|(E - P_2)P_1 h\|^2 \leqq$$
$$\leqq \|(E - P_1)h\|^2 + \|P_1 h\|^2 = \|h\|^2.$$

This inequality shows that the aperture of two linear manifolds does not exceed 1:

$$\Theta(\mathrm{M}_1, \mathrm{M}_2) \leqq 1.$$

In addition to this, we see that the aperture is certainly equal to 1 if one of these manifolds contains a nonzero vector which is orthogonal to the other manifold. This fact implies the following criterion for the equality of the dimensions of two linear manifolds.

THEOREM: *If the aperture of the linear manifolds* M_1 *and* M_2 *is less than* 1, *then*

$$\dim \mathrm{M}_1 = \dim \mathrm{M}_2,$$

i.e., the two linear manifolds have the same dimension.

Proof: It is sufficient to prove that the inequality

$$\dim \mathrm{M}_2 > \dim \mathrm{M}_1$$

implies the existence of a nonzero vector in $\bar{\mathrm{M}}_2$ which is orthogonal to the manifold $\bar{\mathrm{M}}_1$. With this purpose in mind, we project $\bar{\mathrm{M}}_1$ on $\bar{\mathrm{M}}_2$. We get the subspace

$$\mathrm{G} = P_2 \bar{\mathrm{M}}_1,$$

the dimension of which evidently does not exceed the dimension of the subspace $\bar{\mathrm{M}}_1$ and, consequently, is less than the dimension of the subspace $\bar{\mathrm{M}}_2$. Therefore, $\bar{\mathrm{M}}_2 \ominus \mathrm{G}$ contains a nonzero vector, i.e., $\bar{\mathrm{M}}_2$ contains a nonzero vector which is orthogonal to G. This vector is orthogonal to the whole subspace $\bar{\mathrm{M}}_1$ because the subspace $\bar{\mathrm{M}}_1 \ominus \mathrm{G}$ is orthogonal to $\bar{\mathrm{M}}_2$.

The aperture of two linear manifolds has the equivalent definition,

$$(2) \qquad \Theta(\mathrm{M}_1, \mathrm{M}_2) = \max\{ \sup_{f \in \bar{\mathrm{M}}_2, \|f\| = 1} \|(E - P_1)f\|, \sup_{g \in \bar{\mathrm{M}}_1, \|g\| = 1} \|(E - P_2)g\| \}.$$

The quantity

$$\|(E - P_1)f\| = D[f, \bar{\mathrm{M}}_1]$$

represents the distance between the point f and the manifold $\bar{\mathrm{M}}_1$. The significance of this formula consists in the fact that it can be used to define the aperture of two linear manifolds not only in a Hilbert Space, but in any Banach space.[4]

[4] M. G. Krein, M. A. Krasnoselski, and D. P. Milman [1].

We now proceed with the proof of formula (2). According to the original definition of aperture and formula (1),

$$(3) \qquad \Theta(\mathrm{M}_1, \mathrm{M}_2) = \sup_{h \in \mathrm{H}} \frac{\| (P_2 - P_1) h \|}{\| h \|} = $$
$$= \sup_{h \in \mathrm{H}} \frac{\sqrt{\| P_2 (E - P_1) h \|^2 + \| (E - P_2) P_1 h \|^2}}{\| h \|} .$$

Consider the equation obtained from (3) by restricting the vector h to the subspace $\bar{\mathrm{M}}_1$. Then the right member is either unchanged or decreased, so that

$$\Theta(\mathrm{M}_1, \mathrm{M}_2) \geqq \sup_{h \in \bar{\mathrm{M}}_1} \frac{\sqrt{\| P_2 (E - P_1) h \|^2 + \| (E - P_2) P_1 h \|^2}}{\| h \|} =$$
$$= \sup_{h \in \bar{\mathrm{M}}_1} \frac{\| (E - P_2) h \|}{\| h \|} = \rho_2 .$$

In the same way it is proved that

$$\Theta(\mathrm{M}_1, \mathrm{M}_2) \geqq \sup_{h \in \bar{\mathrm{M}}_2} \frac{\| (E - P_1) h \|}{\| h \|} = \rho_1 .$$

Thus,

$$\Theta(\mathrm{M}_1, \mathrm{M}_2) \geqq \max \{\rho_1, \rho_2\}$$

and it remains to prove that

$$\Theta(\mathrm{M}_1, \mathrm{M}_2) \leqq \max \{\rho_1, \rho_2\} .$$

With this purpose in mind we remark that, by the definition of ρ_2,

$$(4) \qquad \| (E - P_2) P_1 h \|^2 \leqq \rho_2^2 \| P_1 h \|^2 .$$

On the other hand

$$\begin{aligned} \|P_2(E - P_1)h\|^2 &= (P_2 \{E - P_1\} h, P_2 \{E - P_1\} h) = \\ &= (P_2 \{E - P_1\} h, \{E - P_1\} h) = \\ &= (P_2 \{E - P_1\} h, \{E - P_1\}^2 h) = \\ &= (\{E - P_1\} P_2 \{E - P_1\} h, \{E - P_1\} h) \leqq \\ &\leqq \| (E - P_1) P_2 (E - P_1) h \| \cdot \| (E - P_1) h \|, \end{aligned}$$

and, consequently, according to the definition of ρ_1,

$$\| P_2 (E - P_1) h \|^2 \leqq \rho_1 \| P_2 (E - P_1) h \| \cdot \| (E - P_1) h \|,$$

so that

$$(5) \qquad \| P_2 (E - P_1) h \| \leqq \rho_1 \| (E - P_1) h \| .$$

The inequalities (4) and (5) imply that

$$\begin{aligned} \|(E - P_2) P_1 h \|^2 + \| P_2 (E - P_1) h \|^2 &\leqq \rho_2^2 \|P_1 h\|^2 + \rho_1^2 \| (E - P_1) h \|^2 \leqq \\ &\leqq \max \{\rho_1^2, \rho_2^2\} [\|P_1 h \|^2 + \| (E - P_1) h \|^2] = \\ &= \| h \|^2 \max \{\rho_1^2, \rho_2^2\}, \end{aligned}$$

so that formula (3) yields

$$\Theta(\mathrm{M}_1, \mathrm{M}_2) \leqq \max \{\rho_1, \rho_2\} .$$

35. Unitary Operators

In three-dimensional Euclidean space the simplest operation after that of projection is rotation of the space, which changes neither the lengths of vectors nor the angles between pairs of them. We now consider an analogous operation in Hilbert space.

DEFINITION: *The operator* U *with domain* H ($\mathrm{D}_U = \mathrm{H}$) *and range* H ($\Delta_U = \mathrm{H}$) *is unitary if*

$$(1) \qquad (Uf, Ug) = (f, g)$$

for $f, g \in \mathrm{H}$.

We emphasize that the given definition does not require that the operator be linear.

We prove first that the unitary operator has an inverse operator, which is also unitary. Recall that an operator T has an inverse if and only if $Tf = Tg$ implies that $f = g$ (cf. Section 14). Therefore, assume that $Uf = Ug$. Then

$$\begin{aligned} 0 &= (Uf - Ug, Uf - Ug) = \\ &= (Uf, Uf) - (Uf, Ug) - (Ug, Uf) + (Ug, Ug) = \\ &= (f, f) - (f, g) - (g, f) + (g, g) = (f - g, f - g), \end{aligned}$$

so that $f = g$. Thus, the operator U^{-1} exists. Since $\mathrm{D}_{U^{-1}} = \Delta_U$ and $\Delta_{U^{-1}} = \mathrm{D}_U$ the operator U^{-1} is defined in the whole space and maps it onto the whole space. Choose $f', g' \in \mathrm{H}$ arbitrarily and let $f = U^{-1}f'$, $g = U^{-1}g'$. Then

$$Uf = f', \; Ug = g',$$

and the substitution of these equations in (1) yields

$$(f', g') = (U^{-1}f', U^{-1}g')$$

so that U^{-1} is unitary.

From the proof just given it follows that, for $f, g \in \mathrm{H}$,

$$(2) \qquad (Uf, g) = (f, U^{-1}g).$$

Indeed, let $U^{-1}g = g'$, so that $g = Ug'$. Then, since U is unitary,

$$(Uf, Ug') = (f, g'),$$

which is equivalent to (2).

We prove now that a unitary operator is necessarily linear. Let

$$f = \alpha_1 f_1 + \alpha_2 f_2.$$

Then, from (2),

$$\begin{aligned} (Uf, g) &= (f, U^{-1}g) = \alpha_1(f_1, U^{-1}g) + \alpha_2(f_2, U^{-1}g) = \\ &= \alpha_1(Uf_1, g) + \alpha_2(Uf_2, g) = (\alpha_1 Uf_1 + \alpha_2 Uf_2, g). \end{aligned}$$

Since g is arbitrary,

$$Uf = \alpha_1 Uf_1 + \alpha_2 Uf_2.$$

Therefore, every unitary operator is linear. Equation (2) shows that for a unitary operator, the adjoint operator coincides with its inverse:

$$U^* = U^{-1}.$$

Frequently the following simple proposition is useful: if a linear operator T satisfies the condition

$$(Tf, Tf) = (f, f) \tag{3}$$

and if $D_T = \Delta_T = H$, then T is unitary.

Proof: By condition (3),

$$(T\{f + \alpha g\}, T\{f + \alpha g\}) = (f + \alpha g, f + \alpha g).$$

Since T is linear,

$$(Tf, Tf) + \alpha(Tg, Tf) + \bar{\alpha}(Tf, Tg) + |\alpha|^2(Tg, Tg) =$$
$$= (f, f) + \alpha(g, f) + \bar{\alpha}(f, g) + |\alpha|^2(g, g).$$

Therefore, again by (3),

$$\alpha(Tg, Tf) + \bar{\alpha}(Tf, Tg) = \alpha(g, f) + \bar{\alpha}(f, g).$$

Since α is arbitrary,

$$(Tf, Tg) = (f, g),$$

so that T is a unitary operator.

36. Isometric Operators

Let H_1 and H_2 be two Hilbert spaces. We shall indicate the scalar product in H_1 by the index 1, and the scalar product in H_2 by the index 2.

DEFINITION: *The operator V with domain H_1 ($D_V = H_1$) and range H_2 ($\Delta_V = H_2$) is isometric if*

$$(Vf, Vg)_2 = (f, g)_1 \tag{1}$$

for $f, g \in H_1$.

In particular, H_1 and H_2 can be subspaces of a space H. In this case the indices on the scalar products are superfluous. Often the term isometric operator applies only in this special case, and the phrase isometric mapping is used in the general case.

A unitary operator in H is a special case of an isometric operator for which $H_1 = H_2 = H$. Many properties of unitary operators carry over to arbitrary isometric operators. We list now some of these properties, omitting those proofs which do not differ essentially from the proofs of corresponding properties for unitary operators.

1° Each isometric operator has an inverse operator which is also isometric.

2° If the operator V is linear, and maps all the space H_1 onto the space H_2 and if

$$(Vf, Vf)_2 = (f,f)_1$$

for $f \in H_1$, then V is an isometric operator.

3° Every isometric operator is linear.

Indeed, let $f', f'' \in H_1$, and $f = \alpha' f' + \alpha'' f''$. Then, for each $g \in H_1$,

$$\begin{aligned}(Vf, Vg)_2 = (f,g)_1 = \alpha'(f',g)_1 + \alpha''(f'',g)_1 = \\ = \alpha'(Vf', Vg)_2 + \alpha''(Vf'', Vg)_2 = (\alpha' Vf' + \alpha'' Vf'', Vg)_2.\end{aligned}$$

Since $\Delta_V = H_2$, it follows that

$$Vf = \alpha' Vf' + \alpha'' Vf'';$$

hence, the operator V is linear.

Isometric operators were introduced implicitly in Section 9, where we introduced the concept of isomorphism of two Hilbert spaces.

To complete the present paragraph, we introduce an important concept which will be used repeatedly in what follows.

DEFINITION: *Let T_1 and T_2 be linear operators defined, respectively, in spaces H_1 and H_2, so that $D_{T_1} \subset H_1$, $\Delta_{T_1} \subset H_1$, $D_{T_2} \subset H_2$, $\Delta_{T_2} \subset H_2$. (In particular, the spaces H_1, H_2 may coincide). The operators T_1 and T_2 are called isomorphic or unitarily equivalent if there exists an isometric operator V, which maps H_1 onto H_2 and D_{T_1} onto D_{T_2}, such that $VT_1 f = T_2 V f$ for each $f \in D_{T_1}$. In other words, T_1 and T_2 are unitarily equivalent if*

$$D_{T_2} = V D_{T_1}$$

and

$$T_1 = V^{-1} T_2 V.$$

37. The Fourier-Plancherel Operator

The object of this section is the proof of the Plancherel theorem, which may be stated as follows. *Let $g(t)$ be any function in $L^2(-\infty, \infty)$ and let*

$$\frac{1}{\sqrt{2\pi}} \frac{d}{dt} \int_{-\infty}^{\infty} \frac{e^{-ist} - 1}{-is} g(s)\, ds = h(t). \tag{1}$$

This formula defines a function $h(t)$ for almost all real t. Furthermore, $h(t) \in L^2(-\infty, \infty)$. The operator $\mathfrak{F}$ defined by $\mathfrak{F} g = h$, where h is given in terms of g by (1), is unitary. Its inverse operator, $\mathfrak{F}^{-1}$, can be expressed by the formula

$$g(t) = \mathfrak{F}^{-1} h(t) = \frac{1}{\sqrt{2\pi}} \frac{d}{dt} \int_{-\infty}^{\infty} \frac{e^{ist} - 1}{is} h(s)\, ds.$$

The operator $\mathfrak{F}$ is called the *Fourier-Plancherel operator.* If it is assumed that the function $g(t)$ is absolutely integrable on the entire real axis, then formula (1) can be replaced by

$$h(t) = \frac{1}{\sqrt{2\pi}} \int_{-\infty}^{\infty} e^{-ist} g(s)\, ds. \tag{2}$$

Thus, in this case $h(t)$ is a *Fourier integral* in the elementary sense.

The set of absolutely integrable functions in $L^2(-\infty, \infty)$ is mapped by the Fourier-Plancherel operator, expressed for this case either by (1) or (2), onto a linear manifold L which is dense in $L^2(-\infty, \infty)$. The value of the Plancherel theorem consists in the fact that it gives an extension of the elementary Fourier integral operator to an operator defined on all of $L^2(-\infty, \infty)$.

This approach to the Fourier-Plancherel operator leads to an equivalent definition. Let $g(t)$ be an arbitrary function in $L^2(-\infty, \infty)$ and let

$$g_N(t) = \begin{cases} g(t) & (-N \leqq t \leqq N) \\ 0 & (|t| > N). \end{cases}$$

Since the function $g_N(t)$ is absolutely integrable,

$$\mathfrak{F} g_N(t) = \frac{1}{\sqrt{2\pi}} \int_{-\infty}^{\infty} e^{-ist} g_N(s)\, ds = \frac{1}{\sqrt{2\pi}} \int_{-N}^{N} e^{-ist} g(s)\, ds.$$

Since the operator $\mathfrak{F}$ is bounded,

$$\lim_{N\to\infty} \| \mathfrak{F} g - \mathfrak{F} g_N \| = 0$$

or

$$h(t) = \mathfrak{F} g(t) = \underset{N\to\infty}{\text{l.i.m.}} \frac{1}{\sqrt{2\pi}} \int_{-N}^{N} e^{-ist} g(s)\, ds. \tag{3}$$

Hence, this alternate definition of the Fourier-Plancherel operator follows from (1).

It is easily proved that, in turn, definition (1) follows from definition (3). Indeed, for any element $f(t) \in L^2(-\infty, \infty)$,

$$\lim_{N\to\infty} (\mathfrak{F} g_N, f) = (\mathfrak{F} g, f). \tag{4}$$

Substituting

$$f(t) = \begin{cases} 1 & (0 \leqq t \leqq \tau), \\ 0 & (t < 0 \text{ and } t > \tau) \end{cases}$$

in (4), we obtain

$$\lim_{N\to\infty} \int_0^{\tau} dt \frac{1}{\sqrt{2\pi}} \int_{-N}^{N} e^{-ist} g(s)\, ds = \int_0^{\tau} \mathfrak{F} g(t)\, dt.$$

After integration, passing to the limit and replacing τ by t, the last equation yields

$$\frac{1}{\sqrt{2\pi}} \int_{-\infty}^{\infty} \frac{e^{-ist}-1}{-is} g(s)\, ds = \int_{0}^{t} \mathfrak{F}g(t)\, dt.$$

This relation implies that, for almost all real t,

$$\mathfrak{F}g(t) = \frac{1}{\sqrt{2\pi}} \frac{d}{dt} \int_{-\infty}^{\infty} \frac{e^{-ist}-1}{-is} g(s)\, ds.$$

Thus, formula (3) implies formula (1).

There exist various proofs of the theorem of Plancherel. The two essential elements of each proof are that the operator $\mathfrak{F}_0$, which is defined by equation (2) on the set $L \subset L^2(-\infty, \infty)$, does not change the norms of functions, and that it maps the set L onto a set which is dense in $L^2(-\infty, \infty)$. Extending the operator $\mathfrak{F}_0$ by continuity onto the whole space $L^2(-\infty, \infty)$, we get the operator $\mathfrak{F}$ which is defined by formula (3). Thus, the domain of definition of this operator is the whole space $L^2(-\infty, \infty)$. Since the operator $\mathfrak{F}$ does not change the norms of functions and since its range is dense in L^2, its range must also be the whole space.[5] To complete the proof it is sufficient to verify that the inverse operator results from the replacement of i by $-i$ in formula (1).

From the standpoint of the geometry of Hilbert space, it is particularly instructive to consider the set L_0 of functions

$$f(t) = P(t)e^{-\frac{t^2}{2}} \tag{5}$$

where $P(t)$ runs through the collection of all polynomials. The set L_0 is dense in $L^2(-\infty, \infty)$. Each function (5) can be represented in the form

$$f(t) = \alpha_0 \varphi_0(t) + \alpha_1 \varphi_1(t) + \dots + \alpha_n \varphi_n(t) \tag{6}$$

where the $\varphi_k(t)$ $(k = 0, 1, 2, \dots)$ are the Tchebysheff-Hermite functions. By the orthogonality of these functions (cf. Section 11)

$$\int_{-\infty}^{\infty} |f(t)|^2 dt = \sqrt{\pi} \sum_{k=0}^{n} 2^k k!\, |\alpha_k|^2.$$

Now we apply the operator $\mathfrak{F}_0$ to the function $f(t)$. For this purpose we make use of the relation

$$\frac{1}{\sqrt{2\pi}} \int_{-\infty}^{\infty} e^{-ist}(-1)^k e^{\frac{s^2}{2}} \frac{d^k}{ds^k} e^{-s^2}\, ds = i^k e^{\frac{t^2}{2}} \frac{d^k}{dt^k} e^{-t^2} \tag{7}$$

[5]Assuming the contrary, we could extend the inverse operator $\mathfrak{F}^{-1}$ onto the whole space $L^2(-\infty, \infty)$. The isometric operator thus obtained would take the identical values at at least two distinct points, which is impossible.

$$(k = 0, 1, 2, \ldots),$$

which will be proved later. The relation (7) implies that

$$\mathfrak{F}_0 \varphi_k = (-i)^k \varphi_k \qquad (k = 0, 1, 2, \ldots).$$

Therefore,

$$(6') \quad h(t) \equiv \mathfrak{F}_0 f(t) = \alpha_0 \varphi_0(t) + (-i)^1 \alpha_1 \varphi_1(t) + \ldots + (-i)^n \alpha_n \varphi_n(t)$$

which yields

$$\int_{-\infty}^{\infty} |h(t)|^2 dt = \int_{-\infty}^{\infty} |f(t)|^2 dt.$$

Thus, the operator $\mathfrak{F}_0$ does not change the norms of functions in the set L_0. Moreover, it follows from our considerations that the range of $\mathfrak{F}_0$ contains L_0, which is dense in $L^2(-\infty, \infty)$. Comparison of (6) and (6′) proves also that in order to change $\mathfrak{F}_0$ to $\mathfrak{F}^{-1}$ it is necessary only to change i to $-i$.

It remains to prove relation (7). Here is the proof:

$$\frac{1}{\sqrt{2\pi}} \int_{-\infty}^{\infty} e^{-ist} (-1)^k e^{\frac{s^2}{2}} \frac{d^k}{ds^k} e^{-s^2} ds = \frac{1}{\sqrt{2\pi}} \int_{-\infty}^{\infty} e^{-s^2} \frac{d^k}{ds^k} (e^{-ist + \frac{s^2}{2}}) ds =$$

$$= \frac{1}{\sqrt{2\pi}} e^{\frac{t^2}{2}} \int_{-\infty}^{\infty} e^{-s^2} \frac{d^k}{ds^k} e^{\frac{(s-it)^2}{2}} ds =$$

$$= \frac{1}{\sqrt{2\pi}} e^{\frac{t^2}{2}} \int_{-\infty}^{\infty} i^k e^{-s^2} \frac{d^k}{dt^k} e^{\frac{(s-it)^2}{2}} ds =$$

$$= \frac{i^k}{\sqrt{2\pi}} e^{\frac{t^2}{2}} \frac{d^k}{dt^k} \int_{-\infty}^{\infty} e^{-\frac{s^2}{2} - \frac{t^2}{2} - ist} ds = i^k e^{\frac{t^2}{2}} \frac{d^k}{dt^k} e^{-t^2}.$$

Examples of operators defined on the space $L^2(0, \infty)$ are $\mathfrak{F}_c$ and $\mathfrak{F}_s$ where

$$\mathfrak{F}_c g(t) = \sqrt{\frac{2}{\pi}} \frac{d}{dt} \int_0^{\infty} \frac{\sin st}{s} g(s) ds = \sqrt{\frac{2}{\pi}} \operatorname*{l.i.m.}_{N \to \infty} \int_0^N g(s) \cos st \, ds,$$

$$\mathfrak{F}_s g(t) = \sqrt{\frac{2}{\pi}} \frac{d}{dt} \int_0^{\infty} \frac{1 - \cos st}{s} g(s) ds = \sqrt{\frac{2}{\pi}} \operatorname*{l.i.m.}_{N \to \infty} \int_0^N g(s) \sin st \, ds.$$

As an exercise we recommend that the reader verify that $\mathfrak{F}_c$ and $\mathfrak{F}_s$ are unitary. Since these operators satisfy the relations

$$\mathfrak{F}_c^* = \mathfrak{F}_c^{-1} = \mathfrak{F}_c,$$
$$\mathfrak{F}_s^* = \mathfrak{F}_s^{-1} = \mathfrak{F}_s,$$

they are self-adjoint.

Chapter IV

GENERAL CONCEPTS AND PROPOSITIONS IN THE THEORY OF LINEAR OPERATORS

38. Closed Operators

In chapter II, a general definition of a linear operator was given, but in the subsequent presentation we considered only bounded, i.e., continuous, operators defined everywhere in H. In the present chapter we begin a study of linear operators which are not necessarily continuous. The related but less restrictive requirement that given operators be *closed* is quite sufficient for many of our purposes.

DEFINITION: *An operator* T (*not necessarily linear*) *is closed if the relations*

$$f_n \in \mathrm{D}_T, \ \lim_{n\to\infty} f_n = f, \ \lim_{n\to\infty} Tf_n = g$$

imply that

$$f \in \mathrm{D}_T, \qquad Tf = g.$$

Thus, the difference between closedness and continuity consists of the following: if the operator T is continuous, then the existence of $\lim_{n\to\infty} f_n$ ($f_n \in \mathrm{D}_T$) implies the existence of $\lim_{n\to\infty} Tf_n$; but if the operator T is only closed, then the convergence of the sequence

$$f_1, f_2, f_3, \dots \qquad (f_n \in \mathrm{D}_T) \tag{1}$$

does not imply the convergence of the sequence

$$Tf_1, Tf_2, Tf_3, \dots . \tag{2}$$

However, if T is closed then, in particular, it has the property that two sequences of the type (2) cannot converge to different limits if the corresponding sequences (1) converge to the same limit.

An operator T having the property mentioned in the preceding sentence may not be closed; but it has *closed extensions*. Among these is the so-called *minimal closed extension*, which is contained in every closed extension of the operator T. The minimal closed extension is uniquely defined for each operator T. It is denoted by $\bar{T}$ and is called the *closure* of T.[1] In

[1] Translator's Note: The reader should verify that the operator $\bar{T}$ defined in the next sentence is closed. To do this, first prove that $\bar{\bar{T}} = \bar{T}$. The minimality of $\bar{T}$ is then easily established.

order to obtain $\bar{T}$, it is sufficient to adjoin to D_T all those elements $f \in \bar{D}_T$ which are limits of sequences (1) generating convergent sequences (2), and to require that

$$\bar{T}f = \lim_{N\to\infty} Tf_n.$$

It is not difficult to show (we leave this to the reader) the truth of the following assertion: if the operator T is closed, then each operator $T - \lambda E$ is closed, and if the inverse operator T^{-1} exists then it is closed.

39. The General Definition of an Adjoint Operator

In Section 22, in the definition of the adjoint operator for a given bounded operator A defined everywhere in H, we started from the fact that each element $g \in$ H uniquely determines an element g^* such that

$$(Af, g) = (f, g^*)$$

for all $f \in$ H. Letting T denote an arbitrary operator, we consider once again the scalar product

$$(Tf, g) \tag{1}$$

where f runs through D_T. We can no longer assert that for every element g the expression (1), as a function of the vector $f \in D_T$, is representable in the form

$$(f, g^*).$$

However, in general, there exist some pairs g and g^* for which

$$(Tf, g) = (f, g^*) \tag{2}$$

for $f \in D_T$. Indeed, this equation holds at least for $g = g^* = 0$.

The existence of vectors g and g^*, for which (2) holds for each $f \in D_T$ is not sufficient to enable us to define an operator T^* which is adjoint to T. It is also necessary that the element g^* be determined uniquely by the element g. This last requirement is fulfilled if and only if D_T is dense in H. Indeed, if D_T is not dense in H then there is a nonzero element h which is orthogonal to D_T; then equation (2) implies that

$$(Tf, g) = (f, g^* + h)$$

for $f \in D_T$. On the other hand, if D_T is dense in H and if, for each $f \in D_T$,

$$(Tf, g) = (f, g_1^*),$$
$$(Tf, g) = (f, g_2^*),$$

then, for each $f \in D_T$,

$$(f, g_1^* - g_2^*) = 0.$$

This implies that $g_1^* = g_2^*$.

Thus, if D_T is dense in H, then the operator T has an adjoint operator T^*. Its domain D_{T^*} is defined as follows: $g \in D_{T^*}$ if and only if there exists a vector g^* such that (2) is satisfied for $f \in D_T$. For each such pair g and g^*,

$$T^*g = g^*.$$

We now list several simple propositions concerning adjoint operators, the proofs of which follow immediately from the definition.

1° The operator T^* is linear.

2° If $S \subset T$, then $S^* \supset T^*$.

3° The operator T^* is closed whether or not T is closed.

4° If the operator T has a closure $\bar{T}$, then $(\bar{T})^* = T^*$.

5° If the operator T^{**} exists, then

$$T \subset T^{**}.$$

The last proposition shows that a necessary condition for the existence of the operator T^{**} is the possibility of closing the operator T. The question as to whether this condition is sufficient for the existence of T^{**} we leave open until Section 46. There we shall consider the problem of the possibility of generalizing to the case of arbitrary operators the equation $T^{**}=T$, which was proved in Section 22 for bounded operators defined everywhere in H.

To complete the present paragraph, we consider the case of a linear operator T which has an inverse operator T^{-1}. Assume that D_T and $D_{T^{-1}}$ are dense in H. Then the operators T^* and $(T^{-1})^*$ exist. We shall prove that

$$(T^*)^{-1} = (T^{-1})^*.$$

To begin with let us assume that f runs through D_T and g runs through $D_{(T^{-1})^*}$. Then

$$(f,g) = (T^{-1}Tf, g) = (Tf, (T^{-1})^*g).$$

But this equation shows that

$$(T^{-1})^*g \in D_{T^*}$$

and

$$T^*(T^{-1})^*g = g. \tag{3}$$

On the other hand, if f runs through $D_{T^{-1}}$ and h runs through D_{T^*}, then

$$(f,h) = (TT^{-1}f, h) = (T^{-1}f, T^*h).$$

It follows that

$$T^*h \in D_{(T^{-1})^*}$$

and

$$(T^{-1})^*T^*h = h. \tag{4}$$

Relations (3) and (4) imply that

$$(T^*)^{-1} = (T^{-1})^*.$$

40. Eigenvectors, Invariant Subspaces and Reducibility of Linear Operators

A complex number λ is called an *eigenvalue* of the linear operator T if there exists a vector $f \neq 0$ such that

$$Tf = \lambda f. \tag{1}$$

Each such vector f is called an *eigenvector* of the operator T (more precisely, f is an eigenvector which belongs to the eigenvalue λ).

For each fixed eigenvalue λ of T the set consisting of all vectors which satisfy equation (1) is a linear manifold which contains at least one non-zero vector; each such set is called an *eigenmanifold* of T. Thus, the eigenmanifold corresponding to a given eigenvalue λ consists of the zero vector and all of the eigenvectors belonging to λ. The multiplicity of an eigenvalue λ is defined as the dimension (finite or infinite) of the corresponding eigenmanifold. If the operator T is closed then each eigenmanifold is closed, so that each eigenmanifold is a subspace.

More general than the concept of a closed eigenmanifold is the concept of an *invariant subspace*. A subspace $\mathrm{H}_1 \subset \mathrm{H}$ is called an invariant subspace of the operator T if every element of D_T belonging to H_1 is mapped by the operator T into an element also belonging to H_1, i.e., if the inclusion relation

$$f \in \mathrm{D}_T \cap \mathrm{H}_1$$

implies the inclusion relation

$$Tf \in \mathrm{H}_1.$$

The operator T determines an operator T_1 defined in the subspace H_1 such that

$$\mathrm{D}_{T_1} = \mathrm{D}_T \cap \mathrm{H}_1, \qquad T_1 \subset T.$$

This operator T_1 is called the *restriction* of the operator T to H_1.

If H_1 is an invariant subspace of the operator T, then its orthogonal complement $\mathrm{H} \ominus \mathrm{H}_1$ may or may not be an invariant subspace of this operator. For the moment, let us suppose that both H_1 and $\mathrm{H}_2 = \mathrm{H} \ominus \mathrm{H}_1$ are invariant subspaces of the operator T, and let T_1 and T_2 be the restrictions of the operator T to H_1 and H_2, respectively. Does the study of the operator T reduce in this case to the study of the two operators T_1 and T_2? The answer is evidently in the affirmative if the operator T is defined everywhere in H. Indeed, for each element $h \in \mathrm{H}$ we have a unique representation

$$h = h_1 + h_2$$

where $h_1 \in \mathrm{H}_1$ and $h_2 \in \mathrm{H}_2$, from which it follows that

$$Th = T_1 h_1 + T_2 h_2.$$

If the operator T is not defined everywhere in H, then the conclusion remains valid only under the additional condition that the projection P_{H_1} on H_1 does not map elements of D_T outside of D_T. Thus, we have the following proposition.

THEOREM 1: *If* H_1 *and its orthogonal complement* H_2 *are invariant subspaces of the operator* T, *and if* $P_{H_1}D_T \subset D_T$ *then, for each* $f \in D_T$,

$$Tf = T_1 f_1 + T_2 f_2,$$

where T_1 *and* T_2 *are the restrictions of* T *to* H_1 *and* H_2, *and* f_1 *and* f_2 *are the projections of* f *on* H_1 *and* H_2.

DEFINITION: *If the subspace* H_1 *satisfies the conditions of Theorem* 1, *then we say that it reduces the operator* T.

It is easy to see that if the subspace H_1 reduces the operator T, then its orthogonal complement H_2 also reduces T. Trivial subspaces which reduce T are the subspace $\{0\}$ and the space H itself. If the operator T does not have any other reducing subspaces, then it is said to be *irreducible*.

THEOREM 2: *Let* P *be the operator of projection on a given subspace* G. *Then* G *reduces* T *if and only if*

$$1) \quad Pf \in D_T \qquad \text{and} \qquad 2) \quad PTf = TPf$$

for $f \in D_T$, *i.e., if the operators* T *and* P *commute.*

Proof: We show first the necessity of the condition of the theorem. If the subspace G reduces T, then $f \in D_T$ implies that $Pf \in D_T$ so that 1) is proved. In order to establish 2), we suppose that

$$f = g + h,$$

where

$$g = Pf.$$

Since G reduces T,

$$Tf = Tg + Th$$

where $Tg \in G$ and $Th \in H \ominus G$. Therefore,

$$PTf = PTg = Tg$$

so that

$$PTf = TPf.$$

In much the same way, the sufficiency of the condition is easily proved.

We say that the projection operator P *reduces* T if the subspace G on which P projects reduces T.

The reduction of the study of the structure of the operator T to the investigation of its reducing subspaces and the restrictions of the operator T to them is based on the following proposition.

THEOREM 3: *Let the subspaces* H_k $(k = 1, 2, 3, \ldots, n;\ n \leqq \infty)$ *be pairwise orthogonal and let*

$$H = \sum_k \oplus H_k.$$

Let T be a linear operator which is reduced by each of the subspaces H_k. *Assume T closed if* $n = \infty$. *Finally, let* P_k *be the operator of projection on* H_k, *and* T_k *the restriction of T to* H_k. *Then* $f \in D_T$ *if and only if*

$$P_k f \in D_{Tk} \text{ and } \sum_{k=1}^{n} \| T_k P_k f \|^2 < \infty. \tag{2}$$

In this case

$$Tf = \sum_{k=1}^{n} T_k P_k f. \tag{3}$$

Proof: Let $f \in D_T$. Since H_k reduces T, we have $P_k f \in D_T$, which implies that

$$P_k f \in D_T \cap H_k = D_{Tk}.$$

In addition, $P_k Tf = TP_k f$. Therefore,

$$Tf = \sum_{k=1}^{n} P_k Tf = \sum_{k=1}^{n} TP_k f = \sum_{k=1}^{n} T_k P_k f.$$

In case $n = \infty$, this result implies the convergence of the series

$$\sum_{k=1}^{\infty} \| T_k P_k f \|^2.$$

Now let us assume conditions (2). If $n < \infty$, then the linearity of D_T implies $f \in D_T$ and equation (3). If $n = \infty$, then the linearity of D_T implies that each of the vectors

$$\sum_{k=1}^{r} P_k f \qquad (r = 1, 2, 3, \ldots)$$

belongs to D_T. Furthermore, the convergence of the series

$$\sum_{k=1}^{\infty} \| T_k P_k f \|^2$$

implies the convergence of the series

$$\sum_{k=1}^{\infty} T_k P_k f = \lim_{r \to \infty} T\left(\sum_{k=1}^{r} P_k f\right).$$

Since the operator T is closed,

$$f \in D_T \text{ and } Tf = \sum_{k=1}^{\infty} T_k P_k f.$$

Remark: The theorem remains valid in the more general case in which the space H is decomposed into an uncountable family of subspaces H_a. This follows from the fact that each of the vectors f and Tf has at most a countable set of nonzero projections on the subspaces H_a (cf. Section 9).

To complete the present paragraph, we consider a typical example. Let the space H be separable and let $\{e_k\}_{k=-\infty}^{\infty}$ be an orthonormal basis. We consider the linear operator U_0 which is defined for these elements by

$$U_0 e_k = e_{k+1} \qquad (-\infty < k < \infty)$$

and then extended to all of H by linearity and continuity; U_0 is a unitary operator. The closure G of the linear envelope of the set of vectors $\{e_k\}_{k=q}^{\infty}$, where q is fixed arbitrarily such that $q > -\infty$, is an invariant subspace of U_0. But G does not reduce U_0. Indeed, if P is the operator of projection on G, then

$$U_0 P e_{q-1} = 0, \quad P U_0 e_{q-1} = P e_q = e_q,$$

so that

$$U_0 P \neq P U_0.$$

The operator U_0 is an example of an operator having no eigenvectors. Indeed if

$$U_0 f = \lambda f, \quad f = \sum_{k=-\infty}^{\infty} a_k e_k \neq 0$$

then

$$\sum_{k=-\infty}^{\infty} a_k e_{k+1} = \lambda \sum_{k=-\infty}^{\infty} a_k e_k$$

whence

$$a_k = \lambda a_{k+1} \qquad (-\infty < k < \infty).$$

Because

$$(f,f) = (U_0 f, U_0 f) = (\lambda f, \lambda f) = |\lambda|^2 (f,f)$$

and $f \neq 0$, we have $|\lambda| = 1$. Therefore,

$$|a_k| = |a_0| \qquad (\pm k = 1, 2, 3, \ldots),$$

which contradicts the fact that

$$0 < \sum_{k=-\infty}^{\infty} |a_k|^2 < \infty.$$

Since U_0 has no eigenvectors, there does not exist a finite subspace reducing U_0. Later (cf. Section 48) we establish the existence of infinite subspaces which reduce U_0.

41. Symmetric Operators

A linear operator A is said to be symmetric[2] *if:*

(a) *its domain* D_A *is dense in* H; *and*

(b) *for* $f, g \in D_A$,

$$(Af, g) = (f, Ag).$$

From this definition it follows that the scalar product (Af, f) is real for $f \in D_A$. It may happen that

$$(Af, f) \geqq 0$$

for each $f \in D_A$. In this case a symmetric operator is said to be *positive*. A *negative* symmetric operator is defined analogously.

If a symmetric operator is bounded, then its extension by continuity is defined everywhere in H and is evidently a bounded self-adjoint operator (cf. Section 22). But if a symmetric operator is not bounded, then by the theorem in Section 25 its domain cannot be the whole space H.

If A is a symmetric operator, then evidently

$$A \subset A^*.$$

Since the adjoint operator is closed, this relation shows that a symmetric operator always has a closure.

If B is a symmetric extension of the operator A, then $B \subset A^*$, i.e., every symmetric extension of the operator A is a restriction of the adjoint operator A^*. Indeed, from $B \supset A$ it follows that $B^* \subset A^*$, and it remains merely to recall that $B \subset B^*$. An operator A which coincides with its adjoint ($A = A^*$) is said to be *self-adjoint*; it does not have a symmetric proper extension. A symmetric operator A not having symmetric proper extensions and not coinciding with its adjoint ($A \subsetneqq A^*$), is called a *maximal* symmetric operator.

THEOREM 1: *A symmetric operator A such that its range Δ_A is all of* H *is self-adjoint.*

Proof: It is sufficient to verify that every element $g \in D_{A^*}$ also belongs to D_A. Thus, let $g \in D_{A^*}$ and $A^*g = g^*$. Since $\Delta_A = H$, there exists an element $h \in D_A$ such that $Ah = g^*$. Consequently, for each $f \in D_A$,

$$(Af, g) = (f, g^*) = (f, Ah) = (Af, h).$$

Again since $\Delta_A = H$, we have $g = h$. So $g \in D_A$ and the theorem is proved.

COROLLARY: *If a bounded self-adjoint operator A has an inverse operator A^{-1} then A^{-1} is self-adjoint* (*bounded or unbounded*).

[2] The word "*Hermitian*" is often used in place of the word "symmetric."

Proof: It suffices to prove that the domain of A^{-1}, i.e., the range of A, is dense in H. Assume, therefore, that the range of A is not dense in H. Then there exists a vector $h \neq 0$ orthogonal to Δ_A, so that $(f, Ah) = (Af, h) = 0$, for every $f \in$ H. It follows that $Ah = 0$, which contradicts the existence of the inverse operator.

THEOREM 2: *The eigenvalues of a symmetric operator are real.*

Proof: If

$$Af = \lambda f \quad (f \neq 0),$$

then

$$\lambda(f,f) = (\lambda f,f) = (Af,f) = (f,Af) = (f,\lambda f) = \bar{\lambda}(f,f),$$

whence

$$\lambda = \bar{\lambda}.$$

THEOREM 3: *Eigenvectors f_1 and f_2 belonging to the two different eigenvalues λ_1 and λ_2 of a symmetric operator are orthogonal.*

Proof: Letting

$$Af_1 = \lambda f_1, \qquad Af_2 = \lambda_2 f_2$$

where $\lambda_1 \neq \lambda_2$, we obtain

$$\lambda_1(f_1,f_2) = (\lambda_1 f_1,f_2) = (Af_1,f_2) = (f_1,Af_2) = (f_1,\lambda_2 f_2) = \lambda_2(f_1,f_2).$$

Therefore

$$(\lambda_1 - \lambda_2)(f_1,f_2) = 0$$

and

$$(f_1,f_2) = 0.$$

THEOREM 4: *If* G *is an invariant subspace of the symmetric operator A and if the projection P on* G *satisfies the relation* $PD_A \subset D_A$ *then the subspace* G *reduces the operator A.*

Proof: In view of Theorem 1 of Section 40, the only thing to prove is that H $\ominus$ G is an invariant subspace of the operator A. According to the conditions of the present theorem, if $f \in D_A \cap (H \ominus G)$ and $g \in D_A$, then

$$(Af, Pg) = 0$$

so that

$$(PAf, g) = 0.$$

Since D_A is dense in H, this implies that

$$PAf = 0$$

and

$$Af \in H \ominus G,$$

so that the theorem is proved.

In conclusion we present the following lemma, which will be needed later (in Appendix II).

LEMMA: *In order that a linear manifold* D ($D_A \subset D \subset D_{A^*}$) *be the domain of a self-adjoint extension of a given symmetric operator A, it is necessary and sufficient that* D *be the set of elements* $f \in D_{A^*}$ *which satisfy the condition*

$$(A^*f, g) = (f, A^*g)$$

for each $g \in D$.

The proof of this lemma follows immediately from the considerations of the present section.

42. More about Isometric and Unitary Operators

In this section we consider isometric operators in the narrow sense, i.e., the domain D_V and the range Δ_V of the operator V are subspaces of the same space H. An isometric operator is said to be *maximal* if it does not have an isometric proper extension.

THEOREM 1:[3] *Each eigenvalue of an isometric operator V has absolute value one.*

Proof: Let

$$Vf = \lambda f,$$

where $f \neq 0$. Then

$$(f,f) = (Vf, Vf) = (\lambda f, \lambda f) = |\lambda|^2 (f,f),$$

whence $|\lambda| = 1$, because $(f,f) \neq 0$.

THEOREM 2: *Eigenvectors f_1 and f_2 belonging to two different eigenvalues λ_1 and λ_2 of an isometric operator V are orthogonal.*

Proof: Let

$$Vf_1 = \lambda_1 f_1, \qquad Vf_2 = \lambda_2 f_2,$$

where $\lambda_1 \neq \lambda_2$. Then

$$(f_1, f_2) = (Vf_1, Vf_2) = (\lambda_1 f_1, \lambda_2 f_2) = \lambda_1 \bar{\lambda}_2 (f_1, f_2)$$

so that

$$(1 - \lambda_1 \bar{\lambda}_2)(f_1, f_2) = 0.$$

Then

$$(f_1, f_2) = 0,$$

since $1 - \lambda_1 \bar{\lambda}_2 \neq 0$.

THEOREM 3: *In order that a subspace* G *reduce a unitary operator U, it is necessary and sufficient that* G *be an invariant subspace of each of the operators U and U^{-1}.*

Proof: Let G reduce U. Then $H \ominus G$ is an invariant subspace of the operator U and, for each $f \in H \ominus G$ and each $g \in G$,

$$(Uf, g) = 0,$$

[3] In essence this fact was established already in Section 40.

$$(f, U^{-1}g) = 0.$$

This implies that $U^{-1}g \in \mathrm{G}$, so that G is an invariant subspace of U^{-1}. Conversely, let G be invariant with respect to the operator U^{-1}. Then, for $f \in \mathrm{H} \ominus \mathrm{G}$ and $g \in \mathrm{G}$,

$$(Uf, g) = (f, U^{-1}g) = 0.$$

It follows that the subspace $\mathrm{H} \ominus \mathrm{G}$ is invariant with respect to the operator U. By an argument similar to that in Theorem 4 of Section 41, G is invariant with respect to U. Consequently, G reduces the operator U.

43. The Concept of the Spectrum (Particularly of a Self-Adjoint Operator)

In linear algebra, the spectrum of a quadratic form is defined as the set of its eigenvalues. Analogously, in the elementary theory of integral equations the spectrum of an equation is defined as the set of eigenvalues of the equation, i.e., the set of eigenvalues of the corresponding integral operator. In this case it turns out that certain nonhomogeneous equations (vector or functional) containing a parameter λ are uniquely solvable for any right member if λ does not belong to the spectrum. These equations are not usually solvable if λ belongs to the spectrum.

We turn now to some general considerations. Suppose that we are given a closed linear operator T defined on a manifold D_T which is dense in H. Let λ denote a parameter which can assume any complex value and consider the operator equation,

$$Tf - \lambda f = g.$$

The study of this equation reduces to an investigation of the linear manifold $\Delta_T(\lambda)$ which consists of the vectors $(T - \lambda E)f$ where f runs through D_T. Thus, $\Delta_T(\lambda)$ is the range of the operator $T - \lambda E$. It can be expressed in the form $\Delta_T(\lambda) = (T - \lambda E)\mathrm{D}_T$. The operator $T - \lambda E = T_\lambda$ defines a correspondence (not necessarily one-to-one) between D_T and $\Delta_T(\lambda)$. If this correspondence is one-to-one, then the operator $T - \lambda E$ has an inverse operator $(T - \lambda E)^{-1}$ with domain $\Delta_T(\lambda)$ and range D_T.

DEFINITION 1: *If* $(T - \lambda E)^{-1}$ *exists and is a bounded operator defined everywhere in* H ($\Delta_T(\lambda) = \mathrm{H}$), *then* λ *is called a* regular value (*or* regular point) *of the operator* T. *All other points of the complex plane comprise the* spectrum *of the operator* T.

In each of the cases mentioned above, which are related to linear algebra and the elementary theory of integral equations, the spectrum of an operator consists of all of its eigenvalues. But, in more general situations, the collection of eigenvalues does not exhaust the spectrum. Indeed, Theorem 1 stated below characterizes the eigenvalues of an operator T as the complex numbers λ for which either the operator $T - \lambda E$ does not have an inverse, or $(T - \lambda E)^{-1}$ exists but is not a bounded operator defined on all of H.

THEOREM 1: *The correspondence between* D_T *and* $\Delta_T(\lambda)$ *determined by the operator* $T - \lambda E$ *is one-to-one if and only if* λ *is not an eigenvalue of the operator* T.

Proof: If the operator $T - \lambda E$ does not determine a one-to-one correspondence between D_T and $\Delta_T(\lambda)$, then there exist $f_1, f_2 \in D_T$ such that $f_1 \neq f_2$ and

$$Tf_1 - \lambda f_1 = g,\ Tf_2 - \lambda f_2 = g.$$

Consequently,

$$Tf = \lambda f,$$

where $f = f_1 - f_2 \neq 0$, so that λ is an eigenvalue of the operator T. The proof of the converse assertion is also simple and is left to the reader.

Instead of considering the general case, which includes all possible hypotheses concerning the operator $(T - \lambda E)^{-1}$ and the domain $\Delta_T(\lambda)$, we restrict ourselves to the important special case in which the original operator is self-adjoint (we shall denote this operator not by T, but by A.)

THEOREM 2: *The number* λ *is an eigenvalue of the self-adjoint operator* A *if and only if*

$$\overline{\Delta_A(\lambda)} \neq H.$$

Proof: Let λ be an eigenvalue of A, so that

$$Af = \lambda f \qquad (f \neq 0).$$

Then, for each $h \in D_A$,

$$(f, \{A - \lambda E\}\, h) = (Af - \lambda f, h) = 0,$$

which implies that

$$f \perp \Delta_A(\lambda).$$

But this is possible only when $\overline{\Delta_A(\lambda)} \neq H$. We suppose now that $\overline{\Delta_A(\lambda)} \neq H$. Then there exists a nonzero vector f which is orthogonal to the manifold $\Delta_A(\lambda)$. Therefore, for each $h \in D_A$,

$$(f, \{A - \lambda E\}\, h) = 0.$$

It follows that $f \in D_{A^*}$ and

$$A^* f = \bar{\lambda} f.$$

But $A^* = A$, so that

$$Af = \bar{\lambda} f,$$

i.e., $\bar{\lambda}$ is an eigenvalue of the operator A. Finally, $\bar{\lambda} = \lambda$ because the eigenvalues of a self-adjoint operator are real.

We note that in the course of the proof of Theorem 2, we have proved the following proposition.

THEOREM 2*: *The eigenmanifold of an operator A corresponding to the eigenvalue* λ *is the orthogonal complement of the linear manifold* $\Delta_A(\lambda) = (A - \lambda E)D_A$.

THEOREM 3: *Nonreal points* λ *in the complex plane are regular points of a self-adjoint operator A.*

Proof: The number $\lambda = \xi + i\eta$ $(\eta \neq 0)$ cannot be an eigenvalue of the operator A. Therefore, on the basis of Theorem 1, the operator $(A - \lambda E)^{-1}$ exists. Letting

$$(A - \lambda E)f = g,$$

we obtain

$$\begin{aligned}\|g\|^2 &= (\{A - \xi E\}f - i\eta f, \{A - \xi E\}f - i\eta f) = \\ &= \|\{A - \xi E\}f\|^2 + i\eta(\{A - \xi E\}f, f) - i\eta(f, \{A - \xi E\}f) + \eta^2\|f\|^2 = \\ &= \|\{A - \xi E\}f\|^2 + \eta^2\|f\|^2,\end{aligned}$$

whence

$$\|f\| \leqq \frac{1}{|\eta|}\|g\|,$$

i.e.,

$$\|(A - \lambda E)^{-1}g\| \leqq \frac{1}{|\eta|}\|g\|.$$

Since this relation is valid for every $g \in \Delta_A(\lambda)$, the operator $(A - \lambda E)^{-1}$ is bounded. By Theorem 2 and the fact that λ is not an eigenvalue of the operator A,

$$\overline{\Delta_A(\lambda)} = H.$$

It remains to show that the manifold $\Delta_A(\lambda)$ is closed. Supposing that $\Delta_A(\lambda) \neq \overline{\Delta_A(\lambda)}$ we extend the bounded operator $(A - \lambda E)^{-1}$ by continuity to $\overline{\Delta_A(\lambda)}$. This extension coincides with the closure of the operator $(A - \lambda E)^{-1}$, which therefore is not closed. But this is impossible because the closedness of the operator A implies the closedness of the operator $(A - \lambda E)^{-1}$.

COROLLARY 1: *The spectrum of a self-adjoint operator is a subset of the real axis.*

COROLLARY 2: *A regular point of the self-adjoint operator A can be defined as a value of the parameter* λ *for which* $\Delta_A(\lambda) = H$.

Proof: If λ is nonreal then it is regular by Theorem 3. If λ is real and $\Delta_A(\lambda) = H$ then, by Theorem 2, λ is not an eigenvalue of the operator A. Therefore, by Theorem 1, there exists the inverse operator $(A - \lambda E)^{-1}$ defined everywhere in H. This operator is self-adjoint and consequently (cf. the beginning of Section 41) it is bounded, so that Definition 1 applies.

Now, without contradicting Definition 1, we can use the following definition.

DEFINITION 2: *If A is a self-adjoint operator, then the point λ is a regular point of A if $\Delta_A(\lambda) = \mathrm{H}$ and λ is a point of the spectrum if $\Delta_A(\lambda) \neq \mathrm{H}$.*

A further refinement of the concept of the spectrum of a self-adjoint operator is given by the following definition.

DEFINITION 3: *We say that the point λ belongs to the point (discrete) spectrum of the self-adjoint operator A if $\overline{\Delta_A(\lambda)} \neq \mathrm{H}$ and λ belongs to the continuous spectrum if $\Delta_A(\lambda) \neq \overline{\Delta_A(\lambda)}$.*

The possibility that λ belongs to the point spectrum and also to the continuous spectrum (for $\overline{\Delta_A(\lambda)} \neq \Delta_A(\lambda) \neq \mathrm{H}$) is not excluded.

On the basis of the theorem 2, the point spectrum of a self-adjoint operator coincides with the set of its eigenvalues.

To complete the present paragraph, we prove the following proposition.

THEOREM 4: *The spectrum of a self-adjoint operator is a closed set.*

Proof: It is sufficient to show that the set of regular points of a self-adjoint operator A is open. Let λ_0 be a regular point. Then there exists a number $k > 0$ such that

$$\|Af - \lambda_0 f\| \geqq k\,\|f\|$$

for $f \in \mathrm{D}_A$. If $0 < \delta \leqq \frac{k}{2}$ then, for $|\lambda - \lambda_0| \leqq \delta$ and $f \in \mathrm{D}_A$,

$$\|Af - \lambda f\| \geqq \|Af - \lambda_0 f\| - \delta\,\|f\| \geqq \frac{1}{2} k\,\|f\|.$$

Therefore, in the first place, λ is not an eigenvalue of the operator A, so that $\overline{\Delta_A(\lambda)} = \mathrm{H}$, and, in the second place, the inverse operator $(A - \lambda E)^{-1}$ is bounded. The equation $\Delta_A(\lambda) = \overline{\Delta_A(\lambda)}$ is a result of the fact that A is closed. Thus, every point λ such that $|\lambda - \lambda_0| \leqq \delta$ is regular, and the theorem is proved.

44. The Resolvent

As in the preceding Section, we begin with an arbitrary closed linear operator T, the domain of which is dense in H, and specialize later to a self-adjoint operator A. The operator $R_\lambda = (T - \lambda E)^{-1}$, which depends on the parameter λ, is called the *resolvent* of the operator T, and is defined for all the values λ for which it exists and for which its domain of definition, i.e., $\Delta_T(\lambda)$, is dense in H.

For each regular point of the operator T the resolvent R_λ is a bounded operator defined on the whole space. The operator R_λ determines a one-to-one correspondence between $\Delta_T(\lambda)$ and D_T. In particular, if λ is a regular point of the operator T then $R_\lambda h = 0$ if and only if $h = 0$.

THEOREM 1: *For any two regular points λ and μ of the operator T the equation,*

$$R_\mu - R_\lambda = (\mu - \lambda) R_\mu R_\lambda,$$

is satisfied. (This is the so-called Hilbert relation.)

Proof: Since λ and μ are regular points of the operator T we have, for each $h \in H$,

$$R_\lambda h = R_\mu (T - \mu E) R_\lambda h,$$
$$R_\mu h = R_\mu (T - \lambda E) R_\lambda h.$$

Subtracting the first equation from the second, we get the desired relation.

From the Hilbert relation follows the commutativity of the resolvents corresponding to any pair of regular values μ, λ:

$$R_\lambda R_\mu = R_\mu R_\lambda.$$

This is a special case of the following general proposition.

THEOREM 2: *In order that the operator T commute with a given bounded operator S which is defined everywhere in H, it is necessary that S commute with the resolvent $R_\lambda = (T - \lambda E)^{-1}$ for each regular value λ, and it is sufficient that S and R_λ commute for at least one regular value λ.*

Proof: We suppose first that the operators T and S commute, i.e., that

$$TSf = STf$$

for $f \in D_T$. If λ is a regular point of the operator T and

$$f = R_\lambda h,$$

then f runs through D_T as h runs through H. But, the commutativity of the operators T and S implies that

$$(T - \lambda E)Sf = S(T - \lambda E)f$$

for each $f \in D_T$, which yields

$$R_\lambda(T - \lambda E) Sf = R_\lambda S(T - \lambda E)f.$$

Since $R_\lambda(T - \lambda E) = E$ and $(T - \lambda E)f = h$, we obtain

$$SR_\lambda h = R_\lambda Sh.$$

Since proof of the second part of the assertion is quite simple, it is omitted.

We turn now to an arbitrary self-adjoint operator A. We shall define its resolvent also for eigenvalues (i.e., for points of the discrete spectrum), after which the resolvent of A will be defined for all points of the λ-plane.

With this aim, we assume that λ' is an eigenvalue of the operator A

and we denote by $G_{\lambda'}$ the corresponding eigenmanifold. As we know, $G_{\lambda'}$ reduces the operator A. Let A' be the restriction of the operator A to $H \ominus G_{\lambda'} = H'$. It is easy to see that A' is a self-adjoint operator in H' for which λ' is not an eigenvalue. We define $R_{\lambda'}$ by

$$R_{\lambda'} = (A' - \lambda' E)^{-1}.$$

The domain of the operator $R_{\lambda'}$ is the manifold $\Delta_{A'}(\lambda')$, which is dense in H'. The manifold $\Delta_{A'}(\lambda')$ consists of all vectors which can be represented in the form

$$(A' - \lambda' E)f' = (A - \lambda' E)f',$$

with $f' \in D_{A'}$. But it is easy to see that this manifold also consists of all vectors of the form

$$(A - \lambda' E)f$$

for $f \in D_A$. Thus,

$$\Delta_{A'}(\lambda') = \Delta_A(\lambda').$$

The range of the operator $R_{\lambda'}$ is obtained if D_A is projected on the orthogonal complement of the eigenmanifold corresponding to λ'.

The manner in which the resolvent R_λ of a self-adjoint operator depends on the value of the parameter λ is shown in the following table:

λ is a regular point of the the operator.	R_λ is a bounded operator defined everywhere in H.
λ belongs to the point spectrum, but does not belong to the continuous spectrum.	R_λ is a bounded operator defined on a set which is not dense in H.
λ belongs to the continuous spectrum but does not belong to the point spectrum.	R_λ is an unbounded operator defined on a set which is dense in H.
λ belongs both to the point spectrum and to the continuous spectrum.	R_λ is an unbounded operator defined on a set which is not dense in H.

To complete the present section, we show that

$$(R_\lambda)^* = R_{\bar{\lambda}} \tag{1}$$

if and only if λ does not belong to the point spectrum of the operator (if it does, then the operator R_λ does not have an adjoint). If λ is real, then $R_\lambda = (A - \lambda E)^{-1}$ is a self-adjoint operator, and (1) is obvious. Suppose λ is not real. Then, for $f, g \in H$,

$$(R_\lambda f, g) = (R_\lambda f, \{A - \bar{\lambda} E\} R_{\bar{\lambda}} g) = (\{A - \lambda E\} R_\lambda f, R_{\bar{\lambda}} g) = (f, R_{\bar{\lambda}} g),$$

which implies formula (1).

45. Conjugation Operators[4]

A *conjugation operator* is an operator I defined on H such that

1) $(If, Ig) = \overline{(f, g)}$,
2) $I^2 f = f$

for $f, g \in$ H. From 2) it follows that the range of the operator I is the whole space H. In fact, each vector $h \in$ H can be represented in the form $h = Ig$ merely by taking $g = Ih$. Instead of the usual linearity, the operator I has the following property, which is sometimes called conjugate linearity:

$$I(\alpha f + \beta g) = \bar{\alpha} If + \bar{\beta} Ig.$$

Indeed, letting

$$g = Ih$$

in 1), we get

$$(If, h) = \overline{(f, Ih)}.$$

It follows that

$$(I(\alpha f + \beta g), h) = \overline{(\alpha f + \beta g, Ih)} = \bar{\alpha}\overline{(f, Ih)} + \bar{\beta}\overline{(g, Ih)} =$$
$$= \bar{\alpha}(If, h) + \bar{\beta}(Ig, h) = (\bar{\alpha} If + \bar{\beta} Ig, h),$$

and the assertion is proved.

An example of a conjugation operator in L^2 is the operation of transition to the complex conjugate function:

$$I\varphi(t) = \overline{\varphi(t)}.$$

For each conjugation operator in a separable space it is possible to select an orthonormal basis $\{e_k\}_1^\infty$ such that if

$$f = \sum_{k=1}^{\infty} x_k e_k,$$

then

$$If = \sum_{k=1}^{\infty} \bar{x}_k e_k.$$

The proof of this simple fact we leave to the reader.

DEFINITION: *A symmetric operator A is said to be real with respect to a given conjugation operator I, if the operators A and I commute, i.e., if $f \in \mathrm{D}_A$ implies that $If \in \mathrm{D}_A$ and*

$$IAf = AIf.$$

THEOREM: *If R_λ is the resolvent of a self-adjoint operator A which is real with respect to a given conjugation operator I, then for each nonreal λ*

$$R_\lambda = IR_\lambda^* I. \tag{1}$$

[4] The results of the present section will be used only in Appendix II.

Proof: We apply the conjugation operator to both members of the equation

$$(A - \lambda E)R_\lambda = E$$

to get

$$I(A - \lambda E)R_\lambda = I.$$

Since A is real with respect to I,

$$(A - \bar{\lambda} E)\, IR_\lambda = I.$$

Now we apply to both members the operator $IR_{\bar{\lambda}}$. This gives

$$R_\lambda = IR_{\bar{\lambda}}\, I.$$

Since $R_{\bar{\lambda}} = R_\lambda^*$, the proof is complete.

If T is a linear operator with domain dense in H and I is a conjugation operator, then, by analogy with the terminology of the theory of matrices, the operator $IT^*I = T'$ is called the transposition with T. Now (1) can be written in the form

$$R_\lambda' = R_\lambda.$$

46. The Graph of an Operator

We consider the set of *ordered pairs* $\{f, g\}$ where the *abscissa* f and the *ordinate* g run through the Hilbert space H. A vector space is defined on this set by means of the equations

$$\alpha\{f, g\} = \{\alpha f, \alpha g\}, \quad \{f_1, g_1\} + \{f_2, g_2\} = \{f_1 + f_2, g_1 + g_2\}$$

(the zero element is $\{0, 0\}$). With the scalar product defined by

$$(\{f_1, g_1\}, \{f_2, g_2\}) = (f_1, f_2) + (g_1, g_2)$$

this vector space becomes a Hilbert space, which we denote by **H**.

Let T be an operator in H. Then the set $\mathbf{M}(T)$ of all points of the form $\{f, Tf\}$ is called the *graph* of the operator T. Every point of the set $\mathbf{M}(T)$ is determined uniquely by its abscissa. Conversely, if all the points of a set **M** in **H** are determined uniquely by their abscissae, then **M** is the graph of some operator in **H**.

Whether or not an operator is closed is reflected in a very simple manner by its graph. In fact, an operator T in H is closed if and only if its graph $\mathbf{M}(T)$ is a closed subset of **H**. Starting from the fact that every subset of **H** has a closure one might be tempted to infer that every operator T has a closure. The fault in this reasoning is that the closure $\overline{\mathbf{M}(T)}$ of the graph of T may contain points which are not determined uniquely by their abscissae. In this case, $\overline{\mathbf{M}(T)}$ is not the graph of any operator. However, if the set $\overline{\mathbf{M}(T)}$ does not contain two distinct points with identical abscissae, then the operator T has a closure $\bar{T}$ and $\mathbf{M}(\bar{T}) = \overline{\mathbf{M}(T)}$.

It is easily seen that if the operator T is linear, then the set $\mathbf{M}(T)$ is a linear manifold in H.

We now define an operator $\boldsymbol{U}$ on $\mathbf{H}$ by

$$\boldsymbol{U}\{f,g\} = \{ig, -if\}.$$

The operator $\boldsymbol{U}$ is unitary since its range is the whole space $\mathbf{H}$ and

$$(\boldsymbol{U}\{f_1,g_1\}, \boldsymbol{U}\{f_2,g_2\}) = (\{ig_1, -if_1\}, \{ig_2, -if_2\}) =$$
$$= (g_1,g_2) + (f_1,f_2) = (\{f_1,g_1\}, \{f_2,g_2\}).$$

We remark also that $\boldsymbol{U}^2 = \boldsymbol{E}$, where $\boldsymbol{E}$ is the identity operator in $\mathbf{H}$.

We now consider the questions raised in Section 39 concerning the existence of T^{**} and the validity of the equation $T^{**} = \bar{T}$. The graph of T will be used in the analysis of these questions. To begin with, we consider the equation

$$(Tf,g) - (f,g^*) = -i(\{iTf, -if\}, \{g,g^*\}) = -i(\boldsymbol{U}\{f,Tf\}, \{g,g^*\}),$$

which holds for each $f \in \mathrm{D}_T$ and any pair of elements g and g^* in H. This equation has the following consequences:

1° In order that the elements $g, g^* \in \mathrm{H}$ satisfy the equation,

$$(Tf,g) = (f,g^*),$$

for each $f \in \mathrm{D}_T$, it is necessary and sufficient that the element $\{g, g^*\}$ of the space $\mathbf{H}$ be orthogonal to the image $\boldsymbol{U}\mathbf{M}(T)$ under $\boldsymbol{U}$ of the graph of the operator T.

2° If the operator T^* exists, then its graph is given by

$$\mathbf{M}(T^*) = \mathbf{H} \ominus \overline{\boldsymbol{U}\mathbf{M}(T)}.$$

3° The operator T^* exists if and only if each point of the set

$$\mathbf{H} \ominus \overline{\boldsymbol{U}\mathbf{M}(T)}$$

is determined uniquely by its abscissa.

Consequence 3° is another criterion for the existence of the adjoint operator (the first criterion, the density of the manifold D_T in H, was established in Section 39).

THEOREM 1: *If the linear operator T with domain dense in* H *has a closure then the operator T^{**} exists and is the closure* [5] *of the operator T:*

$$T^{**} = \bar{T}.$$

(*In particular, if the operator T is closed and $\bar{\mathrm{D}}_T = \mathrm{H}$, then $T = T^{**}$.*)

Proof: We assume, at first, that the operator T is closed. Then the set $\mathbf{M}(T)$ is closed, so that $\boldsymbol{U}\mathbf{M}(T)$ also is closed. Therefore, the relation in consequence 2° can be written in the form

$$\mathbf{H} = \boldsymbol{U}\mathbf{M}(T) \oplus \mathbf{M}(T^*).$$

---

[5] Thus, Theorem 1 gives a method for finding the closure of an operator. This method is used sometimes in applications (cf. Section 49).

Hence, applying the operator U, we get

$$\mathbf{H} = \mathbf{M}(T) \oplus U\mathbf{M}(T^*) \tag{1}$$

which implies that

$$\mathbf{H} \ominus U\mathbf{M}(T^*) = \mathbf{M}(T). \tag{2}$$

Since the points of the graph $\mathbf{M}(T)$ are determined uniquely by their abscissae, it follows by means of consequence 3° that the operator T^{**} exists. In view of (2) and consequence 2°, T^{**} coincides with T. So the theorem is proved for the case in which the operator T is closed. Let us assume now that the operator T is not closed but has a closure $\bar{T}$. Then, by what has been proved,

$$(\bar{T})^{**} = \bar{T}.$$

But

$$(\bar{T})^{**} = [(\bar{T})^*]^* = (T^*)^* = T^{**}$$

and, hence,

$$T^{**} = \bar{T},$$

which proves the theorem.

To complete the present paragraph, we prove by the method of graphical representation another remarkable proposition.

THEOREM 2: *If T is a closed linear operator with domain dense in* H, *then the product T^*T is a self-adjoint (and, therefore, positive) operator.*

Proof: First, we note that for $f, g \in D_{T^*T}$,

$$(T^*Tf, g) = (Tf, Tg) = (f, T^*Tg)$$

and

$$(T^*Tf, f) = (Tf, Tf) \geqq 0.$$

Thus, T^*T is a positive operator. Let h be an arbitrary element of H. Since T is closed, relation (1) holds and, therefore, the element $\{h, 0\} \in \mathbf{H}$ has a unique representation in the form

$$\{h, 0\} = \{f_0, Tf_0\} + U\{g_0, T^*g_0\}$$

or equivalently,

$$\{h, 0\} = \{f_0, Tf_0\} + i\{T^*g_0, -g_0\}.$$

It follows that

$$h = f_0 + iT^*g_0, \qquad 0 = Tf_0 - ig_0$$

whence

$$h = (E + T^*T)f_0.$$

Thus, for every $h \in$ H, the equation

$$(E + T^*T)f = h \tag{3}$$

is solvable (uniquely). From this it follows that D_{T^*T} is dense in H. Indeed, let h be any vector which is orthogonal to D_{T^*T}. Now, h can be represented in the form (3). Hence, for each $g \in D_{T^*T}$,

$$0 = (h, g) = ((E + T^*T)f, g) = (f, (E + T^*T)g).$$

Let $g = f$ to get

$$0 = (f,f) + (f, T^*Tf) = (f,f) + (Tf, Tf).$$

This equation yields $f = 0$, which implies that $h = 0$. Therefore, D_{T^*T} is dense in H. It follows that T^*T and, hence, $E + T^*T$ are symmetric operators. But, according to what has been proved, the range of the operator $E + T^*T$ is the whole space H. Therefore (cf. Section 41), $E + T^*T$ is a self-adjoint operator. Since

$$T^*T = (E + T^*T) - E,$$

T^*T is self-adjoint. This completes the proof. Under the conditions of the theorem, it is proved in the same way that TT^* is a positive self-adjoint operator.

47. Matrix Representations of Unbounded Symmetric Operators

The present paragraph is essentially an extension of Section 26. We assume again that the space H is separable and concern ourselves with the question of the matrix representation of an operator A. The operator A is now unbounded, symmetric and closed.

As in Section 26, we select an orthonormal basis $\{e_k\}_1^\infty$ in H, which is no longer arbitrary but must belong to the dense subset D_A. Let

$$(1) \qquad Ae_k = c_k \qquad (k = 1, 2, 3, \ldots)$$

and

$$(2) \qquad (Ae_k, e_i) = a_{ik} \qquad (i, k = 1, 2, 3, \ldots).$$

We shall attempt to recover the operator A from its values for the vectors e_k or, what is the same thing, from the matrix (a_{ik}).

With this purpose in mind, we introduce the linear envelope $L = L(e_1, e_2, e_3, \ldots)$ of the set of all vectors e_k $(k = 1, 2, 3, \ldots)$. Let B denote the linear operator defined for e_k, $k \geqq 1$, by

$$(3) \qquad Be_k = c_k, \qquad (k = 1, 2, 3, \ldots)$$

and extended by linearity to L. Thus, the domain D_B of B is the linear manifold L. Since $\bar{a}_{ik} = a_{ki}$, B is symmetric. In view of (1) and (3), A is a closed extension of B. Therefore, the closure (the minimal closed extension) $\bar{B}$ of B satisfies the relation

$$\bar{B} \subset A.$$

If a closed linear operator has the matrix representation (a_{ik}) with respect to the basis $\{e_k\}_1^\infty$ then this operator must be $\bar{B}$ rather than any of its closed extensions, which also satisfy (1) and (2). It is possible that $\bar{B} = A$. In this case we say that the operator A is represented by the matrix (a_{ik}) with respect to the basis $\{e_k\}_1^\infty$. A change of basis $\{e_k\}_1^\infty$ changes the matrix (a_{ik}) and the operator $\bar{B}$. Therefore, the question arises: given a closed symmetric operator A is it possible to find an orthonormal basis $\{e_k\}_1^\infty$ with respect to which $\bar{B} = A$, i.e., with respect to which the operator A has a matrix representation? Below (Theorem 3) we answer this question in the affirmative.

DEFINITION: *An orthonormal basis $\{e_k\}_1^\infty$ is called a basis for a matrix representation for a closed symmetric operator A if:*

1. *the elements of this basis belong to* D_A; and
2. *there is a minimal closed linear operator which assumes the value Ae_k at e_k, $k \geqq 1$.*

In contrast with Section 26 we have not considered yet the question of the construction of the components of Af from those of f. The following two theorems are devoted to this question.

THEOREM 1: *Let A be a closed symmetric operator; let $\{e_k\}$ be an arbitrary orthonormal basis, the elements of which belong to D_A; and, finally, let*

$$(Ae_k, e_i) = a_{ik} \qquad (i, k = 1, 2, 3, \ldots).$$

Then the value of Af for each $f \in D_A$ is given by the formulas

$$(4_1) \qquad Af = \sum_{i=1}^{\infty} y_i e_i,$$

$$(4_2) \qquad y_i = \sum_{k=1}^{\infty} a_{ik} x_k \qquad (i = 1, 2, 3, \ldots)$$

if

$$(5) \qquad f = \sum_{k=1}^{\infty} x_k e_k.$$

Proof: The proof follows from the equations,

$$y_i = (Af, e_i) = (f, Ae_i) = \sum_{k=1}^{\infty} (f, e_k)(e_k, Ae_i) =$$

$$= \sum_{k=1}^{\infty} (f, e_k)(Ae_k, e_i) = \sum_{k=1}^{\infty} a_{ik} x_k \quad (i = 1, 2, 3, \ldots).$$

THEOREM 2: *Let A be a closed symmetric operator; let $\{e_k\}_1^\infty$ be a basis for a matrix representation of A; and let*

$$a_{ik} = (Ae_k, e_i) \qquad (i, k = 1, 2, 3, \ldots).$$

Finally, define the operator T by the relations

$$Tf = \sum_{i=1}^{\infty} z_i e_i,$$

$$z_i = \sum_{k=1}^{\infty} a_{ik} x_k$$

on the set D_T *of all vectors*

$$f = \sum_{k=1}^{\infty} x_k e_k,$$

for which

$$\sum_{i=1}^{\infty} \left| \sum_{k=1}^{\infty} a_{ik} x_k \right|^2 < \infty.$$

Then $T = A^*$, *i.e.,* T *is the adjoint of* A.

Proof: We prove first that

$$(6_1) \qquad A^* \subset T.$$

Let $g \in D_{A^*}$ and $A^*g = g^*$. Letting

$$g = \sum_{k=1}^{\infty} x_k e_k, \quad g^* = \sum_{i=1}^{\infty} z_i e_i,$$

we obtain

$$z_i = (g^*, e_i) = (g, Ae_i) = \sum_{k=1}^{\infty} (g, e_k)(e_k, Ae_i) =$$

$$= \sum_{k=1}^{\infty} (g, e_k)(Ae_k, e_i) = \sum_{k=1}^{\infty} a_{ik} x_k.$$

We also have

$$\sum_{i=1}^{\infty} |z_i|^2 = \|g^*\|^2 < \infty.$$

Therefore, the vector g belongs to D_T and $Tg = g^*$, so that relation (6_1) is proved. In the proof we did not use the hypothesis that the basis $\{e_k\}_1^{\infty}$ is a basis for the matrix representation of A. We now prove that

$$(6_2) \qquad T \subset A^*.$$

Letting $g \in D_T$ and

$$g = \sum_{k=1}^{\infty} x_k e_k$$

we obtain

$$(Ae_i, g) = \sum_{k=1}^{\infty} \bar{x}_k (Ae_i, e_k) = \sum_{k=1}^{\infty} a_{ki} \bar{x}_k.$$

But since

$$(Tg, e_i) = \sum_{k=1}^{\infty} a_{ik} x_k = \sum_{k=1}^{\infty} \bar{a}_{ki} x_k$$

we also obtain

$$(Ae_i, g) = \overline{(Tg, e_i)} = (e_i, Tg).$$

It follows that

$$(7) \qquad (Af, g) = (f, Tg)$$

for each f in the linear envelope of the vectors e_i $(i = 1, 2, 3, \ldots)$. But, since $\{e_k\}_1^{\infty}$ is a basis for a matrix representation of the operator A, equation

(7) holds for $f \in D_A$. Hence, $g \in D_{A^*}$ and $A^*g = Tg$, so that relation (6_2) is established. Since relations (6_1) and (6_2) yield $T = A^*$, the theorem is proved.

We remark that Theorem 2 yields the equation

$$\sum_{i=1}^{\infty}\Big(\sum_{k=1}^{\infty} a_{ik}x_k\Big)\bar{y}_i = \sum_{k=1}^{\infty} x_k \overline{\Big(\sum_{i=1}^{\infty} a_{ki}y_i\Big)} \tag{8}$$

where it is assumed that the vector $\sum_{k=1}^{\infty} x_k e_k$ belongs to D_A and that the vector $\sum_{i=1}^{\infty} y_i e_i$ belongs to D_{A^*}. Equation (8) can be expressed in the form

$$\sum_{i=1}^{\infty}\Big(\sum_{k=1}^{\infty} a_{ik}x_k\Big)\bar{y}_i = \sum_{k=1}^{\infty}\Big(\sum_{i=1}^{\infty} a_{ik}\bar{y}_i\Big)x_k.$$

If the reversal of order of summation which this equation indicates is valid for every vector in D_{A^*}, then the operator A^* is symmetric. In this and only this case A is a self-adjoint operator.

The above theorems indicate why matrix representations of symmetric operators cannot be based on formulas of the form (4_1), (4_2) and (5), as was done in Section 26 for bounded operators.

The possibility of the matrix representation of an operator is equivalent to the property that the operator can be recovered from the matrix.

THEOREM 3: *There exists a basis for matrix representation of each closed symmetric operator A.*

Proof: We prove first that there exists a sequence $\{f_k\}_1^\infty \subset D_A$ such that, for each $f \in D_A$, there is a subsequence $\{f_{k_i}\}_{i=1}^\infty$ for which

$$\lim_{i\to\infty} f_{k_i} = f, \quad \lim_{i\to\infty} Af_{k_i} = Af.$$

After this, in order to complete the proof of the theorem it remains only to orthogonalize the sequence $\{f_k\}_1^\infty$. As a preliminary to the construction of the sequence $\{f_k\}_1^\infty$, we choose an arbitrary sequence $\{h_k\}_1^\infty$ which is dense in H. If for a triple of integers, (m, n, p), there exist elements $f \in D_A$ satisfying the inequalities

$$\|h_m - f\| \leq \frac{1}{p}, \qquad \|h_n - Af\| \leq \frac{1}{p},$$

then we associate with this triple any one such element f and denote it by $f_{m,n,p}$. Since $\{h_k\}_1^\infty$ is dense in H, this defines an infinite sequence $\{f_{m,n,p}\}$. The possibility that a triple (m, n, p) might not have a corresponding element, or that the same element might correspond to different triples will not matter in what follows. We enumerate the elements $f_{m,n,p}$ in order to obtain the desired sequence $\{f_k\}_1^\infty$. For the proof we choose an arbitrary element of $f \in D_A$, an arbitrary number $\varepsilon > 0$, and any integer $p' \geqq \frac{2}{\varepsilon}$. Since the sequence $\{h_k\}_1^\infty$ is dense in H there exist integers m', n' such that

$$(9_1) \qquad \| h_{m'} - f \| \leqq \frac{1}{p'}, \; \| h_{n'} - Af \| \leqq \frac{1}{p'} .$$

In view of (9_1) the triple (m', n', p') has an associated element $f_{m',n',p'}$ such that

$$(9_2) \qquad \| h_{m'} - f_{m',n',p'} \| \leqq \frac{1}{p'}, \; \| h_{n'} - Af_{m',n',p'} \| \leqq \frac{1}{p'} .$$

From (9_1) and (9_2) it follows that

$$\| f_{m',n',p'} - f \| \leqq \varepsilon, \; \| Af_{m',n',p'} - Af \| \leqq \varepsilon.$$

Since $\varepsilon > 0$ is arbitrary, the existence of the required subsequence $\{f_{k_i}\}_{i=1}^\infty$ is established, so that the theorem is proved.

We have proved that to each closed symmetric operator there corresponds a (Hermitian) matrix which represents the operator in terms of some basis. However, not every Hermitian matrix defines a symmetric operator.

THEOREM 4: *If the Hermitian matrix (a_{ik}) satisfies the relations*

$$(10) \qquad \sum_{i=1}^{\infty} | a_{ik} |^2 < \infty \qquad (k = 1, 2, 3, \ldots),$$

then, in terms of each orthonormal basis, it defines a closed symmetric operator.

Proof: It suffices to let

$$Ae_k = \sum_{i=1}^{\infty} a_{ik} e_i \quad (k = 1, 2, 3, \ldots)$$

and then to construct the operator $\bar{B}$ by the method indicated at the beginning of the present section.

The Hermitian matrices satisfying condition (10), but which are not bounded, are called *unbounded* Hermitian matrices. An unbounded Hermitian matrix (a_{ik}) does not admit, in general, a transformation by an arbitrary unitary matrix by means of the formula

$$(\mathring{a}_{ik}) = (u_{ir}^*) \cdot (a_{rs}) \cdot (u_{sk}).$$

If the corresponding infinite series converge so that such a transformation is formally possible, then it can happen, nevertheless, that the transformed Hermitian matrix $(\mathring{a}_{ik})$ does not satisfy condition (10) and hence does not determine an operator in H. Moreover, even if the transformed matrix satisfies condition (10) then the operator $\mathring{A}$ determined by it may not coincide with A. (It is remarkable that the intersection $D_A \cap D_{\mathring{A}}$ may turn out to be empty.)

These various possibilities form the foundations for the so-called pathological properties of unbounded Hermitian matrices.[6] These properties make matrices unsuitable for the study of unbounded symmetric operators.

48. The Operation of Multiplication by the Independent Variable

If $[a, b]$ is a finite interval, then the operator Q of multiplication by independent variable is defined for each function $\varphi = \varphi(t) \in L^2(a, b)$ by the equation

$$Q\,\varphi(t) = t\,\varphi(t). \tag{1}$$

In this case Q is a bounded self-adjoint operator, the norm of which is equal to the larger of the numbers $|a|$ and $|b|$. If $[a, b]$ is an infinite interval, then the multiplication operator Q is defined by formula (1) on the manifold D_Q of functions $\varphi(t) \in L^2(a, b)$ such that $t\,\varphi(t) \in L^2(a, b)$. This manifold D_Q is dense in $L^2(a, b)$ since it contains the set D of all functions in $L^2(a, b)$ each of which vanishes outside of a finite interval (the interval depending on the function). If $[a, b]$ is an infinite interval then it is evident that Q is an unbounded symmetric operator. We show now that Q is self-adjoint.

Let $\psi \in D_{Q^*}$ and $\psi^* = Q^*\psi$. Then for each $\varphi \in D_Q$,

$$(Q\,\varphi, \psi) = (\varphi, \psi^*).$$

Equivalently,

$$\int_a^b t\,\varphi(t)\overline{\psi(t)}\,dt = \int_a^b \varphi(t)\overline{\psi^*(t)}\,dt$$

or

$$\int_a^b \varphi(t)\,\{t\,\overline{\psi(t)} - \overline{\psi^*(t)}\}\,dt = 0.$$

The last equation holds, in particular, for any function $\varphi(t)$ in $L^2(a, b)$ which vanishes outside of a finite interval. Therefore, for any finite α and β in the interval (a, b),

$$\int_\alpha^\beta \varphi(t)\,\{t\,\overline{\psi(t)} - \overline{\psi^*(t)}\}\,dt = 0.$$

It follows that, for almost all t in $[a, b]$,

$$\psi^*(t) = t\,\psi(t).$$

Hence

$$t\,\psi(t) \in L^2(a, b),$$

[6] Cf. J. v. Neumann [1].

and

$$\psi \in D_Q, \ \psi^* = Q\psi.$$

This argument shows that $Q^* \subset Q$. But since $Q \subset Q^*$ by the symmetry of Q, we have $Q = Q^*$. From the equation $Q = Q^*$ it follows, in particular, that Q is closed.

If we restrict the multiplication operator to a smaller domain, i.e., if we define an operator Q_1 by

$$Q_1 \varphi(t) = t \varphi(t), \quad D_{Q_1} = D,$$

where $D \subset D_Q$, then the above argument yields

$$Q_1^* = Q.$$

It follows that (cf. Section 46)

$$Q = Q^* = Q^{**} = \bar{Q}_1,$$

i.e., the operator Q defined originally is the closure of the operator Q_1.

The operator of multiplication by the independent variable does not have any eigenfunctions. Since

$$Q \varphi = \lambda \varphi$$

implies that

$$\int_a^b |t - \lambda|^2 |\varphi(t)|^2 dt = 0$$

we have $\varphi(t) = 0$ almost everywhere, i.e., $\varphi = 0$.

All points of the interval $[a, b]$ belong to the continuous spectrum of the operator Q since the manifold

$$\Delta_Q(\lambda) = (Q - \lambda E) D_Q \quad (a < \lambda < b)$$

consists of all functions $\psi(t)$ in $L^2(a, b)$ which remain in $L^2(a, b)$ after division by $t - \lambda$; i.e., $\psi \in \Delta_Q(\lambda)$ if and only if

$$\psi(t) \in L^2(a, b)$$

and

$$\frac{\psi(t)}{t - \lambda} \in L^2(a, b).$$

It is evident that $\Delta_Q(\lambda)$ is dense in $L^2(a, b)$. But it does not coincide with $L^2(a, b)$ because, for instance, it does not contain a function equal to one in a neighborhood of the point $t = \lambda$.

The subspace $M = M_e$ of functions in $L^2(a, b)$ which are equal to zero outside of some fixed point set $e \subset [a, b]$ evidently reduces[7] the operator Q. If $\varphi \in L^2(\alpha, \beta)$, where $\beta - \alpha \leqq \varepsilon$, then the operator Q satisfies the inequality,

---

[7] It is not difficult to prove that all subspaces reducing the operator Q have this form.

$$\| Q\varphi - \lambda_0\varphi \| \leqq \varepsilon \| \varphi \|,$$

for each fixed λ_0 in the interval $[\alpha, \beta]$; i.e., Q differs by not more than ε in norm from the similarity transformation $\lambda_0 E$. Thus, decomposing the interval $[a, b]$ into subintervals of sufficiently small length, we get a decomposition of the space $L^2(a, b)$ into a countable sum of subspaces, each reducing Q and in each of which the operator Q differs by an arbitrarily small amount from a similarity operator.[8]

It is possible to consider the operator of multiplication not only by the independent variable but also by a function of it. We restrict ourselves here to one particular case. Let the operator U be defined on all functions $\varphi(t)$ in $L^2(0, 2\pi)$ by the equation

$$U\varphi(t) = e^{it}\varphi(t).$$

Each element of the orthonormal basis $\left\{\frac{1}{\sqrt{2\pi}} e^{ikt}\right\}_{k=-\infty}^{\infty}$ is transformed by the operator U into the following element:

$$Ue^{ikt} = e^{i(k+1)t} \qquad (-\infty < k < \infty).$$

From this fact it follows that the operator U is unitarily equivalent to the operator U_0 defined in a separable space H in Section 40. Indeed, if we define the isometric operator V which maps H onto $L^2(0, 2\pi)$ by the equation

$$Ve_k = \frac{1}{\sqrt{2\pi}} e^{ikt} \quad (-\infty < k < \infty), \tag{2}$$

then

$$U = VU_0V^{-1}.$$

This representation of U in terms of U_0 will be used to find the (infinitely many) subspaces which reduce U_0.[9] We shall base the derivation on the following general proposition (the proof of which is left to the reader). If the subspace G of the space H reduces a linear operator T, and if the operator T_1 defined on the space H_1 is unitarily equivalent to the operator T $(T_1 = VTV^{-1})$, then the subspace $G_1 = VG$ of the space H_1 reduces the operator T_1. According to this proposition the class of subspaces reducing the operator U_0 includes the set G_e, which consists of all the elements

$$f(t) = V^{-1}\varphi(t)$$

such that $\varphi(t) = 0$ outside of the arbitrary subset e of $[0, 2\pi]$ and the operator V is defined by formula (2).

[8] In Chapter VI we shall prove that every self-adjoint operator has an analogous approximate decomposition into a similarity transformation. This fact plays a fundamental role in all of the theory.

[9] Cf. the end of Section 40.

In conclusion we note that, in place of the multiplication operator Q in the space $L^2(a, b)$, one can consider the operator Q_σ of multiplication by an independent variable in the space $L^2_\sigma(a, b)$. It is not difficult to verify that Q_σ is a self-adjoint operator. We propose to the reader as useful exercises the proofs of the following propositions.

(a) The real regular values of the operator Q_σ are the points of constancy of the function $\sigma(t)$.

(b) The eigenvalues of the operator Q_σ are the points of discontinuity of the function $\sigma(t)$.

(c) The continuous spectrum of the operator Q_σ is the set of all non-isolated points of increase of the function $\sigma(t)$.

The operators Q_σ play a special role in the theory of self-adjoint operators: in Chapter VI we shall see that the study of any self-adjoint operator can be reduced to the study of the operators Q_σ.

49. A Differential Operator

We shall consider the *differential operator* P in $L^2(a, b)$ which is defined by the equation

$$P\varphi = i\frac{d\varphi}{dt}$$

for each function $\varphi(t)$ in its domain D_P. A function $\varphi(t)$ belongs to the domain D_P if certain conditions, given below, are satisfied.

(A) If $\varphi \in D_P$ then the function $\varphi(t)$ is absolutely continuous on each finite subset of the interval $[a, b]$, and both $\varphi(t)$ and $\varphi'(t)$ belong to $L^2(a, b)$. We examine separately the cases of a finite interval $[a, b]$, a semi-axis, and the whole axis. In the first two cases D_P consists of the functions which, besides condition (A), also satisfy certain boundary conditions (B).

1° *Finite interval.* In this case we assume that the interval is $[0, 2\pi]$ and the boundary conditions have the form

$$\text{(B)} \qquad \varphi(0) = \varphi(2\pi) = 0.$$

The set D_P of functions satisfying conditions (A) and (B) is evidently dense in $L^2(0, 2\pi)$. Then P is a symmetric (unbounded) operator, since

$$(P\varphi,\psi) = \int_0^{2\pi} i\varphi'(t)\overline{\psi(t)}dt$$

$$= i\{\varphi(2\pi)\overline{\psi(2\pi)} - \varphi(0)\overline{\psi(0)} + \int_0^{2\pi} \varphi(t)\overline{\{i\psi'(t)\}}\,dt$$

for $\varphi, \psi \in D_P$. This equation implies that

$$(P\varphi, \psi) = (\varphi, P\psi),$$

since the integral expression vanishes. The integral expression vanishes also in case φ belongs to D_P and ψ satisfies only the condition (A). Hence, every function $\psi(t)$ satisfying condition (A) belongs to D_{P^*}, and

$$P^*\psi(t) = i\psi'(t).$$

Conversely, let $\psi \in D_{P^*}$ and $P^*\psi = \psi^*$. Then, for each $\varphi \in D_P$,

$$(P\varphi, \psi) = (\varphi, \psi^*) = \int_0^{2\pi} \varphi(t)\overline{\psi^*(t)}\,dt =$$

$$= -i\int_0^{2\pi} \varphi(t)\frac{d}{dt}\overline{\left\{-\int_0^t i\psi^*(s)ds + C\right\}}dt,$$

where C is an arbitrary constant. Integrating by parts we get

$$(1) \qquad (P\varphi, \psi) = \int_0^{2\pi} i\varphi'(t)\overline{\left\{-\int_0^t i\psi^*(s)\,ds + C\right\}}\,dt,$$

since $\varphi(t)$ vanishes at the end-points of the integral. From (1) it follows that, for $\varphi(t) \in D_P$,

$$(2) \qquad \int_0^{2\pi} \varphi'(t)\overline{\left\{\psi(t) + \int_0^t i\psi^*(s)\,ds - C\right\}}\,dt = 0.$$

We define C by the equation

$$\int_0^{2\pi}\left\{\psi(t) + \int_0^t i\psi^*(s)\,ds - C\right\}dt = 0$$

and substitute for $\varphi(t)$ the function

$$\varphi_0(t) = \int_0^t\left\{\psi(t) + \int_0^t i\psi^*(s)\,ds - C\right\}dt,$$

which evidently belongs to D_P. Then (2) assumes the form

$$\int_0^{2\pi} |\,\psi(t) + \int_0^t i\psi^*(s)\,ds - C\,|^2\,dt = 0$$

which yields

$$\psi(t) + \int_0^t i\psi^*(s)\,ds - C = 0.$$

Therefore, for almost all t,

$$i\psi'(t) = \psi^*(t).$$

We have proved that the domain of the operator P^* is the set of all functions $\psi(t)$ satisfying relation (A), and that

$$P^*\psi(t) = i\psi'(t).$$

From the proof of these facts it follows that the symmetric operator P is not self-adjoint; in fact, the functions in D_P must satisfy two conditions (A) and (B), whereas the functions in D_{P^*} must satisfy only the condition (A).

We prove now that the operator P is closed. In place of a direct proof, we proceed, first, to prove that (cf. Section 46)

$$P^{**} = P.$$

From the relation

$$P \subset P^*$$

it follows that

$$P^{**} \subset P^*$$

Therefore the functions $\chi(t)$ in $D_{P^{**}}$ satisfy condition (A) and

$$P^{**}\chi(t) = i\chi'(t).$$

Hence, for each $\psi \in D_{P^*}$,

$$\int_0^{2\pi} \psi(t)\overline{i\chi'(t)}\,dt = (\psi, P^{**}\chi) = (P^*\psi, \chi) = \int_0^{2\pi} i\psi'(t)\overline{\chi(t)}\,dt =$$

$$= i[\psi(2\pi)\overline{\chi(2\pi)} - \psi(0)\overline{\chi(0)}] + \int_0^{2\pi} \psi(t)\,\overline{i\chi'(t)}\,dt,$$

which implies that

$$\psi(2\pi)\overline{\chi(2\pi)} - \psi(0)\overline{\chi(0)} = 0.$$

Since the values of $\psi(0)$ and $\psi(2\pi)$ are arbitrary, this equation is satisfied if and only if

$$\chi(0) = \chi(2\pi) = 0,$$

i.e., if $\chi(t) \in D_P$ We have proved that

$$P^{**} \subset P.$$

Since the reverse relation

$$P \subset P^{**}$$

always holds, we have

$$P = P^{**}.$$

This equation implies that the operator P is closed.

If we replace (A) and (B) by more rigid conditions, requiring, for example, that the functions in the domain of the operator be repeatedly (or even infinitely often) differentiable and that these derivatives vanish for $t = 0$ and $t = 2\pi$, then the closure of the operator is P. In order to be convinced of this fact one may verify that the adjoint of the adjoint of the new operator coincides with P.

However, it is possible to strengthen (A) and (B) to such an extent that the closure of the operator obtained does not coincide with the operator P. For example, if we leave condition (A) unchanged, and replace condition (B) by

$$\varphi(0) = \varphi(\pi) = \varphi(2\pi) = 0,$$

then the operator P_0 obtained turns out to be closed, but $P_0 \neq P$ (of course, $P_0 \subset P$).

In this connection we remark that the subspace of functions in $L^2(0, 2\pi)$ which vanish for $\pi < t \leqq 2\pi$ (we can identify this subspace with $L^2(0, \pi)$) reduces the operator P_0 but does not reduce the operator P.

We now look for the symmetric extensions of the operator P. Let $\tilde{P}$ be one of them. Since $\tilde{P} \subset P^*$ the functions in $D_{\tilde{P}}$ satisfy relation (A). Hence, for all functions $\varphi, \psi \in D_{\tilde{P}}$ we have

$$(\tilde{P}\varphi, \psi) = \int_0^{2\pi} i\varphi'(t)\overline{\psi(t)}\,dt = i[\varphi(2\pi)\overline{\psi(2\pi)} - \varphi(0)\overline{\psi(0)}] + \int_0^{2\pi} \varphi(t)\overline{i\psi'(t)}\,dt =$$

$$= i[\varphi(2\pi)\overline{\psi(2\pi)} - \varphi(0)\overline{\psi(0)}] + (\varphi, \tilde{P}\psi).$$

Since the operator $\tilde{P}$ is symmetric, it must satisfy the relation

$$\varphi(2\pi)\overline{\psi(2\pi)} - \varphi(0)\overline{\psi(0)} = 0. \tag{3}$$

Since $\tilde{P} \neq P$, there exists a function $\psi_0(t)$ in $D_{\tilde{P}}$ not satisfying relation (B); assume, for the sake of definiteness, that $\psi_0(2\pi) \neq 0$. Letting $\psi(t) = \psi_0(t)$ in (3), we obtain for each function $\varphi(t) \in D_{\tilde{P}}$ the relation

$$\varphi(2\pi) = \theta\varphi(0) \tag{$\tilde{B}$}$$

where

$$\theta = \frac{\overline{\psi_0(0)}}{\overline{\psi_0(2\pi)}}.$$

Since condition ($\tilde{B}$) holds for $\varphi(t) = \psi_0(t)$, we must have $|\theta| = 1$.

Our result may be stated as follows: all functions $\varphi(t)$ in $D_{\tilde{P}}$ must satisfy conditions (A) and ($\tilde{B}$), where for each extension $\tilde{P}$ the constant θ has absolute value one.

We now prove the converse: every function $\psi(t)$ which satisfies conditions (A), ($\tilde{B}$) belongs to $D_{\tilde{P}}$. With this purpose in mind we select a constant α such that

$$\psi(0) - \alpha\psi_0(0) = 0$$

and we let

$$\varphi(t) = \psi(t) - \alpha\,\psi_0(t).$$

It is easy to see that $\varphi(t)$ satisfies conditions (A) and (B) and, hence, belongs to D_P. But, since $D_P \subset D_{\tilde{P}}$, we have $\varphi(t) \in D_{\tilde{P}}$, so that the function

$$\psi(t) = \varphi(t) + \alpha\,\psi_0(t)$$

also belongs to $D_{\tilde{P}}$.

Thus, the symmetric extensions $\tilde{P}$ of the operator P are characterized by conditions (A) and ($\tilde{B}$). The extension of the operator P has led to a weakening of condition (B).

In view of the fact that each extension is determined by a number θ ($|\theta| = 1$), which appears in ($\tilde{B}$), we shall write P_θ instead of $\tilde{P}$. It is not difficult to verify that $D_{P_\theta^*}$ consists of the functions $\psi(t)$ in D_{P^*} such that

$$\varphi(2\pi)\overline{\psi(2\pi)} - \varphi(0)\overline{\psi(0)} = 0$$

for each $\varphi(t) \in D_{P_\theta}$. Hence it follows that D_{P_θ} and $D_{P_\theta^*}$ coincide, so that every extension P_θ of the operator P is a self-adjoint operator.

We now determine the spectrum of the operator P_θ (for simplicity only for $\theta = 1$). The equation

$$P_1\varphi = \lambda\varphi$$

implies that

$$i\varphi'(t) = \lambda\varphi(t)$$

and

$$\varphi(2\pi) = \varphi(0).$$

Hence

$$\lambda = \lambda_k = k$$

$$\varphi(t) = \varphi_k(t) = e^{-ikt} \qquad (\pm k = 0, 1, 2, \ldots).$$

It is not difficult to verify that for $\lambda \neq \lambda_k$ the equation

$$(P_1 - \lambda E)f = g$$

or, equivalently, the equation

$$if'(t) - \lambda f(t) = g(t) \qquad [f(2\pi) = f(0)]$$

is solvable for each function $g(t) \in L^2(0, 2\pi)$. Hence, P_1 has no continuous spectrum.

2° *Semi-axis* $[0, \infty)$. If the function $\varphi(t)$ satisfies condition (A) in the case of a semi-axis, then the product $\varphi(t)\varphi'(t)$ is absolutely integrable on this interval. The formula

$$\int_0^t \varphi(s)\overline{\varphi'(s)}\,ds = |\varphi(t)|^2 - |\varphi(0)|^2 - \int_0^t \varphi'(s)\overline{\varphi(s)}\,ds$$

shows that $|\varphi(t)|$ has a limit as $t \to \infty$. But, since $\varphi(t) \in L^2(0, \infty)$, we must have

$$\lim_{t\to\infty} \varphi(t) = 0.$$

As we see, the boundary condition on the right end of the semi-axis is automatically fulfilled.

For the second condition in the definition of the domain D_P we shall require that

(B) $$\varphi(0) = 0.$$

For every $\varphi, \psi \in D_P$ we have

$$(P\varphi, \psi) = i\int_0^\infty \varphi'(t)\overline{\psi(t)}\,dt = \int_0^\infty \varphi(t)\,\overline{i\psi'(t)}\,dt = (\varphi, P\psi).$$

Hence P is a symmetric operator.

As in the case of a finite interval it is not difficult to show that D_{P^*} is the set of all functions satisfying the single condition (A), and that

$$P^*\psi = i\psi'.$$

Thus, $P \neq P^*$, so that the operator P is not self-adjoint.

In contrast with the case of a finite interval the differential operator on the semi-axis does not have a symmetric proper extension. In fact the domain of definition of such an extension $\tilde{P}$ would have to contain a function $\varphi_0(t)$ which is different from zero for $t = 0$. But then we would have

$$(\tilde{P}\psi_0, \psi_0) = i\int_0^\infty \psi_0'(s)\overline{\psi_0(s)}\,ds = i\,|\psi_0(0)|^2 + \int_0^\infty \psi_0(s)\overline{i\psi_0'(s)}\,ds \neq (\psi_0, \tilde{P}\psi_0)$$

which is impossible. Thus, the differential operator on the semi-axis is a maximal symmetric operator. Later (cf. Section 82) we shall prove that it is irreducible.

3° *The whole real axis.* For each function $\varphi(t)$ satisfying condition (A) the boundary conditions

$$\lim_{t\to\infty} \varphi(t) = \lim_{t\to-\infty} \varphi(t) = 0$$

are automatically fulfilled. Therefore, the domain of D_P is defined by the single requirement (A) and it is proved without difficulty that P is a self-adjoint operator. The operator P has no eigenvalues because the equation

$$i\frac{d\varphi}{dt} = \lambda\varphi$$

does not have a nontrivial solution in $L^2(-\infty, \infty)$.

We now establish a connection between the operator Q (of multiplication by the independent variable) and the operator P. From this connection it will follow that every point of the real axis belongs to the continuous spectrum of the operator P. The connection between the two operators amounts to the fact that they are unitarily equivalent. To show this we shall use the relations

$$\psi(t) = \frac{1}{\sqrt{2\pi}} \int_{-\infty}^{\infty} \varphi(s) e^{-ist} ds,$$

$$i\psi'(t) = \frac{1}{\sqrt{2\pi}} \int_{-\infty}^{\infty} s\, \varphi(s) e^{-ist} ds.$$

Thus, multiplication of the function $\varphi(s)$ by s corresponds to differentiation of the function $\psi(t)$.

THEOREM: *The operators P and Q are related by*

$$P = \mathfrak{F} Q \mathfrak{F}^{-1}$$

where $\mathfrak{F}$ is the Fourier-Plancherel operator.

Proof: The proof consists of two parts. First, we shall show that $h \in D_Q$ implies that

$$\mathfrak{F}h \in D_P, \; P\mathfrak{F}h = \mathfrak{F}Qh$$

and then we shall show that $g \in D_P$ implies that

$$\mathfrak{F}^{-1}g \in D_Q, \quad \mathfrak{F}^{-1}Pg = Q\mathfrak{F}^{-1}g.$$

Let $h \in D_Q$. Then

$$(4) \quad \mathfrak{F} Q h(t) = \frac{d}{dt}\frac{1}{\sqrt{2\pi}} \int_{-\infty}^{\infty} \frac{e^{-ist}-1}{-is} s h(s) ds = i\frac{d}{dt}\frac{1}{\sqrt{2\pi}} \int_{-\infty}^{\infty} \{e^{-ist} - 1\} h(s) ds$$

and, since $h \in D_Q$,

$$\int_{-\infty}^{\infty} |h(s)| ds < \infty.$$

Therefore,

$$\mathfrak{F} h(t) = \frac{1}{\sqrt{2\pi}} \int_{-\infty}^{\infty} e^{-ist} h(s) ds,$$

and relation (4) can be expressed in the form

$$\mathfrak{F} Q h(t) = i\frac{d}{dt}\frac{1}{\sqrt{2\pi}} \int_{-\infty}^{\infty} e^{-ist} h(s) ds$$

which implies both the inclusion $\mathfrak{F} h \in D_P$ and the equation

$$\mathfrak{F}Qh = P\mathfrak{F}h.$$

Let us assume now that $g \in \mathrm{D}_P$. Then

$$\mathfrak{F}^{-1}Pg(t) = \frac{d}{dt}\frac{1}{\sqrt{2\pi}}\int_{-\infty}^{\infty}\frac{e^{ist}-1}{is}\,ig'(s)\,ds =$$

$$= -\frac{d}{dt}\frac{1}{\sqrt{2\pi}}\int_{-\infty}^{\infty}\left(\frac{e^{ist}\,ist-(e^{ist}-1)}{s^2}\right)g(s)\,ds =$$

$$= \frac{d}{dt}\frac{t}{\sqrt{2\pi}}\int_{-\infty}^{\infty}\frac{e^{ist}-1}{is}\,g(s)\,ds + \frac{d}{dt}\frac{1}{\sqrt{2\pi}}\int_{-\infty}^{\infty}\frac{e^{ist}-1-ist}{s^2}\,g(s)\,ds =$$

$$= t\frac{d}{dt}\frac{1}{\sqrt{2\pi}}\int_{-\infty}^{\infty}\frac{e^{ist}-1}{is}\,g(s)\,ds = t\,\mathfrak{F}^{-1}g(t).$$

Since the left member belongs to $L^2(-\infty, \infty)$, so does the right member. Hence, $\mathfrak{F}^{-1}g \in \mathrm{D}_Q$ and

$$\mathfrak{F}^{-1}Pg = Q\mathfrak{F}^{-1}g.$$

Since every real point belongs to the continuous spectrum of Q, the equation

$$(Q-\lambda E)g = f$$

is not always solvable. Hence, the equation,

$$(Q-\lambda E)\,\mathfrak{F}^{-1}h = f,$$

is not always solvable. Neither is

$$\mathfrak{F}(Q-\lambda E)\,\mathfrak{F}^{-1}h = \mathfrak{F}f$$

which can be reduced to the form

$$(P-\lambda E)h = f_1.$$

Since P has no eigenvalues, every real point belongs to its continuous spectrum.

Using the unitary equivalence of the operators Q and P it is easy to determine the subspaces which reduce P. Each such subspace consists of the functions

$$\psi(t) = \mathfrak{F}\varphi(t)$$

where $\varphi(t)$ is equal to zero outside of some fixed set on the real axis.

50. The Inversion of Singular Integrals

To begin with let us assume the following important theorem[10] from the theory of functions: *for each function* $g(x) \in L^2(-\infty, \infty)$, *the equation*

$$g(x) = \frac{1}{\pi} \int_{-\infty}^{\infty}{}' \frac{f(y)}{x-y}\, dy \tag{1}$$

(where the prime signifies that the integral is considered as the principal value in the sense of Cauchy) has a solution[11]; *the solution belongs to* $L^2(-\infty, \infty)$, *and is given by the formula*

$$f(x) = \frac{1}{\pi} \int_{-\infty}^{\infty}{}' \frac{g(y)}{y-x}\, dy. \tag{2}$$

Moreover,

$$\int_{-\infty}^{\infty} |f(x)|^2\, dx = \int_{-\infty}^{\infty} |g(x)|^2\, dx.$$

Therefore, the singular integrals in (1) *and* (2) *define unitary operators in* $L^2(-\infty, \infty)$.

Equations (1) and (2) form a pair of inversion formulas. If we assume that the first of the functions $f(x)$ and $g(x)$ is even, we find without difficulty that the second is odd. This leads to the following pair of inversion formulas in the space $L^2(0, \infty)$:

$$g(x) = \frac{2}{\pi} \int_{0}^{\infty}{}' \frac{f(y)\, x}{x^2 - y^2}\, dy,$$

$$f(x) = \frac{2}{\pi} \int_{0}^{\infty}{}' \frac{g(y)\, y}{y^2 - x^2}\, dy$$

where

$$\int_{0}^{\infty} |f(x)|^2 dx = \int_{0}^{\infty} |g(x)|^2\, dx.$$

Letting

$$x^2 = s,\; y^2 = t,\; f(x) = F(s),\; g(x) = \sqrt{s}\, G(s),$$

---

[10] Cf. Titchmarsh [1], page 120.

[11] When we say that a function is the solution of the equation (1) we mean that, after the substitution of this function, the right member exists and equals the left member for almost all x.

we get

$$\text{(3)} \qquad \begin{cases} G(s) = \dfrac{1}{\pi} \displaystyle\int_0^{\infty}{}' \frac{1}{\sqrt{t}} \frac{F(t)}{s-t}\, dt, \\[2ex] F(s) = \dfrac{1}{\pi} \displaystyle\int_0^{\infty}{}' \sqrt{t} \frac{G(t)}{t-s}\, dt. \end{cases}$$

Therefore,

$$\text{(4)} \qquad \int_0^{\infty} \frac{1}{\sqrt{t}} |F(t)|^2 dt = \int_0^{\infty} \sqrt{t}\, |G(t)|^2 dt.$$

If we agree to denote the space $L_\sigma^2(a, b)$, where $\sigma'(x) = p(x)$, by the symbol $L^2(p(x); a, b)$, then we can say that formulas (3) give an isometric mapping of $L^2\left(\frac{1}{\sqrt{x}}; 0, \infty\right)$ onto $L^2(\sqrt{x}; 0, \infty)$.

Now we can easily derive inversion formulas for a finite interval. With this aim in mind we let

$$s = \frac{1+x}{1-x}, \quad t = \frac{1+y}{1-y}$$

in (3) and (4), to get the formulas

$$\text{(5)} \qquad \begin{cases} g(x) = \dfrac{1}{\pi} \displaystyle\int_{-1}^{1}{}' \sqrt{\frac{1-y}{1+y}} \frac{f(y)}{x-y}\, dy, \\[2ex] f(x) = \dfrac{1}{\pi} \displaystyle\int_{-1}^{1}{}' \sqrt{\frac{1+y}{1-y}} \frac{g(y)}{y-x}\, dy, \end{cases}$$

$$\text{(6)} \qquad \int_{-1}^{1} \sqrt{\frac{1-x}{1+x}} |f(x)|^2 dx = \int_{-1}^{1} \sqrt{\frac{1+x}{1-x}} |g(x)|^2 dx.$$

This gives an isometric mapping of the space

$$L^2\left(\sqrt{\frac{1-x}{1+x}}; -1, 1\right)$$

onto the space

$$L^2\left(\sqrt{\frac{1+x}{1-x}}; -1, 1\right).$$

These formulas lead to an inversion in which, instead of one interval of the real axis, there is a system of such intervals. We confine ourselves here to one simple example.[12] Let E denote a point set which consists of the finite intervals

$$(a_1, b_1), (a_2, b_2), \dots (a_n, b_n) \quad (a_1 < b_1 < a_2 < b_2 < \dots < b_n)$$

and let

$$p(x) = \sqrt{-\frac{(x-b_1)(x-b_2)\dots(x-b_n)}{(x-a_1)(x-a_2)\dots(x-a_n)}}, \quad q(x) = \frac{1}{p(x)}.$$

Then the desired formulas have the form

$$\begin{cases} g(x) = \dfrac{1}{\pi} \displaystyle\int_E' p(y) \frac{f(y)}{y-x} \, dy, \\ f(x) = \dfrac{1}{\pi} \displaystyle\int_E' q(y) \frac{g(y)}{x-y} \, dy, \end{cases}$$

and

$$\int_E p(x) \, |f(x)|^2 \, dx = \int_E q(x) \, |g(x)|^2 \, dx.$$

The isometric mapping of the space $L^2(p(x); E)$ on the space $L^2(q(x); E)$, defined by these formulas, has a curious property. It maps polynomials into polynomials of the same power. The proofs and further generalizations and examples the reader will find in the works cited above.

---

[12] N. I. Akhieser [1], N. I. Muschelishvili [1].

Chapter V

SPECTRAL ANALYSIS OF COMPLETELY CONTINUOUS OPERATORS

51. A Lemma

The present chapter is devoted to the spectral theory of certain classes of completely continuous operators. Since it is a direct and easily surveyed generalization of the corresponding sections of linear algebra and of the elementary theory of integral equations, the spectral theory of completely continuous operators represents the most natural introduction to the general spectral theory of operators in Hilbert space.

In the formulation of the spectral theory of completely continuous operators, the completeness of the space, as we shall see below, is not generally exploited. Moreover, by the elimination of the requirement of completeness, the domain of application of the theory is extended. Therefore, in the present chapter, together with propositions referring to operators in a Hilbert space H, there is established a series of propositions concerning operators in an arbitrary inner product space R. A number of these propositions are related to the lemma to which the present section is devoted.

LEMMA: *If* $\{g_k\}_0^\infty$ *is an infinite orthonormal sequence of vectors in* R *and*

$$Ag_k = \beta_{k0}g_0 + \beta_{k1}g_1 + \ldots + \beta_{kk}g_k \qquad (k = 1, 2, 3, \ldots),$$

where A *is a completely continuous operator in* R, *then*

$$\lim_{k\to\infty} \beta_{kk} = 0.$$

Proof: Let $n > m$. Then

$$\| Ag_n - Ag_m\|^2 = \|\beta_{nn}g_n + \ldots + \beta_{n,m+1}g_{m+1} + (\beta_{nm} - \beta_{mm})g_m + \ldots + (\beta_{n0} - \beta_{m0})g_0\|^2 =$$
$$= |\beta_{nn}|^2 + \ldots + |\beta_{n,m+1}|^2 + |\beta_{nm} - \beta_{mm}|^2 + \ldots + |\beta_{n0} - \beta_{m0}|^2 \geqq |\beta_{nn}|^2.$$

Suppose that β_{kk} does not tend to zero as $k \to \infty$. Then there exists an infinite sequence of positive integers

$$n_1 < n_2 < n_3 < \ldots,$$

such that

$$|\beta_{n_jn_j}| \geqq \delta > 0 \qquad (j = 1, 2, 3, \ldots)$$

for some δ. Therefore,

$$\| Ag_{n_k} - Ag_{n_i} \|^2 \geqq \delta^2 > 0$$

which implies that the infinite sequence of vectors $\{Ag_{n_j}\}_{j=1}^{\infty}$ does not contain a convergent subsequence, and this contradicts the compactness of the set of vectors $\{Ag_k\}_0^{\infty}$; i.e., it contradicts the complete continuity of the operator A.

52. Properties of the Eigenvalues of Completely Continuous Operators in R

The propositions which we establish in the present and following sections are generalizations of well-known propositions in the theory of integral equations. In the present section, as its title indicates, operators are considered in an inner product space R. In the following section, we shall assume that the system R is complete, so that R becomes a Hilbert space.

THEOREM 1: *If A is a completely continuous operator in* R *and $\rho > 0$, then A has only a finite number of linearly independent eigenvectors such that the corresponding eigenvalues exceed ρ in modulus.*

Proof: Supposing the contrary, we assume that there exists an infinite sequence of linearly independent vectors f_n $(n = 1, 2, 3, \ldots)$, for which

$$Af_n = \lambda_n f_n, \quad |\lambda_n| > \rho > 0 \qquad (n = 1, 2, 3, \ldots).$$

Orthogonalizing $\{f_n\}_1^{\infty}$, we get the orthonormal sequence of vectors

$$\begin{aligned} g_1 &= \alpha_{11} f_1, \\ g_2 &= \alpha_{21} f_1 + \alpha_{22} f_2, \\ &\cdots\cdots\cdots \\ g_k &= \alpha_{k1} f_1 + \alpha_{k2} f_2 + \ldots + \alpha_{kk} f_k, \\ &\cdots\cdots\cdots\cdots\cdots \end{aligned}$$

Then

$$Ag_k = \alpha_{k1} Af_1 + \alpha_{k2} Af_2 + \ldots + \alpha_{kk} Af_k = \alpha_{k1}\lambda_1 f_1 + \alpha_{k2}\lambda_2 f_2 + \ldots + \alpha_{kk}\lambda_k f_k,$$

and, consequently,

$$\begin{aligned} Ag_k - \lambda_k g_k &= \alpha_{k1}(\lambda_1 - \lambda_k) f_1 + \ldots + \alpha_{k,k-1}(\lambda_{k-1} - \lambda_k) f_{k-1} = \\ &= \beta_{k1} g_1 + \beta_{k2} g_2 + \ldots + \beta_{k,k-1} g_{k-1}. \end{aligned}$$

Thus,

$$Ag_k = \beta_{k1} g_1 + \beta_{k2} g_2 + \ldots + \beta_{k,k-1} g_{k-1} + \lambda_k g_k.$$

By the lemma of the preceding section, $\lambda_k \to 0$ as $k \to \infty$, which contradicts the assumption that $|\lambda_k| > \rho > 0$ for every k.

COROLLARY 1: *Only the point zero* (0) *can be a limit point of the eigenvalues of a completely continuous operator in* R.

COROLLARY 2: *To each nonzero eigenvalue of a completely continuous operator in* R *belong only a finite number of linearly independent eigenvectors. In other words, each nonzero eigenvalue of a completely continuous operator in* R *has finite multiplicity.*

COROLLARY 3: *Each completely continuous operator in* R *has only countably many linearly independent eigenvectors belonging to nonzero eigenvalues.*

THEOREM 2: *If A is a completely continuous operator in* R *and if, for a fixed $\lambda \neq 0$, the equation*

$$Af - \lambda f = h \tag{1}$$

has a solution for each $h \in$ R, *then the equation*

$$Af - \lambda f = 0 \tag{2}$$

has the unique solution $f = 0$, i.e., λ is not an eigenvalue of the operator A.

Proof: Supposing the contrary, we assume that equation (2) has a solution $f_0 \neq 0$. Thus,

$$Af_0 = \lambda f_0.$$

Solving equation (1) with $h = f_0$, we get a certain vector f_1:

$$Af_1 - \lambda f_1 = f_0.$$

Analogously, we find a vector f_2 such that

$$Af_2 - \lambda f_2 = f_1.$$

Continuing this process, we find an infinite sequence of vectors $\{f_k\}_1^\infty$ such that

$$Af_k - \lambda f_k = f_{k-1} \qquad (k = 1, 2, 3, \ldots).$$

We also have

$$Af_0 - \lambda f_0 = 0.$$

We show next that the vectors f_k $(k = 0, 1, 2, \ldots)$ are linearly independent. Supposing the contrary, we assume that in the sequence of vectors

$$f_0, f_1, f_2, \ldots, f_k, \ldots$$

the first one which is a linear combination of the preceding vectors is f_n, so that

$$f_n = \alpha_0 f_0 + \alpha_1 f_1 + \ldots + \alpha_{n-1} f_{n-1}. \tag{3}$$

Applying the operator A to both members of this equation, we obtain

$$\lambda f_n + f_{n-1} = \alpha_0 \lambda f_0 + \alpha_1 (\lambda f_1 + f_0) + \ldots + \alpha_{n-1} (\lambda f_{n-1} + f_{n-2}).$$

This equation and (3) yield

$$f_{n-1} = \alpha_1 f_0 + \alpha_2 f_1 + \ldots + \alpha_{n-1} f_{n-2},$$

which contradicts our assumption. Therefore, the vectors f_k $(k = 0, 1, 2, \ldots)$ are linearly independent. Orthogonalizing this sequence, we get the orthonormal sequence

$$g_0 = a_{00} f_0,$$
$$g_1 = a_{10} f_0 + a_{11} f_1,$$
$$\ldots\ldots\ldots\ldots$$
$$g_k = a_{k0} f_0 + a_{k1} f_1 + \ldots + a_{kk} f_k,$$
$$\ldots\ldots\ldots\ldots\ldots\ldots$$

It follows that

$$\begin{aligned} Ag_k &= a_{k0} \lambda f_0 + a_{k1}(\lambda f_1 + f_0) + \ldots + a_{kk}(\lambda f_k + f_{k-1}) = \\ &= a_{k1} f_0 + a_{k2} f_1 + \ldots + a_{kk} f_{k-1} + \lambda g_k = \\ &= \beta_{k0} g_0 + \beta_{k1} g_1 + \ldots + \beta_{k,k-1} g_{k-1} + \lambda g_k. \end{aligned}$$

Since this equation contradicts the lemma of the preceding section, our theorem is proved.

COROLLARY 4: *If for a fixed $\lambda \neq 0$ equation (1) is solvable for each $h \in R$, then, given $h \in R$, this equation has a unique solution and, consequently, the operator $A - \lambda E$ has an inverse operator $(A - \lambda E)^{-1}$.*

THEOREM 3: *Let A be a completely continuous operator in R and fix $\lambda \neq 0$. There exists a constant L, depending only on A and λ, such that if the equation*

$$Af - \lambda f = h, \qquad (1)$$

with a given right member h, has a solution f, then, for at least one of its solutions,

$$\|f\| \leqq L \|h\|.$$

Proof: Fix h and assume that (1) has a solution f^*. If λ is an eigenvalue of A, let

$$f_1, f_2, \ldots, f_k$$

be a complete system of linearly independent eigenvectors of the operator A belonging to λ. In this case, the general solution of equation (1) has the form

$$f = f^* + \alpha_1 f_1 + \alpha_2 f_2 + \ldots + \alpha_k f_k,$$

where $\alpha_1, \alpha_2, \ldots, \alpha_k$ are arbitrary complex numbers. We select these numbers so that the norm $\| f \|$ of the vector f is a minimum. Let $\overset{0}{f}$ denote the solution of (1) with minimum norm. If λ is regular then $\overset{0}{f} = f^*$. Now let h run through the set M of all vectors for which the equation (1) is solvable. To each vector $h \in \mathrm{M}$, there corresponds a minimal solution $\overset{0}{f}$. We must show that[1]

$$\sup_{h \in \mathrm{M}} \frac{\| \overset{0}{f} \|}{\| h \|} < \infty.$$

We suppose the contrary. Then there exists a sequence of vectors $\{h_k\}_1^\infty$ such that, as $k \to \infty$

$$\frac{\| \overset{0}{f}_k \|}{\| h_k \|} \to \infty$$

where $\overset{0}{f}_k$ is the minimal solution of (1) with right member h_k. We divide both sides of the equation

$$A\overset{0}{f}_k - \lambda \overset{0}{f}_k = h_k \qquad (k = 1, 2, 3, \ldots)$$

by $\| \overset{0}{f}_k \|$ to get the equation

$$A\overset{0}{f'}_k - \lambda \overset{0}{f'}_k = h'_k \qquad (k = 1, 2, 3, \ldots)$$

where

$$h'_k = \frac{h_k}{\| \overset{0}{f}_k \|}, \; \| \overset{0}{f'}_k \| = 1 \qquad (k = 1, 2, 3, \ldots).$$

Thus, the minimal solution $\overset{0}{f'}_k$ of equation (1) has norm one if the right member is h'_k. Since the operator A is completely continuous, there exists a subsequence

$$\overset{0}{f'}_{n_1}, \overset{0}{f'}_{n_2}, \ldots$$

of $\{\overset{0}{f'}_k\}$ for which the limit

$$\lim_{i \to \infty} A\overset{0}{f'}_{n_i}$$

exists. Since

$$h'_k \to 0$$

as $k \to \infty$, the limit

$$\lim_{i \to \infty} \overset{0}{f'}_{n_i} = g$$

---

[1] Translator's Note: In the displayed inequality $\overset{0}{f}$ is determined by h.

also exists and, consequently,

$$Ag - \lambda g = 0$$

where $\| g \| = 1$. Thus, g is an eigenvector of the operator A.

Both the vector $\overset{0}{f'_{n_i}} - g$, and the vector $\overset{0}{f'_{n_i}}$ are solutions of equation (1) with the right member h'_{n_i}. But, because the minimum norm of a solution of this equation is one, we have, for each i,

$$\| \overset{0}{f'_{n_i}} - g \| \geqq 1.$$

Since this is impossible, the theorem is proved.

53. Further Properties of Completely Continuous Operators

THEOREM 1: *If λ is a nonzero eigenvalue of the completely continuous operator A in the Hilbert space* H, *then $\bar{\lambda}$ is an eigenvalue of the operator A^*.*

Proof: Let the vector h run through H. Then the vector

$$g = Ah - \lambda h$$

does not run through the whole space, but only through some linear manifold G $\subset$ H, because the equation

$$Af - \lambda f = g$$

is not solvable for every right member g. It is not difficult to see that the manifold G is closed, so that G is a subspace. Indeed, if $g_n \in$ G $(n = 1, 2, 3, \ldots)$ then, according to Theorem 3 of the preceding section, there exist a constant L and vectors $\overset{0}{h}_n$ for which

$$A\overset{0}{h}_n - \lambda \overset{0}{h}_n = g_n \qquad (n = 1, 2, 3, \ldots)$$

and

$$\| \overset{0}{h}_n \| \leqq L \| g_n \| \qquad (n = 1, 2, 3, \ldots).$$

If $g_n \to g$, then the sequence of vectors $\{\overset{0}{h}_n\}_{n=1}^{\infty}$ is bounded and, therefore, it has a subsequence $\{\overset{0}{h}_{n_i}\}_{i=1}^{\infty}$ such that

$$\lim_{i \to \infty} A\overset{0}{h}_{n_i}$$

exists. Consequently,

$$h = \lim_{i \to \infty} \overset{0}{h}_{n_i}$$

exists, and we obtain

$$Ah - \lambda h = g.$$

Hence, $g \in$ G, so that the manifold G is closed. Since the subspace G does not coincide with H, there exists a nonzero vector f which is orthogonal to G. Then, for each $h \in$ H,

$$(Ah - \lambda h, f) = 0$$

so that

$$(Ah, f) = (\lambda h, f)$$

or

$$(Ah, f) = (h, \bar{\lambda} f).$$

This relation shows that $f \in \mathrm{D}_{A^*}$ and

$$A^* f = \bar{\lambda} f.$$

The theorem is proved.

Now we can strengthen Theorem 2 of the preceding section somewhat. This theorem asserts that if $\lambda \neq 0$ is an eigenvalue of the completely continuous operator A in R, then the equation

$$Af - \lambda f = g \tag{1}$$

is not solvable for every $g \in$ R. Supposing that A is now defined in the Hilbert space H, we shall determine the set of vectors g for which equation (1) is solvable.

THEOREM 2: *Let A be a completely continuous operator in* H *and fix $\lambda \neq 0$. Then equation* (1) *has a solution if and only if the vector g is orthogonal to the eigenmanifold* F *of the operator A^* belonging to $\bar{\lambda}$. If $\bar{\lambda}$ is not an eigenvalue of the operator A^*, then* F *is understood to be the null subspace, i.e., in this case, equation* (1) *is solvable for every right member $g \in$* H.

Proof: Let G be the set of all vectors g having a representation

$$g = Ah - \lambda h,$$

for some $h \in$ H. We must show that $\mathrm{H} \ominus \mathrm{G}$ coincides with the eigenmanifold F of the operator A^* belonging to $\bar{\lambda}$. Let the vector f be orthogonal to G. Then, repeating the corresponding part of the proof of Theorem 1, we find that

$$A^* f = \bar{\lambda} f.$$

Thus, $f \in$ F, which implies that

$$\mathrm{H} \ominus \mathrm{G} \subset \mathrm{F}.$$

In particular, if $\bar{\lambda}$ is not an eigenvalue of the operator A^*, then $\mathrm{H} \ominus \mathrm{G} = 0$, so that G $=$ H and equation (1) is solvable for every right member $g \in$ H. It remains to show that if $\bar{\lambda}$ is an eigenvalue of the operator A^*, then

$$\mathrm{F} \subset \mathrm{H} \ominus \mathrm{G}. \tag{2}$$

Let $f \in$ F and let g be any vector of the form

$$g = Ah - \lambda h.$$

Then

$$(f, g) = (f, Ah - \lambda h) = (f, Ah) - \bar{\lambda}(f, h) = (A^*f - \bar{\lambda} f, h) = 0.$$

This implies that $f \perp$ G. Therefore, F $\perp$ G and relation (2) is proved. At the same time, the theorem is proved.

The reader familiar with the elementary theory of integral equations will recognize that the propositions established above are generalizations of two theorems of Fredholm.

A generalization of a third theorem of Fredholm is also valid. It is formulated as follows: *the eigenmanifolds of the completely continuous operators A and A^* belonging to the characteristic values λ and $\bar{\lambda}$ have the same dimension.* We shall not prove this theorem.

54. The Existence Theorem for Eigenvectors of Completely Continuous Self-Adjoint Operators

The fundamental theorem concerning completely continuous self-adjoint operators asserts: *every completely continuous self-adjoint nonzero operator A in* R *has at least one eigenvector e which belongs to a nonzero eigenvalue λ.*

We shall give two different proofs of this theorem.

First proof: Let

$$M = \sup_{\|g\|=1} |(Ag, g)| = \sup_{\|g\|=1} \| Ag \|.$$

Clearly, $M > 0$. It follows from the definition of the supremum that there exists a sequence of normalized vectors $\{g_n\}_1^\infty$ for which

$$\lim_{n\to\infty} (Ag_n, g_n)$$

exists, and is equal to $+M$ or $-M$. We denote this nonzero limit by λ. Since A is completely continuous, the bounded sequence $\{g_n\}_1^\infty$ has a subsequence $\{g_{n_i}\}_{i=1}^\infty$ for which

$$\lim_{i\to\infty} Ag_{n_i} = h \tag{1}$$

exists. From the equation

$$\| Ag_{n_i} - \lambda g_{n_i} \|^2 = \|Ag_{n_i}\|^2 - 2\lambda\, (Ag_{n_i}, g_{n_i}) + \lambda^2$$

it follows that

$$\lim_{i\to\infty} \| Ag_{n_i} - \lambda g_{n_i} \|^2 = \| h \|^2 - 2\lambda^2 + \lambda^2 = \| h \|^2 - \lambda^2. \tag{2}$$

Since

$$\| Ag_{n_i} \| \leqq M \| g_{n_i} \| = M = | \lambda |$$

equation (1) yields

$$\| h \| \leqq | \lambda | .$$

But, since the left member of equation (2) is non-negative, $\| h \| = | \lambda |$, so that (2) becomes

(3) $$\lim_{i \to \infty} \| Ag_{n_i} - \lambda g_{n_i} \| = 0.$$

It follows from (1) and (3) that $\lim_{i \to \infty} g_{n_i}$ exists and is equal to $\frac{h}{\lambda}$. Introducing the vector $e = \frac{h}{\lambda}$, which has norm one, we obtain from (1) and (3) the equation

$$Ae - \lambda e = 0.$$

This completes the first proof.

Second proof: We choose an arbitrary vector f_0 for which $Af_0 \neq 0$. It is not difficult to prove that $A^n f_0 \neq 0$, $n = 1, 2, 3, \ldots$. Indeed, assume that $A^n f_0 = 0$ for some n. If n is even and $n = 2k$ then

$$0 = (A^n f_0, f_0) = (A^k f_0, A^k f_0).$$

If n is odd and $n = 2k - 1$ then

$$0 = (A^n f_0, A f_0) = (A^k f_0, A^k f_0).$$

Thus, for each n, if $A^n f_0 = 0$, then $A^k f_0 = 0$, where $k = \left[\frac{n}{2} \right]$. Repeating this argument, we get finally that $Af_0 = 0$, which is impossible.[2] From this result it follows that there exist two infinite sequences of nonzero vectors

$$\{f_k\}_0^\infty, \{f'_k\}_0^\infty$$

defined by means of the equations

$$f'_k = \frac{f_k}{\| f_k \|}, \quad f_{k+1} = Af'_k \qquad (k = 0, 1, 2, \ldots).$$

From these definitions it follows that

(4) $$\| f_k \| \leqq \| f_{k+1} \| \qquad (k = 0, 1, 2, \ldots)$$

and

(5) $$\| f_{k-1} \| \cdot \| f_k \| = (f_{k-1}, f_{k+1}) = (f_{k+1}, f_{k-1}) \qquad (k = 1, 2, 3, \ldots).$$

Indeed, for $k \geqq 1$,

$$\begin{aligned} \| f_k \| &= (f_k, f'_k) = (Af'_{k-1}, f'_k) = (f'_{k-1}, Af'_k) = \\ &= (f'_{k-1}, f_{k+1}) \leqq \| f'_{k-1} \| \cdot \| f_{k+1} \| = \| f_{k+1} \|, \end{aligned}$$

[2] Translator's Note: The proof is slightly easier if n is chosen as the *smallest* integer for which $A^n f_0 = 0$. Since $Af_0 \neq 0$ by hypothesis, $n > 1$. By the argument given above, $A^k f_0 = 0$ for some $k < n$. This is a contradiction.

which proves the inequality (4) and

$$(f'_{k-1}, f_{k+1}) = \| f_k \|,$$

which implies (5).

Again denote the norm of the operator A by M. Then

$$\| A f'_{k-1} \| \leqq M$$

and, consequently,

$$\| f_k \| \leqq M.$$

Therefore, $\{ \| f_k \| \}_1^\infty$ is a bounded nondecreasing sequence of positive numbers, so that it has a finite (positive) limit:

$$\lim_{k \to \infty} \| f_k \| = \lambda. \tag{6}$$

Moreover, since A is completely continuous, there exists a subsequence $\{f'_{n_i}\}_{i=1}^\infty$ for which the sequence of vectors

$$f_{n_i+1} = A f'_{n_i} \qquad (i = 1, 2, 3, \ldots)$$

has a limit as $i \to \infty$:

$$f_{n_i+1} \to g.$$

Noting that

$$f_{n_i+2} = A f'_{n_i+1} = \frac{A f_{n_i+1}}{\| f_{n_i+1} \|},$$

we infer without difficulty that the sequence $\{f_{n_i+2}\}_{i=1}^\infty$ converges. Similarly, the convergence of the sequence $\{f_{n_i+3}\}_{i=1}^\infty$ is established. We suppose that, as $i \to \infty$,

$$f_{n_i+2} \to h, \ f_{n_i+3} \to h'$$

and evaluate $\| h' - g \|^2$. Using relations (5) and (6), we get

$$\| h' - g \|^2 = \lim_{i \to \infty} \| f_{n_i+3} - f_{n_i+1} \|^2 =$$

$$= \lim_{i \to \infty} \{ \| f_{n_i+3} \|^2 + \| f_{n_i+1} \|^2 - (f_{n_i+3}, f_{n_i+1}) - (f_{n_i+1}, f_{n_i+3}) \} = 0.$$

Therefore, $h' = g$. On the other hand, since

$$h = \frac{Ag}{\lambda}, \qquad h' = \frac{Ah}{\lambda}$$

we have

$$Ag = \lambda h, \qquad Ah = \lambda g,$$

so that

$$A(h + g) = \lambda(h + g), \ A(h - g) = -\lambda(h - g).$$

The vector g is not equal to zero because $\| g \| = \lambda$. So at least one of the vectors $h + g$ and $h - g$ is different from zero. This nonzero vector is an eigenvector of the operator A belonging to an eigenvalue which is either λ or $-\lambda$.

55. The Spectrum of a Completely Continuous Self-Adjoint Operator in R

In the preceding section we proved that each nonzero completely continuous self-adjoint operator A in R has at least one nonzero eigenvalue λ. In this section we construct a complete system of nonzero eigenvalues of the operator A.

THEOREM: *Each nonzero completely continuous self-adjoint operator A has a finite or infinite orthonormal sequence of eigenvectors $e_1, e_2, e_3, \ldots$ belonging to its nonzero eigenvalues $\lambda_1, \lambda_2, \lambda_3, \ldots$ ($|\lambda_1| \geqq |\lambda_2| \geqq |\lambda_3| \geqq \ldots$) such that for each vector f of the form $f = Ah$ the Parseval relation*

$$\|f\|^2 = \sum_k |(f, e_k)|^2$$

holds.

Proof: On the basis of the theorem of the preceding section there exists a vector e_1 such that $\| e_1 \| = 1$ and

$$Ae_1 = \lambda_1 e_1$$

where

$$\lambda_1 = \pm \sup_{\|g\|=1} |(Ag, g)|.$$

For later convenience we re-denote our linear system R by R_1, and the operator A by A_1. Let

$$R_2 = R_1 \ominus e_1.$$

It is clear that R_2 is also a linear metrizable system. Moreover, if $f \in R_2$, then $A_1 f \in R_2$, since the equation $(f, e_1) = 0$ implies that

$$(A_1 f, e_1) = (f, A_1 e_1) = (f, \lambda_1 e_1) = \lambda_1 (f, e_1) = 0.$$

The restriction A_2 of the operator A_1 to R_2 is also a completely continuous self-adjoint operator. If the operator A_2 is not identically zero then, by the theorem of the preceding section, there exists a vector $e_2 \in R_2$ such that

$$A_2 e_2 = \lambda_2 e_2 \qquad (\| e_2 \| = 1).$$

Since $e_2 \in R_2$ we have $(e_2, e_1) = 0$. Furthermore,

$$|\lambda_2| = \sup_{\|f\|=1, f \in R_2} |(A_1 f, f)| \leqq \sup_{\|g\|=1} |(A_1 g, g)| = |\lambda_1|.$$

In the same manner we define the linear system

$$R_3 = R_2 \ominus e_2$$

and let A_3 denote the restriction of A_1 to R_3. If $A_3 \neq 0$, we select a normalized eigenvector e_3 and the corresponding eigenvalue λ_3. This process comes to an end at the nth step only if the restriction A_n of the operator

A_1 to R_n turns out to be identically zero. In that case we get a finite orthonormal sequence of vectors

$$e_1, e_2, \ldots, e_{n-1},$$

belonging to the nonzero eigenvalues

$$\lambda_1, \lambda_2, \ldots \lambda_{n-1},$$

where

$$|\lambda_1| \geqq |\lambda_2| \geqq \ldots \geqq |\lambda_{n-1}|$$

and

$$|\lambda_k| = \sup_{\|f\|=1, f \in R_k} |(A_1 f, f)|.$$

If the process does not come to an end after a finite number of steps, then we get an infinite orthonormal sequence $\{e_k\}_1^\infty$ such that the corresponding sequence of eigenvalues converges to zero: $\lambda_k \to 0$ as $k \to \infty$. Let m be the numbers of elements of the sequence $\{e_k\}$ if the sequence is finite, and let m be an arbitrary fixed integer otherwise. Now we choose $h \in H$ arbitrarily and let

$$g = h - \sum_{k=1}^{m} (h, e_k) e_k.$$

Since

$$(g, e_k) = 0 \qquad (k = 1, 2, 3, \ldots, m)$$

we have $g \in R_{m+1}$. Therefore,

$$\| Ag \|^2 \leqq \| A_{m+1} \|^2_{R_{m+1}} \cdot \| g \|^2$$

or

$$\| Ah - \sum_{k=1}^{m} (h, e_k) Ae_k \|^2 \leqq \| A_{m+1} \|^2_{R_{m+1}} \cdot \| g \|^2. \tag{1}$$

Noting that

$$(h, e_k) Ae_k = (h, e_k) \lambda_k e_k = (h, \lambda_k e_k) e_k = (h, Ae_k) e_k = (Ah, e_k) e_k$$

and also that

$$\| g \| \leqq \| h \|$$

and letting $f = Ah$, we obtain from (1) the inequality

$$\| f - \sum_{k=1}^{m} (f, e_k) e_k \|^2 \leqq \| A_{m+1} \|^2_{R_{m+1}} \cdot \| h \|^2.$$

If the sequence

$$e_1, e_2, e_3, \ldots$$

is finite, then $A_{m+1} = 0$, so that (2) yields

$$f = \sum_{k=1}^{m} (f, e_k) e_k.$$

If the sequence $\{e_k\}$ is infinite, then it follows from (2) that

$$\|f-\sum_{k=1}^{m}(f, e_k)e_k\|^2 \leqq \lambda_{m+1}^2\|h\|^2$$

which yields

$$\|f\|^2-\sum_{k=1}^{m}|(f, e_k)|^2 \leqq \lambda_{m+1}^2\|h\|^2.$$

Letting $m \to \infty$, we get

$$\|f\|^2=\sum_{k=1}^{\infty}|(f, e_k)|^2.$$

56. Completely Continuous Normal Operators

Let S denote a completely continuous normal operator in the linear system R. We shall consider the operator $A = S^*S = SS^*$, which is completely continuous, self-adjoint, and nonnegative:

$$(Af, f) \geqq 0$$

for every $f \in$ R. The nonnegativity of A follows from the fact that

$$(Af, f)=(S^*Sf, f) = (Sf, Sf).$$

This property implies that all the eigenvalues of A are nonnegative. We denote them by

$$\rho_1^2 \geqq \rho_2^2 \geqq \rho_3^2 \geqq \ldots.$$

On the basis of the preceding section the operator A has a complete orthonormal system of eigenvectors g_k:

$$Ag_k = \rho_k^2 g_k \qquad (k = 1, 2, 3, \ldots).$$

We choose a positive eigenvalue of the operator A (let it be ρ^2) and suppose that its multiplicity is r. Let

$$g^{(1)}, g^{(2)}, \ldots, g^{(r)},$$

$$(1) \qquad (g^{(i)}, g^{(k)}) = \delta_{ik} \qquad (i, k = 1, 2, 3, \ldots, r),$$

comprise a complete orthonormal system of eigenvectors of the operator A belonging to the eigenvalue ρ^2. We define $h^{(i)}$ such that

$$S^*g^{(i)} = \rho h^{(i)} \qquad (i = 1, 2, 3, \ldots, r).$$

Then

$$\rho Sh^{(i)} = SS^*g^{(i)} = Ag^{(i)} = \rho^2 g^{(i)}$$

so that

$$Sh^{(i)} = \rho g^{(i)} \qquad (i = 1, 2, 3, \ldots, r).$$

Since

$$(h^{(i)}, h^{(k)}) = \frac{1}{\rho^2}(S^*g^{(i)}, S^*g^{(k)}) = \frac{1}{\rho^2}(Ag^{(i)}, g^{(k)}) = (g^{(i)}, g^{(k)})$$

the vectors $h^{(i)}$ comprise an orthonormal system:

$$(2) \qquad (h^{(i)}, h^{(k)}) = \delta_{ik} \qquad (i, k = 1, 2, 3, \ldots, r).$$

Furthermore,

$$Ah^{(i)} = \frac{1}{\rho} SS^*S^*g^{(i)} = \frac{1}{\rho} S^*Ag^{(i)} = \rho S^*g^{(i)} = \rho^2 h^{(i)} \qquad (i = 1, 2, 3, \ldots, r).$$

So $h^{(i)}$ is an eigenvector of the operator A which belongs to the eigenvalue ρ^2. Hence, $h^{(i)}$ is a linear combination of the vectors

$$g^{(1)}, g^{(2)}, \ldots, g^{(r)},$$

i.e.,[3]

$$h^{(i)} = \alpha_{i1} g^{(1)} + \alpha_{i2} g^{(2)} + \ldots + \alpha_{ir} g^{(r)}. \tag{3}$$

We consider the n-dimensional space G spanned by the vectors $g^{(i)}$ ($i = 1, 2, 3, \ldots, r$). Let U denote the linear operator in G which is defined by

$$Ug^{(i)} = h^{(i)} \qquad (i = 1, 2, 3, \ldots, r).$$

Relations (1) and (2) imply that U is a unitary operator. We proceed to determine the eigenvectors of U. If f is an eigenvector and

$$f = x_1 g^{(1)} + x_2 g^{(2)} + \ldots + x_r g^{(r)},$$

then from the equation

$$Uf = \theta f$$

it follows that

$$\begin{aligned} \alpha_{11}x_1 + \alpha_{21}x_2 + \ldots + \alpha_{r1}x_r &= \theta x_1, \\ \alpha_{12}x_1 + \alpha_{22}x_2 + \ldots + \alpha_{r2}x_r &= \theta x_2, \\ \cdots\cdots\cdots\cdots\cdots\cdots \\ \alpha_{1r}x_1 + \alpha_{2r}x_2 + \ldots + \alpha_{rr}x_r &= \theta x_r . \end{aligned}$$

Thus, we obtain the equation

$$\begin{vmatrix} \alpha_{11} - \theta & \alpha_{21} & \ldots & \alpha_{r1} \\ \alpha_{12} & \alpha_{22} - \theta & \ldots & \alpha_{r2} \\ \cdots & \cdots & \cdots & \cdots \\ \alpha_{1r} & \alpha_{2r} & \ldots & \alpha_{rr} - \theta \end{vmatrix} = 0.$$

Conversely, each root θ of this equation corresponds to an eigenvector of the operator U. Since U is unitary, each root has absolute value one. Let $\theta^{(1)}$ be one of these roots and let $f^{(1)}$ be the corresponding eigenvector. Consider the subspace G_2 consisting of all vectors of the space $G \equiv G_1$ which are orthogonal to $f^{(1)}$. Obviously, U is a unitary operator in G_2. Therefore, repeating the foregoing argument, we find that the operator U has an eigenvector $f^{(2)}$ in G_2 with corresponding eigenvalue $\theta^{(2)}$. Continuing this "splitting off" process, we construct an orthonormal system of r vectors

$$f^{(1)}, f^{(2)}, \ldots, f^{(r)}.$$

[3] Translator's Note: Since the $g^{(i)}$ are linearly independent, the coefficients in (3) are unique.

These vectors satisfy the equations

$$Uf^{(i)} = \theta^{(i)} f^{(i)} \qquad (i = 1, 2, 3, \ldots, r).$$

The vectors $f^{(i)}$ also span the space G. If $f^{(i)}$ is given in terms of the vectors $g^{(i)}$ by the equation

$$f^{(i)} = x_1^{(i)} g^{(1)} + x_2^{(i)} g^{(2)} + \ldots + x_r^{(i)} g^{(r)}$$

then

$$S^* f^{(i)} = \rho \{ x_1^{(i)} h^{(1)} + x_2^{(i)} h^{(2)} + \ldots + x_r^{(i)} h^{(r)} \} = \rho \, U f^{(i)} = \rho \, \theta^{(i)} f^{(i)}.$$

Therefore, $f^{(i)}$ is an eigenvector of the operator S^* belonging to the eigenvalue $\rho\theta^{(i)}$. Analogously it is proved that $f^{(i)}$ is an eigenvector of the operator S belonging to the eigenvalue $\rho\bar{\theta}^{(i)}$. In the system of eigenvectors g_1, g_2, g_3, ... of the operator A we now replace

$$g^{(1)}, g^{(2)}, \ldots, g^{(r)}$$

by the vectors

$$f^{(1)}, f^{(2)}, \ldots, f^{(r)}.$$

If we proceed in the same way with each eigenvalue of the operator A having multiplicity larger than 1, we get the proof of the first part of the following theorem.

THEOREM: *If S is a completely continuous nonzero normal operator in* R, *then there exists an orthonormal system of vectors $\{e_k\}$ and a system of nonzero (complex) numbers $\{\lambda_k\}$ for which*

$$Se_k = \lambda_k e_k, \qquad S^* e_k = \bar{\lambda}_k e_k \qquad (k = 1, 2, 3, \ldots).$$

This system of vectors is complete in the sense that each element f of the form Sh or S^*h is represented by the series

$$f = \sum_k (f, e_k) e_k. \tag{4}$$

In order to prove the second part of the assertion we shall let $f = Sh$ and consider the vector

$$f' = S^* f = S^* Sh = Ah.$$

The theorem in Section 55 on completely continuous self-adjoint operators and the fact[4] that the vectors e_k form an orthonormal system of eigenvectors of A which is complete in the sense of this theorem yield the Parseval relation,

$$\| f' \|^2 = \sum_{k=1}^{\infty} | (f', e_k) |^2.$$

Thus, f' is the strong limit of the sequence

$$\sum_{k=1}^{n} (f', e_k) e_k \qquad (n = 1, 2, 3, \ldots).$$

[4] We assume for definiteness that they form an infinite set.

Therefore,

$$(f',h) = \lim_{n\to\infty} \sum_{k=1}^{n} (f',e_k)(e_k,h) = \sum_{k=1}^{\infty} (f',e_k)(e_k,h).$$

But since

$$(f',h) = (S^*f,h) = (f,Sh) = (f,f),$$
$$(f',e_k) = (S^*f,e_k) = (f,Se_k) = \bar{\lambda}_k(f,e_k),$$
$$(e_k,h) = \frac{1}{\bar{\lambda}_k}(S^*e_k,h) = \frac{1}{\bar{\lambda}_k}(e_k,Sh) = \frac{1}{\bar{\lambda}_k}(e_k,f)$$

the Parseval relation yields

$$\|f\|^2 = \sum_{k=1}^{\infty} |(f,e_k)|^2.$$

Therefore, equation (4) is proved if $f = Sh$. A similar proof works if $f = S^*h$.

57. Applications to the Theory of Almost Periodic Functions

As was indicated in Section 13, a (complex) continuous function $f(t)$ $(-\infty < t < \infty)$ is *almost periodic* if for each $\varepsilon > 0$ there exists $l = l(\varepsilon) > 0$ such that each real interval of length $l(\varepsilon)$ contains at least one number τ for which

$$|f(t+\tau) - f(t)| < \varepsilon$$

for all t. Each such τ is called a *translation number*.

Starting from this definition, it is not difficult to establish several simple properties of almost periodic functions. We list these properties without proof.[5]

I. Every almost periodic function is bounded and uniformly continuous on the whole real axis.

II. For each almost periodic function $f(t)$ there exists the so-called *mean value*

$$\mathbf{M}\{f(t)\} = \lim_{T\to\infty} \frac{1}{2T} \int_{-T}^{T} f(t)\,dt = \lim_{T\to\infty} \frac{1}{2T} \int_{-T+a}^{T+a} f(t)\,dt.$$

The right hand limit is uniform in a.

III. The product and sum of two almost periodic functions are almost periodic.

IV. If $f(t)$ is an almost periodic function and if

$$\mathbf{M}\{|f(t)|^2\} = 0,$$

then $f(t) \equiv 0$.

[5] The reader will find a detailed presentation in the survey article by B. M. Levitan [1].

The set of all almost periodic functions becomes a metrized linear system R if the scalar product is defined by the formula

$$(f,g) = \mathbf{M}\{f(t)\overline{g(t)}\}. \tag{1}$$

Now property IV has the equivalent form: $(f,f) = 0$ implies $f = 0$.

In Section 13, the scalar product (1) was introduced for all polynomials of the form

$$\sum_k A_k e^{i\lambda_k t}. \tag{2}$$

Being the sum of periodic functions which are, therefore, almost periodic, every such polynomial belongs to our linear system R. In particular, the linear system R contains the continuum of functions, $e^{i\lambda t}$ $(-\infty < \lambda < \infty)$, which form an orthonormal system. Therefore, as was the case with the space B^2 defined in Section 13, the linear system R is nonseparable.

The central place in the theory of almost periodic functions is occupied by the theorem, mentioned in Section 13, to the effect that each almost periodic function is the uniform limit on the whole real axis of polynomials of the form (2). As was first shown by Bohr, it is possible to prove this theorem with the aid of a particular method of harmonic analysis of almost periodic functions. Instead of the usual Fourier coefficients of pure periodic functions, we introduce in the harmonic analysis of the almost periodic function $f(t)$ the function

$$a(\lambda) = \lim_{T\to\infty} \frac{1}{2T} \int_{-T}^{T} f(t) e^{-i\lambda t}\, dt = \mathbf{M}\{f(t) e^{-i\lambda t}\} = (f, e^{i\lambda t}).$$

The existence of $a(\lambda)$ for each real λ follows from properties II and III. On the basis of the remark preceding Theorem 4 in Section 9, for each function $f(t) \in$ R there exist only countably many values of the parameter λ for which $a(\lambda) \neq 0$. We denote these values by $\lambda_1, \lambda_2, \ldots$ and we call them the *Fourier exponents* of the function $f(t)$. We call the corresponding numbers

$$C_k = a(\lambda_k) \qquad (k = 1, 2, 3, \ldots)$$

the *Fourier coefficients* of the function $f(t)$. Thus, to each function $f(t) \in$ R there corresponds a *Fourier series:*

$$f(t) \sim \sum_k C_k e^{i\lambda_k t}.$$

From the general situation of Section 8, it follows that

$$\sum_k |C_k|^2 \leqq \mathbf{M}\{|f(t)|^2\}. \tag{3}$$

This is a generalization of the so-called *Bessel's inequality*.

The fundamental theorem of the Bohr theory asserts that there is equality in (3) for each almost periodic function $f(t)$. We present next a proof of the Bohr theorem. The proof, which is due to H. Weyl, is based on the theory of completely continuous operators. For this purpose, fix $f(t) \in \mathrm{R}$ and let

$$v(s) = \lim_{T\to\infty} \frac{1}{2T} \int_{-T}^{T} f(s-t)\,u(t)\,dt = \underset{t}{\mathbf{M}}\{f(s-t)\,u(t)\}.$$

This formula associates with each function $u(t) \in \mathrm{R}$ a function $v(t)$ which, it is evident, also belongs to R. Thus, we have a linear operator, determined by the function $f(t)$, which maps $u(t)$ into $v(t)$. We denote this operator by A.

We shall prove that A is a normal operator. For this purpose we introduce a second operator B: let $w = Bu$ if

$$w(s) = \underset{t}{\mathbf{M}}\{\overline{f(t-s)}\,u(t)\}.$$

We shall prove that

(a) $(Au_1, u_2) = (u_1, Bu_2)$ for $u_1(t), u_2(t) \in \mathrm{R}$,

(b) $AB = BA$.

As for property (a), we have

$$\begin{aligned}(Au_1, u_2) &= \underset{s}{\mathbf{M}}\{\underset{t}{\mathbf{M}}[f(s-t)\,u_1(t)]\,\overline{u_2(s)}\} = \\ &= \underset{t}{\mathbf{M}}\{u_1(t)\,\underset{s}{\mathbf{M}}[f(s-t)\,\overline{u_2(s)}]\} = \\ &= \underset{t}{\mathbf{M}}\{u_1(t)\,\overline{\underset{s}{\mathbf{M}}[\overline{f(s-t)}\,u_2(s)]}\} = (u_1, Bu_2).\end{aligned}$$

The reversal of the order of the operations $\underset{s}{\mathbf{M}}$ and $\underset{t}{\mathbf{M}}$ is valid because the mean value is a uniform limit according to the alternate definition given in property II above. We now prove property (b). First, we note that

$$\begin{aligned}ABu &= \underset{t}{\mathbf{M}}\{f(s-t)\,\underset{\tau}{\mathbf{M}}[\overline{f(\tau-t)}\,u(\tau)]\} = \\ &= \underset{\tau}{\mathbf{M}}\{u(\tau)\,\underset{t}{\mathbf{M}}[f(s-t)\overline{f(\tau-t)}]\}.\end{aligned}$$

Next, we consider in detail the function of s and τ defined by

$$\varphi = \underset{t}{\mathbf{M}}[f(s-t)\overline{f(\tau-t)}] = \lim_{T\to\infty} \frac{1}{2T} \int_{-T}^{T} f(s-t)\overline{f(\tau-t)}\,dt.$$

After the change of varable from t to s indicated by

$$\tau - t = \sigma - s$$

we get

$$\varphi = \lim_{T\to\infty} \frac{1}{2T} \int_{\tau+s-T}^{\tau+s+T} f(\sigma - \tau)\overline{f(\sigma - s)}\, d\sigma =$$

$$= \lim_{T\to\infty} \frac{1}{2T} \int_{-T}^{T} f(\sigma - \tau)\overline{f(\sigma - s)}\, d\sigma = \underset{\sigma}{M}\, [f(\sigma - \tau)\overline{f(\sigma - s)}].$$

Hence,

$$ABu = \underset{\tau}{\mathbf{M}}\, \{u(\tau)\, \underset{\sigma}{\mathbf{M}}\, [f(\sigma - \tau)\overline{f(\sigma - s)}\,]\} =$$

$$= \underset{\sigma}{\mathbf{M}}\, \{\overline{f(\sigma - s)}\, \underset{\tau}{\mathbf{M}}\, [f(\sigma - \tau)\, u(\tau)]\} = BAu,$$

and property (b) is proved.

Now we are ready to prove that A is a completely continuous operator. Let G denote an infinite set of functions $u(t) \in$ R for which

$$\| u \|^2 = \mathbf{M}\{ | u(t) |^2\} \leqq 1.$$

We must prove that G contains a sequence $\{u_n(t)\}_1^\infty$ such that the corresponding sequence $\{v_n(t)\}_1^\infty$, where $v_n = Au_n$, converges in the norm of R to some function $V(t) \in$ R. The function $u(t) \in$ G is transformed by the operator A into the function

$$v(s) = \lim_{T\to\infty} \frac{1}{2T} \int_{-T}^{T} f(s - t)\, u(t)\, dt \tag{4}$$

so that

$$| v(s) | \leq \lim_{T\to\infty} \sqrt{\frac{1}{2T} \int_{-T}^{T} | f(s - t) |^2 dt}\, \sqrt{\frac{1}{2T} \int_{-T}^{T} | u(t) |^2 dt}$$

or

$$| v(s) | \leqq \sqrt{\mathbf{M}\{ | f(t) |^2\}} = C.$$

Thus, the transformed function $v(s)$ is uniformly bounded. Furthermore, from formula (4) it follows that

$$v(s') - v(s'') = \lim_{T\to\infty} \frac{1}{2T} \int_{-T}^{T} \{f(s' - t) - f(s'' - t)\}\, u(t)\, dt$$

which yields

$$| v(s') - v(s'') | \leqq \sqrt{\underset{t}{\mathbf{M}}\{ | f(s' - t) - f(s'' - t) |^2\}}.$$

This inequality shows that the function $v(s)$ is uniformly continuous. Indeed, $f(t)$ is uniformly continuous, so that for each $\varepsilon > 0$ there exists $\delta = \delta(\varepsilon) > 0$ such that

$$|f(s'-t)-f(s''-t)| \leqq \varepsilon \qquad (-\infty < t < \infty)$$

for $|s'-s''| \leqq \delta$. Hence, for $|s'-s''| \leqq \delta$,

$$|v(s')-v(s'')| \leqq \sqrt{\mathbf{M}\{\varepsilon^2\}} = \varepsilon.$$

According to the well-known theorem of Arzelà the set of functions $v(t)$ contains a sequence $\{v_n(t)\}_1^\infty$ which is uniformly convergent on the whole real axis. We denote the limit function by $V(t)$. It also satisfies the inequality

$$|V(s')-V(s'')| \leqq \sqrt{\mathbf{M}_t\{|f(s'-t)-f(s''-t)|^2\}}.$$

Therefore, if τ is a translation number of the function $f(t)$ corresponding to ε and if $s'-s''=\tau$, then

$$|V(s')-V(s'')| \leqq \varepsilon.$$

From this it is evident that $V(t)$ is an almost periodic function. Since

$$\sqrt{\mathbf{M}\{|w_1(t)-w_2(t)|^2\}} \leqq \sup_{-\infty<t<\infty} |w_1(t)-w_2(t)|$$

the sequence $\{v_n(t)\}_{n=1}^\infty$ converges not only uniformly on the real axis, but also in the norm of R. It follows that the operator A is completely continuous.

Let μ be a nonzero eigenvalue of A and let

$$g_1(t), g_2(t), \ldots, g_n(t) \tag{5}$$

be a complete orthonormal system of eigenfunctions (in other words, eigenvectors) of the operator A belonging to the eigenvalue μ. If $g(t)$ is any one of these eigenfunctions, then

$$\lim_{T\to\infty} \frac{1}{2T}\int_{-T}^{T} f(s-t)g(t)\,dt = \mu g(s).$$

It follows that for each real number σ the function $g(t+\sigma)$ also is an eigenfunction belonging to μ. From this fact we obtain the equations

$$g_r(t+\sigma) = \sum_{k=1}^{n} c_{rk}(\sigma) g_k(t) \qquad (r = 1, 2, 3, \ldots, n). \tag{6}$$

Thus, the almost periodic functions in the orthonormal system (5) satisfy the functional equations (6). By various methods it is possible to prove that each of the functions $g_r(t)$ has the form[6]

$$g_r(t) = \sum_{k=1}^{n} c_r^{(k)} e^{i\lambda_k t}$$

where the $c_r^{(k)}$ are constants. If necessary, one can replace the original eigenfunctions $g_r(t)$ by linear combinations of them to obtain the eigenfunctions

$$e^{i\lambda_1 t}, e^{i\lambda_2 t}, \ldots, e^{i\lambda_n t}.$$

According to the definition of the function $\alpha(\lambda)$,

$$\lim_{T\to\infty} \frac{1}{2T}\int_{-T}^{T} f(s-t)\, e^{i\lambda_r t}\, dt = e^{i\lambda_r s} \lim_{T\to\infty} \frac{1}{2T}\int_{-T}^{T} f(s-t)\, e^{i\lambda_r (t-s)}\, dt = e^{i\lambda_r s}\, \alpha(\lambda_r).$$

Thus, we have found the general form of the eigenfunctions of the operator A and have proved that the eigenfunction $e^{i\lambda_r t}$ corresponds to the eigenvalue which is the Fourier constant

$$\alpha(\lambda_r) = \mathbf{M}\{f(t) e^{-i\lambda_r(t)}\}.$$

Let

$$e^{i\lambda_1 t}, e^{i\lambda_2 t}, e^{i\lambda_3 t}, \ldots$$

be the sequence (in some order) of all eigenfunctions of the operator A and let

$$C_1 = \alpha(\lambda_1),\ C_2 = \alpha(\lambda_2),\ C_3 = \alpha(\lambda_3), \ldots$$

be the corresponding eigenvalues. We consider the series

$$\sum_{k=1}^{\infty} C_k (\Phi, e^{i\lambda_k t})\, e^{i\lambda_k t}, \tag{7}$$

where $\Phi \in \mathrm{R}$. Since

$$\sum_{k=m}^{\infty} | C_k (\Phi, e^{i\lambda_k t}) | \leqq \sqrt{\sum_{k=m}^{\infty} |C_k|^2}\ \sqrt{\sum_{k=m}^{\infty} |(\Phi, e^{i\lambda_k t})|^2} \leqq$$

$$\leqq \sqrt{\sum_{k=m}^{\infty} |C_k|^2}\ \sqrt{\mathbf{M}\{|\Phi(t)|^2\}}$$

---

[6] For this one must first show that the functions $c_{rk}(\sigma)$ are continuously differentiable. Then the system (6) yields the system of differential equations,

$$g_r'(t) = \sum_{k=1}^{n} c_{rk}'(0)\, g_k(t) \qquad (r = 1, 2, 3, \ldots, n).$$

The solutions of these equations are linear combinations of functions of the form

$$(a_0 + a_1 t + \ldots + a_m t^m)\, e^{\beta t}.$$

Since each function $g_k(t)$ is bounded uniformly, β must be pure imaginary and the polynomial factor must reduce to the constant term.

Another proof, which is richer in ideas, is based on the simultaneous reduction to diagonal form of the Abelian group of unitary matrices

$$(c_{rk}(\sigma))_{r,\, k=1}$$

the series (7) converges uniformly in t. Now we examine the function

$$\underset{s}{\mathbf{M}}\{f(t-s)\overline{f(-s)}\}.$$

It is of the form Ag where $g(t) \in \mathrm{R}$. According to the theorem in Section 56, this function can be represented by the series

$$\sum_{k=1}^{\infty} C_k \underset{s}{\mathbf{M}}\{\overline{f(-s)}e^{-i\lambda_k s}\} e^{i\lambda_k t}$$

which converges in the norm of R. By what was proved above[7] this series converges uniformly, so that

$$\underset{s}{\mathbf{M}}\{f(t-s)\overline{f(-s)}\} = \sum_{k=1}^{\infty} C_k \underset{s}{\mathbf{M}}\{\overline{f(-s)}e^{-i\lambda_k s}\} e^{i\lambda_k t}.$$

Hence, for $t = 0$,

$$\mathbf{M}\{|f(-s)|^2\} = \sum_{k=1}^{\infty} C_k \underset{s}{\mathbf{M}}\{\overline{f(-s)}e^{-i\lambda_k s}\}$$

or

$$\mathbf{M}\{|f(t)|^2\} = \sum_{k=1}^{\infty} |C_k|^2,$$

and, finally, the fundamental theorem of Bohr is proved.

[7] The role of the function $\Phi(t)$ is now played by $\overline{f(-t)}$.

BIBLIOGRAPHY

AKHIEZER, N. I. — [1] On some inversion formulas of singular integrals, *Izv. Akad. Nauk SSSR*, **9** (1945), 275–290; [2] Infinite Jacobi matrices and the moment problem, *Uspekhi mat. Nauk*, **9** (1941); [3] Integral operators with Carleman kernels, *Uspekhi mat. Nauk II*, **5** (21) (1947); [4] On the theory of maximal symmetric operators in Hilbert space, *Učenie Zapiski Av. Inst III*, **2** (1940).

AMBROSE, W. — [1] Spectral resolution of groups of unitary operators, *Duke Math. Journal*, **11** (1944), 589–595.

ARONZAJN, N. — [1] Caractérisation métrique de l'espace de Hilbert, etc., *C. R. Acad. Sci. Paris*, **201** (1935), 811–813, 873–875.

BANACH, S. — [1] *Théorie des opérations linéaires*, Warsaw (1932); [2] Sur les fonctionnelles linéaires, *Studia Math.*, **1** (1929), 211–216.

BEURLING, A. — [1] Sur les intégrales de Fourier absolument convergentes et leur application à une transformation fonctionnelle, *IX Congrès des Math. Scandinaves, Helsingfors* (1938), 345–366.

BOCHNER, S. — [1] *Vorlesungen über Fouriersche Integrale*, Leipzig, (1932); [2] Spektraldarstellung linearer Scharen unitärer Operatoren, *Sitzber. Preuss. Akad. Wiss.* (1933), 371–376; [3] Inversion formulae and unitary transformations, *Ann. of Math.*, **35** (1934), 111–115.

BOHNENBLUST, H. F. and A. SOBCZYK — [1] Extensions of functionals on complex linear spaces, *Bulletin Amer. Math. Soc.*, **44** (1938), 91–93.

CARATHEODORY, C. — [1] Bemerkungen zum Riesz-Fischerschen Satz und zur Ergodentheorie, *Abhandl. Math. Sem. d. Hansischen Univ.*, **14** (1941), 351–398.

CARLEMAN, T. — [1] *Sur les équations intégrales singulières à noyau réel symétrique*, Uppsala (1923); [2] Zur Theorie der linearen Integralgleichungen, *Math. Zeitschr.*, **9** (1921), 196–217; [3] Application de la théorie des équations intégrales linéaires aux équations différentielles non linéaires, *Acta Math.*, **59** (1932), 63–87.

COOPER, J. L. B. — [1] The spectral analysis of self-adjoint operators, *Quarterly Jour. of Math.*, **16** (1945), 31–48; [2] One-parameter semigroups of isometric operators in Hilbert space, *Ann. of Math.*, **48** (1947), 827–842; [3] Symmetric operators in Hilbert space, *Proc. London Math. Soc.*, **50** (1948), 11–55.

COURANT, R. — [1] Zur Theorie der linearen Integralgleichungen, *Math. Ann.*, **89** (1923), 161–178.

COURANT, R. and D. HILBERT — [1] *Methoden der mathematischen Physik*, Berlin (1931).

DELSARTE, J. — [1] *Les groupes de transformations linéaires dans l'espace de Hilbert*, Mém. Sci. math., **57** (1923).

DOETSCH, G. — [1] *Theorie und Anwendung der Laplace-Transformation*, Berlin (1937).

DUNFORD, N. — [1] Spectral theory I, Convergence to projections, *Trans. Amer. Math. Soc.*, **54** (1943), 185–217; [2] Spectral theory, *Bulletin Amer. Math. Soc.*, **49** (1943), 637–651.

EBERLEIN, W. F. — [1] A note on the spectral theorem, *Bulletin Amer. Math. Soc.*, **52** (1946), 328–331.

EVANS, G. C. — [1] *Functionals and their applications.* Selected topics, including integral equations, Amer. Math. Soc. Coll. Publ. V, New York (1918).

FRECHET, M. — [1] Sur les opérations linéaires, *Trans. Amer. Math. Soc.*, **8** (1907), 433–446.

FREDHOLM, J. — [1] Les équations intégrales linéaires, *C. R. du Congr. de Stockholm*, (1910), 92–100.

FREUDENTHAL, H. — [1] Über die Friedrichssche Fortsetzung halbbeschränkter Hermitescher Operatoren, *Proc. Acad. Amsterdam*, **39** (1936), 832–833.

FRIEDRICHS, K. — [1] Spektraltheorie halbbeschränkter Operatoren, *Math. Ann.*, **109** (1934), 465–487; 685–713; also **110** (1935), 777–779; [2] Über die ausgezeichnete Randbedingung in der Spektraltheorie der halbbeschränkten gewöhnlichen Differentialoperatoren zweiter Ordnung, *Math. Ann.*, **112** (1935), 1–23; [3] Beiträge zur Theorie der Spektralschar, *Math. Ann.*, **110** (1935), 54–62; [4] Die unitären Invarianten selbstadjungierter Operatoren im Hilbertschen Raum, *Jber. Deutsch. Math.-Vereinig.*, **45** (1935), 79–82; [5] On differential operators in Hilbert spaces, *Amer. J. Math.*, **61** (1939), 523–544.

FUKAMIYA, M. — [1] On one-parameter groups of operators, *Proc. Imp. Acad. Sci. Tokyo*, **16** (1940), 262–265.

GELFAND, I. M. — [1] Sur une lemme de la théorie des espaces linéaires, *Soobšč. Chark. mat. Obšč.* (4), **13** (1936), 35–40; [2] Abstract functions and linear operators, *Mat. Sbornik N. S.*, **4** (1938), 235–284.

GLAZMAN, I. M. — [1] On the theory of singular quasi-differential operators, *Chark.*, (1949); [2] A class of solutions of a classical moment problem, *Zapiski naučno-issl. Inst. mat. i mech. i Chark. mat. Obšč.* (4), **20** (1950); [3] On the theory of singular differential operators, *Uspekhi mat. Nauk*, **5** (part 6) (1950).

GLAZMAN, I. — [1] On the defect index of differential operators, *Doklady Akad. Nauk SSSR*, **64** (#2) (1949).

GODEMENT, R. — [1] Sur une généralisation d'un théorème de Stone, *Comptes Rendus Acad. Sci. Paris*, **218** (1944), 901–903.

GRAFF, A. A. — [1] On the theory of linear differential systems in one-dimensional regions, *Mat. Sbornik*, **60** (1946); also **63** (1947).

HAHN, H. — [1] Über Folgen linearer Operationen, *Monatshefte Math. Phys.*, **32** (1922), 3–88.

HALMOS, P. R. — [1] *Introduction to Hilbert space and the theory of spectral multiplicity*, New York (1951).

HALPERIN, I. — [1] Closures and adjoints of linear differential operators, *Ann. of Math.*, **38** (1937), 880–919.

HAMBURGER, H. — [1] Beiträge zur Konvergenztheorie der Stieltjesschen Kettenbrüche, *Math. Zeitschr.*, **4** (1919), 186–222.

HELLINGER, E. and O. TOEPLITZ — [1] Grundlagen für eine Theorie der unendlichen Matrizen, *Math. Ann.*, **69** (1910), 289–330; [2] Integralgleichungen und Gleichungen mit unendlich vielen Unbekannten, *Enzyklopädie d. Math. Wiss. II C*, **13**, Leipzig (1928).

HILBERT, D. — [1] *Grundzüge einer allgemeinen Theorie der linearen Integralgleichungen*, Leipzig (1912); [2] Wesen und Ziele einer Analysis der unendlich vielen unabhängigen Variablen, *Rend. Circ. Math. Palermo*, **27** (1909), 59–74.

HILDEBRAND, T. H. — [1] Über vollstetige, lineare Transformationen, *Acta Math.*, **51** (1928), 311–318.

HILLE, E. — [1] *Functional analysis and semi-groups*, Amer. Math. Soc. Coll. Publ., New York (1948); [2] Notes on linear transformations, *Trans. Amer. Math. Soc.*, **39** (1936), 131–153; [3] On semi-groups of transformations in Hilbert space, *Proc. Nat. Acad. Sci.*, **24** (1938) 159–161; [4] Analytical semi-groups in the theory of linear transformations, *Neuvième Congrès Math. Scand., Helsingfors* (1938), 135–145.

HILLE, E. and J. D. TAMARKIN — [1] On the theory of linear integral equations I, *Ann. of Math.* (2), **31** (1930), 479–528.

JULIA, G. — [1] *Introduction mathématique aux théories quantiques II*, Paris (1938).

KAKUTANI, S. — [1] Iteration of linear operators in complex Banach spaces, *Proc. Imp. Acad. Tokyo*, **14** (1938), 295–300.

KAZ, I. — [1] On Hilbert spaces generated by monotone Hermitean matrix functions, *Zapiski naučno-issl. Inst. mat. i mech. i Chark. mat. Obšč.*, **22** (1950).

KODAIRA, K. — [1] On some fundamental theorems in the theory of operators in Hilbert space, *Proc. Imp. Acad. Tokyo*, **15** (1939), 207–210.

KRASNOSELSKI, M. A. — [1] On defect numbers of closed operators, *Doklady Akad. Nauk SSSR*, **56** (1947), 559–561; [2] On self-adjoint extensions of Hermitean operators, *Ukrainskij mat. J. A. N. USSR*, **1** (1949), 21–38.

KREIN, M. — [1] On resolvents of Hermitean operators with defect indices (m, m), *Doklady Akad. Nauk SSSR*, **52** (#8) (1946), 657–660; [2] On a general method of decomposition of positive definite kernels into elementary products, *Doklady Akad. Nauk SSSR*, **53** (#1) (1946); [3] The theory of self-adjoint extensions of half-bounded Hermitean operators and their applications, Part I, *Mat. Sbornik*, *20* (*62*): **3** (1947), 431–495; [4] Part II of the above article, *Mat. Sbornik*, *21* (*63*): 3 (1947), 366–404; [5] On Hermitean operators with directed functionals, *Sbornik prac. Inst. Mat. A. N. Ukr. SSR*, **10.**

KREIN, M. and M. A. KRASNOSELSKI — [1] Fundamental theorems on the extensions of Hermitean operators and some of their applications to the theory of orthogonal polynomials and the moment problem, *Uspekhi mat. Nauk*, **2** (Part 3) (1947).

KREIN, M., M. A. KRASNOSELSKI and D. P. MILMAN — [1] On the defect numbers of linear operators in Banach space and on some geometric questions, *Sbornik trudov Inst. A. N. Ukr. SSR*, **11** (1948).

LANDAU, E. — [1] Über einen Konvergenzsatz, *Göttinger Nachrichten*, (1907), 25–27.

LENGYEL, B. — [1] On the spectral theorem of self-adjoint operators, *Acta Sci. Math. Szeged*, **9** (1939), 174–186.

LENGYEL, B. A. and M. H. STONE — [1] Elementary proof of the spectral theorem, *Ann. of Math.*, **37** (1936), 853–864.

LEVITAN, B. M. — [1] *Almost periodic functions*, Moscow (1953); [2] *Entwicklung nach Eigenfunktionen*, Gostechizdat (1950).

LIFSCHITZ, M. S. — [1] An application of the theory of Hermitean operators to the generalized moment problem, *Doklady Akad. Nauk SSSR*, **44** (#1) (1944); [2] A class of linear operators in Hilbert space, *Mat. Sbornik*, **19** (61): **2** (1946), 239–260; [3] On the theory of isometric operators with equal deficiency numbers, *Doklady Akad. Nauk SSSR*, **58** (#1) (1947), 13–15; [4] On the theory of elementary divisors of Hermitean operators, *Doklady Akad. Nauk SSSR*, **60** (#1) (1948), 17–20.

LORCH, E. R. — [1] Functions of self-adjoint transformations in Hilbert space, *Acta Sci. Math. Szeged*, **7** (1934), 136–146; [2] On a calculus of operators in reflexive vector spaces, *Trans. Amer. Math. Soc.*, **45** (1939), 217–234; [3] The spectrum of linear transformations, *Trans. Amer. Math. Soc.*, **52** (1942), 238–248.

LÖWIG, H. — [1] Komplexe euklidischer Räume von beliebiger endlicher oder unendlicher Dimensionszahl, *Acta Sci. Math. Szeged*, **7** (1934), 1–33.

MIMURA, J. — [1] Über Funktionen von Funktionaloperatoren in einem Hilbertschen Raum, *Jap. J. Math.*, **13** (1936), 119–128.

MURRAY, F. J. — [1] *An introduction to linear transformations in Hilbert space*, Ann. of Math. Studies #4, Princeton (1941); [2] Linear transformations between Hilbert spaces and applications of this theory to linear partial differential equations, *Trans. Amer. Math. Soc.*, **37** (1935), 301–338.

MUSCHELISCHVILI, N. I. — [1] *Singular integral equations*, North Holland Pub. Co. (1953).

NAIMARK, M. A. — [1] Spectral functions of a symmetric operator, *Izv. A. N. SSSR*, **4** (#3) (1940), 277–318; [2] On self-adjoint extensions of the second kind of a symmetric operator, *Izv. A. N. SSSR*, **4** (#1) (1940), 53–104; [3] On the square of a closed symmetric operator, *Dokl. Akad. Nauk SSSR*, **26** (1940), 866-870; [4] On a representation of additive operator functions of sets, *Doklady Akad. Nauk SSSR*, **41** (1943), 373–375; [5] Extremal spectral functions of symmetric operators, *Izv. A. N. SSSR*, **11** (#4) (1947).

NAKANO, H. — [1] Zur Eigenwerttheorie normaler Operatoren, *Proc. Phys.-Math. Soc. Jap. III, Series 21* (1939), 315–339; [2] Über Abelsche Ringe von Projektionsoperatoren, *Proc. Phys.-Math. Soc. Jap. III, Series 21* (1939), 357–375; [3] Funktionen mehrerer hypermaximaler normaler Operatoren, *Proc. Phys.-Math. Soc. Jap. III, Series 21* (1939), 713–728; [4] Hypermaximalität normaler Operatoren, *Proc. Phys.-Math. Soc. Jap. III, Series 22* (1940), 259–264; [5] Unitärinvarianten hypermaximaler normaler Operatoren im Hilbertschen Raum, *Ann. of Math.*, **42** (1941), 657–664; [6] Über den Beweis des Stoneschen Satzes, *Ann. of Math.*, **42** (1941), 665–667; [7] Unitärinvarianten im allgemeinen euklidischen Raum, *Math. Ann.*, **118** (1941), 112–133.

NEUMANN, J. v. — [1] *Mathematische Grundlagen der Quantenmechanik*, Berlin (1932); [2] Zur Theorie der unbeschränkten Matrizen, *J. reine angew. Math.*, **161** (1929), 208–236; [3] Allgemeine Eigenwerttheorie Hermitescher Funktionaloperatoren, *Math. Ann.*, **102** (1929), 49–131; [4] Zur Algebra der Funktionaloperationen und Theorie der normalen Operatoren, *Math. Ann.*, **102** (1929), 370–427; [5] Über Funktionen von Funktionaloperatoren, *Ann. of Math.*, **32** (1931), 191–226; [6] Über adjungierte Funktionaloperatoren, *Ann. of Math.* **33** (1932), 294–310; [7] Über einen Satz von M. H. Stone, *Ann. of Math.*, **33**, (1932), 567–573; [8] Zur Operatorenmethode in der klassischen Mechanik, *Ann. of Math.*, **33** (1932), 587–648; [9] On normal operators, *Proc. Acad. Sci. U.S.A.*, **21** (1935), 366–369; [10] Charakterisierung des Spektrums eines Integraloperators, *Actualités Sci. et Ind.*, **229** (1935).

PALEY, R. E. A. C. and N. WIENER — [1] *Fourier transforms in the complex domain*, New York (1934).

PHILLIPS, R. S. — [1] On linear transformations, *Trans. Amer. Math. Soc.*, **48** (1940), 516–541.

PLANCHEREL, M. — [1] Contribution à l'étude de la représentation d'une fonction arbitraire par des intégrales définies, *Rend. Circ. Math. Palermo*, **30** (1910), 289–335; [2] Sur les formules de réciprocité du type de Fourier, *J. London Math. Soc.*, **8** (1933), 220–226.

PLESNER, A. I. — [1] Spectral theory of linear operators I, *Uspekhi mat. Nauk*, **9** (1941); [2] Über halbunitäre Operatoren, *Doklady Akad. Nauk SSSR*, **25** (1939), 710–712; [3] Über Funktionen eines maximalen Operators, *Doklady Akad. Nauk SSSR*, **23** (1939), 327–330; [4] Spectral analysis of maximal operators, *Doklady Akad. Nauk SSSR*, **22** (#5) (1939), 225–228.

PLESNER, A. I. and W. A. ROCHLIN — [1] Spectral theory of linear operators II, *Uspekhi mat. Nauk*, 1 (Part 1) (11) (1946).

POVZNER, A. J. — [1] On the method of directed functionals of M. G. Krein, *Zap. naučno-issl. Inst. mat. i mech. i Chark. mat. Obšč. Series 4*, **20** (1950).

REID, W. T. — [1] Symmetrizable completely continuous linear transformations in Hilbert space, *Duke Math. J.*, **18** (1951), 41–56.

REDLICH, F. — [1] Spektraltheorie in nichtseparablen Räumen, *Math. Ann.*, **110** (1934), 342–356; [2] Störungstheorie der Spektralzerlegung I, *Math. Ann.* **113** (1936), 600–619; II, **113** (1936), 677–685; III, **116** (1939), 555–570; IV, **117** (1940), 356–382; V, **118** (1942), 462–484.

RIESZ, F. — [1] *Les systèmes d'équations linéaires à une infinité d'inconnues*, Paris (1931); [2] Untersuchungen über Systeme integrierbarer Funktionen, *Math. Ann.*, **69** (1910), 449–497; [3] Über die linearen Transformationen des komplexen Hilbertschen Raumes, *Acta Sci. Math. Szeged*, **5** (1930), 23–54; [4] Über Sätze von Stone und Bochner, *Acta Sci. Math. Szeged*, **6** (1933), 184–198; [5] Zur Theorie des Hilbertschen Raumes, *Acta Sci. Math. Szeged*, **7** (1934), 34–38; [6] Sur les fonctions des transformations hermitiennes dans l'espace de Hilbert, *Acta Sci. Math. Szeged*, **7** (1935), 147–159.

RIESZ, F. and E. R. LORCH — [1] The integral representation of unbounded self-adjoint transformations in Hilbert space, *Trans. Amer. Math. Soc.*, **39** (1936), 331–340.

RIESZ, F. and B. SZ.-NAGY — [1] Über Kontraktionen des Hilbertschen Raumes, *Acta Sci. Math. Szeged*, **10** (1943), 202–205; [2] *Functional analysis*, New York (1955).

SCHAUDER, J. — [1] Über lineare vollstetige Funktionaloperationen, *Studia Math.*, **2** (1930), 183–196.

SCHMIDT, E. — [1] Auflösung der allgemeinen linearen Integralgleichung, *Math. Ann.*, **64** (1907), 161–174; [2] Über die Auflösung linearer Gleichungen mit abzählbar unendlich vielen Unbekannten, *Rend Circ. Math. Palermo*, **25** (1908), 53–77.

SMIRNOV, V. I. — [1] Course of higher mathematics V, (1947).

SMITHIES, F. — [1] The Fredholm theory of integral equations, *Duke Math. J.*, **8** (1941), 107–130.

STONE, M. H. — [1] Linear transformations in Hilbert space, *Proc. Nat. Acad. Sci. U.S.A.*, **15** (1929), 198–200, 423–425; **16** (1930), 172–175; [2] On one-parameter unitary groups in Hilbert space, *Ann. of Math.*, **33** (1932), 643–648; [3] *Linear transformations in Hilbert space*, New York (1932).

STRAUS, A. W. — [1] Generalized resolvents of symmetric operators, *Izv. A. N. SSSR*, **18** (#1) (1954).

SUKHOMLINOV, G. — [1] Über Fortsetzung von linearen Funktionalen in linearen komplexen Räumen und linearen Quaternionenräumen, *Mat. Sbornik N. S.*, **3** (1938), 353–358.

SZ.-NAGY, B. — [1] Perturbations des transformations autoadjointes dans l'espace de Hilbert, *Commentarii Math. Helv.*, **19** (1946–47), 347–366; [2] On semi-groups of self-adjoint transformations in Hilbert space, *Proc. Nat. Acad. Sci. U.S.A.*, **24** (1938), 559–560; [3] Expansion theorems of Paley-Wiener type, *Duke Math. J.*, **14** (1947), 975–978; [4] On uniformly bounded linear transformations in Hilbert space, *Acta Sci. Math. Szeged*, **11** (1947), 152–157; [5] Perturbations des transformations linéaires fermées, *Acta Sci. Math. Szeged*, **14** (1951), 125–137; [6] *Spektraldarstellung linearer Transformationen des Hilbertschen Raumes*, Berlin (1942).

TAYLOR, A. E. — [1] Linear operations which depend analytically upon a parameter, *Ann. of Math.*, **39** (1938), 574–593; [2] Spectral theory of closed distributive operators, *Acta. Math.*, **84** (1950), 189–224.

TEICHMÜLLER, O. — [1] Operatoren im Wachschen Raum, *J. reine angew. Math.*, **174** (1935), 73–124.

TITCHMARCH, E. C. — [1] *Introduction to the theory of Fourier integrals*, Oxford Univ. Press (1937).

VIGIER, J. P. — [1] Étude sur les suites infinies d'opérateurs hermitiens, *Thesis, Geneva*, (1946).

VISSER, C. — [1] Note on linear operators, *Proc. Acad. Amsterdam*, **40** (1937), 270–272.

WECKEN, F. J. — [1] Zur Theorie linearer Operatoren, *Math. Ann.*, **110** (1935), 722–725; [2] Unitärinvarianten selbstadjungierter Operatoren, *Math. Ann.*, **116** (1939), 422–455.

WEYL, H. — [1] Über beschränkte quadratische Formen, deren Differenz vollstetig ist, *Rend. Circ. Math. Palermo*, **27** (1909), 373–392; [2] Über gewöhnliche Differentialgleichungen mit Singularitäten, *Math. Ann.*, **68** (1910), 220–269.

WHITTAKER, E. T. and G. N. WATSON — [1] *A course of modern analysis*, Cambridge (1952).

WINTNER, A. — [1] *Spektraltheorie unendlicher Matrizen*, Leipzig (1929).

ZAANEN, A. C. — [1] Über vollstetige symmetrische und symmetrisierbare Operatoren, *Nieuw Archief voor Wiskunde* (2), **22** (1948), 57–80; [2] On linear functional equations, *Nieuw Archief voor Wiskunde*, **22** (1948), 269–282.

ZYGMUND, A. — [1] *Trigonometrical series*, Warsaw (1935).

INDEX

THEORY OF LINEAR OPERATORS IN HILBERT SPACE

VOLUME II

CONTENTS

Chapter VI

THE SPECTRAL ANALYSIS OF UNITARY AND SELF-ADJOINT OPERATORS

58. The Trigonometric Moment Problem

Our next considerations will be based on a series of facts related to the so-called *moment problem.* In general form, the moment problem can be formulated as follows: let there be given a set of functions $u_\alpha(t)$ $(a \leqq t \leqq b)$ and a set of numbers c_α; the parameter α determines a one-to-one correspondence between these sets; it is required to find a nondecreasing function of bounded variation $\sigma(t)$ $(a \leqq t \leqq b)$ satisfying the system of equations

$$\int_a^b u_\alpha(t)\, d\sigma(t) = c_\alpha. \tag{1}$$

This problem consists of several parts, of which the basic one is the determination of conditions for the solvability of the above system of equations in the indicated class of functions $\sigma(t)$.

The earliest moment problem, which concerned the important special case with the given set of functions

$$1, t, t^2, \ldots,$$

(*the algebraic moment problem*) was studied by P. L. Tchebysheff and A. A. Markov in their remarkable works on limiting values of integrals.

In the present section we consider the trigonometric moment problem. In this case the system (1) has the form

$$c_k = \int_0^{2\pi} e^{ikt} d\sigma(t) \quad (\pm k = 0, 1, 2, \ldots; \bar{c}_k = c_{-k}). \tag{2}$$

THEOREM: *For the existence of a nondecreasing function*[1] $\sigma(t)$ *satisfying equations* (2) *it is* (a) *necessary that the non-negativity of the trigonometric sum*

$$\sum_{k=-n}^{n} \xi_k e^{ikt} \qquad (n = 0, 1, 2, \ldots)$$

in the whole interval $[0, 2\pi]$ *implies the inequality*

[1] The boundedness of the variation follows automatically from equation (2) for $k = 0$.

$$\sum_{k=-n}^{n} \xi_k c_k \geqq 0$$

and, (b) *sufficient that for any real number u the expressions*

$$\sum_{k=-n}^{n} \left(1 - \frac{|k|}{n}\right) e^{-iku} c_k \qquad (n = 1, 2, 3, \ldots)$$

be non-negative, which correspond to the particular trigonometric sums

$$\sum_{k=-n}^{n} \left(1 - \frac{|k|}{n}\right) e^{ik(t-u)} = \frac{1}{n} \left| \sum_{k=0}^{n-1} e^{ik(t-u)} \right|^2 .$$

Proof: If the system (2) has a solution $\sigma(t)$, then

$$\sum_{k=-n}^{n} \xi_k c_k = \int_0^{2\pi} \sum_{k=-n}^{n} \xi_k e^{ikt} d\sigma(t),$$

and this yields the necessity of the first condition stated in the theorem. In order to prove the second part of the theorem, we consider a sequence of trigonometric sums

$$\psi_n(u) = \sum_{k=-n}^{n} \left(1 - \frac{|k|}{n}\right) c_k e^{-iku} \qquad (n = 1, 2, 3, \ldots),$$

which are non-negative by hypothesis. Let

$$\sigma_n(t) = \frac{1}{2\pi} \int_0^t \psi_n(u)\, du \qquad (n = 1, 2, 3, \ldots).$$

These are nondecreasing functions for which

$$\sigma_n(0) = 0, \ \sigma_n(2\pi) = c_0 \qquad (n = 1, 2, 3, \ldots).$$

Thus, we have a sequence of functions $\sigma_n(t)$ which are nondecreasing in the interval $[0, 2\pi]$ and satisfy the inequality

$$0 \leqq \sigma_n(t) \leqq c_0 \qquad (n = 1, 2, 3, \ldots).$$

By the first theorem of Helly there exists a nondecreasing function $\sigma(t)$ and a subsequence $\{\sigma_{n_j}(t)\}_{j=1}^{\infty}$ such that

$$\lim_{j \to \infty} \sigma_{n_j}(t) = \sigma(t)$$

at each point of continuity of $\sigma(t)$. By the second theorem of Helly

$$\int_0^{2\pi} e^{ikt} d\sigma(t) = \lim_{j\to\infty} \int_0^{2\pi} e^{ikt} d\sigma_{n_j}(t) = \lim_{j\to\infty} \frac{1}{2\pi} \int_0^{2\pi} e^{ikt} \psi_{n_j}(t)\, dt =$$

$$= \lim_{j\to\infty} \left(1 - \frac{|k|}{n_j}\right) c_k = c_k,$$

and our assertion is completely proved.

We turn now to the question of the number of solutions of the system (2) under the assumption that it is solvable. Let us assume that (2) has two distinct nondecreasing solutions, $\sigma(t)$ and $\sigma^*(t)$. Their difference

$$\omega(t) = \sigma(t) - \sigma^*(t)$$

is a function of bounded variation for which

$$\int_0^{2\pi} e^{ikt} d\,\omega(t) = 0 \qquad (\pm k = 0, 1, 2, \ldots). \tag{3}$$

Now we shall prove that if the (real or even complex) function of bounded variation $\omega(t)$ satisfies (3), then it is constant at all its points of continuity. With this aim, we consider, for $k = \pm 1, \pm 2, \ldots$, the equation

$$\int_0^{2\pi} e^{ikt} d\,\omega(t) = e^{ikt}\omega(t)\Big|_0^{2\pi} - ik\int_0^{2\pi} e^{ikt}\,\omega(t)\,dt =$$

$$= \int_0^{2\pi} d\,\omega(t) - ik\int_0^{2\pi} e^{ikt}\omega(t)\,dt.$$

It follows by means of (3) that

$$\int_0^{2\pi} e^{ikt}\omega(t)\,dt = 0 \qquad (\pm k = 1, 2, 3, \ldots).$$

Letting

$$C = \frac{1}{2\pi}\int_0^{2\pi} \omega(t)\,dt,$$

we find that

$$\int_0^{2\pi} e^{ikt}\{\omega(t) - C\}\,dt = 0 \qquad (\pm k = 0, 1, 2, \ldots).$$

It follows from this equation and a fundamental uniqueness theorem from the theory of Fourier series that

$$\omega(t) - C = 0$$

at each point of continuity of $\omega(t)$.

From what has been proved, it follows, in particular, that, if $\sigma(t)$ and $\sigma^*(t)$ are two distinct solutions of (2), then their difference is constant at all the points where this difference is continuous. We say, therefore, that the solution of (2) is *essentially* unique. If we impose certain normalization conditions on the function $\sigma(t)$, then the solution is unique without reservation, i.e., the italicized word "essentially" is unnecessary. These normalization conditions involve, first, the points of discontinuity of $\sigma(t)$ and, second, the end points of the interval $[0, 2\pi]$. At each point t of discontinuity in the open interval $(0, 2\pi)$, the limits $\sigma(t-0)$ and $\sigma(t+0)$ exist. The value at the point t itself can be any number lying between these limits. We let

$$\sigma(t) = \sigma(t-0),$$

i.e., we require that the function $\sigma(t)$ be continuous from the left. In order to motivate further normalization conditions, we point out that the function $\sigma(t)$ itself is of no interest; rather, the interest is in the integrals

$$\int_0^{2\pi} f(t)\, d\sigma(t),$$

where $f(t)$ is a continuous function with period 2π. Every integral of this kind may be represented in the form

$$\int_0^{2\pi} f(t)\, d\sigma(t) = \int_{+0}^{2\pi-0} f(t)\, d\sigma(t) + $$

$$+ f(2\pi)\{\sigma(2\pi) - \sigma(2\pi-0)\} + f(0)\{\sigma(+0) - \sigma(0)\}.$$

Since

$$f(2\pi) = f(0)$$

we have

$$\int_0^{2\pi} f(t) d\sigma(t) = \int_{+0}^{2\pi-0} f(t) d\sigma(t) + f(0)\{\sigma(2\pi) - \sigma(2\pi-0) + \sigma(+0) - \sigma(0)\} =$$

$$= \int_0^{2\pi-0} f(t)\, d\sigma^*(t) = \int_0^{2\pi} f(t)\, d\sigma^*(t),$$

where

$$\sigma^*(t) = \sigma(t) \qquad (0 < t < 2\pi),$$

$$\sigma^*(2\pi) = \sigma^*(2\pi-0) = \sigma(2\pi-0),$$

$$\sigma^*(+0) - \sigma^*(0) = \sigma(+0) - \sigma(0) + \sigma(2\pi) - \sigma(2\pi-0).$$

Thus, the new function $\sigma^*(t)$ is continuous from the left also at the point 2π. At the point zero it has a well-determined jump from the right (*concentrated mass*) which in special cases can equal zero. Finally, let us define

$$\tilde{\sigma}(t) = \sigma^*(t) - \sigma^*(0).$$

Then the equation

$$\int_0^{2\pi} f(t)\, d\sigma(t) = \int_0^{2\pi} f(t)\, d\tilde{\sigma}(t)$$

holds for any continuous periodic function $f(t)$. Both $\tilde{\sigma}(t)$ and $\sigma^*(t)$ are continuous from the left in the interval $0 < t \leqq 2\pi$. But, in addition, $\tilde{\sigma}(0) = 0$. We will always impose these normalization conditions on the solutions of the trigonometric moment problem. They serve to determine a unique solution.

59. Analytic Functions with Values in a Half-plane

In the present section we consider analytic functions, regular in the interior of a circle or a half-plane, which take on values which lie in some half-plane.

THEOREM 1: *In order that a given function $f(\zeta)$, finite in the circle $|\zeta| < 1$, admit the representation*

$$f(\zeta) = i\beta + \int_0^{2\pi} \frac{e^{it} + \zeta}{e^{it} - \zeta}\, d\sigma(t), \tag{1}$$

where β is a real constant and $\sigma(t)$ is a nondecreasing function[2], *it is necessary and sufficient that $f(\zeta)$ be regular and have a non-negative real part in the circle $|\zeta| < 1$.*

PRELIMINARY REMARK. We have here a particular moment problem with the moment function

$$u_\zeta(t) = \frac{e^{it} + \zeta}{e^{it} - \zeta}.$$

The role of the moments is played by the numbers

$$c_\zeta = f(\zeta) - i\beta,$$

where the parameter ζ runs through a discrete sequence in the circle $|\zeta| < 1$.

We remark also that $\beta = \Im f(0)$.

Proof: The necessity of the condition of the theorem is obvious, since the right member of formula (1) is regular in the region $|\zeta| < 1$ and

$$\Re\left\{\int_0^{2\pi} \frac{e^{it} + \zeta}{e^{it} - \zeta}\, d\sigma(t)\right\} = \int_0^{2\pi} \frac{1 - r^2}{1 - 2r\cos(t - \varphi) + r^2}\, d\sigma(t) \geqq 0,$$

where $r = |\zeta| < 1$ and $\varphi = \arg \zeta$.

---

[2] The boundedness of the variation of $\sigma(t)$ is obtained automatically from the finiteness of the value $f(0)$.

We prove now that the condition is sufficient. Assume that $f(\zeta)$ is regular and has non-negative real part in the circle $|\zeta| < 1$. Then, as is known from the theory of functions, $f(\zeta)$ is given, for $|\zeta| \leqq R < 1$, by

$$f(\zeta) = i\beta + \frac{1}{2\pi}\int_0^{2\pi} \frac{Re^{it} + \zeta}{Re^{it} - \zeta}\, u(Re^{it})\, dt,$$

where

$$u(re^{it}) = \frac{f(re^{it}) + \overline{f(re^{it})}}{2}$$

is the real part of the function $f(\zeta)$. Hence

$$\Re f(0) = \frac{1}{2\pi}\int_0^{2\pi} u(Re^{it})\, dt.$$

The indicated representation can be written in the form

$$f(\zeta) = i\beta + \int_0^{2\pi} \frac{Re^{it} + \zeta}{Re^{it} - \zeta}\, d\,\sigma_R(t), \tag{2}$$

where

$$\sigma_R(t) = \frac{1}{2\pi}\int_0^t u(Re^{is})\, ds.$$

Since $u(Re^{is}) \geqq 0$ by assumption, $\sigma_R(t)$ is a nondecreasing function of t and, for $0 \leqq t \leqq 2$,

$$0 \leqq \sigma_R(t) \leqq \sigma_R(2\pi) = \Re f(0).$$

Thus, the set of functions $\sigma_R(t)$ $(0<R<1)$ is uniformly bounded. By the Helly theorem, there exists a nondecreasing function $\sigma(t)$ and a sequence

$$R_1 < R_2 < R_3 < \ldots \qquad (R_j \to 1),$$

such that

$$\lim_{j\to\infty} \sigma_{R_j}(t) = \sigma(t)$$

at each point of continuity of $\sigma(t)$. Applying the second Helly theorem to (2), we find that

$$f(\zeta) = i\beta + \int_0^{2\pi} \frac{e^{it} + \zeta}{e^{it} - \zeta}\, d\sigma(t) \tag{3}$$

for each ζ in the circle $|\zeta| < 1$. Thus, the theorem is proved.

If we expand $f(\zeta)$ in a Maclaurin series

$$f(\zeta) = c + 2c_{-1}\zeta + 2c_{-2}\zeta^2 + \ldots$$

and let $\frac{c+\bar{c}}{2} = c_0$, then by (3) we obtain the following expressions for the coefficients

$$c_0 = \int_0^{2\pi} d\,\sigma(t)\,,$$

$$c_{-k} = \int_0^{2\pi} e^{-ikt} d\,\sigma(t) \qquad (k = 1, 2, 3, \ldots).$$

Introducing the numbers $c_k = \bar{c}_{-k}(k = 1, 2, 3, \ldots)$, we obtain the equations which appear in the trigonometric moment problem. Therefore, under the normalization conditions of the preceding section, the function $\sigma(t)$ in (3) is uniquely determined by $f(\zeta)$.

This fact can be proved also with the aid of the following well-known inversion formula from the theory of Fourier series.

THEOREM 2: (*R. Nevanlinna*) *In order that a given function* $\varphi(z)$, *finite in the half-plane* $\Im z > 0$, *admit the representation*

$$\varphi(z) = \alpha + \mu z + \int_{-\infty}^{\infty} \frac{1 + tz}{t - z} d\,\sigma(t)$$

(where $\mu \geqq 0$ *and* α *are two real constants, and* $\sigma(t)$ *is a nondecreasing function*[3]*), it is necessary and sufficient that* $\varphi(z)$ *be regular and have non-negative imaginary part in the half-plane* $\Im z > 0$. *Here the Stieltjes integral with infinite limits of integration is to be interpreted as*

$$\lim_{\substack{A \to -\infty \\ B \to \infty}} \int_A^B \frac{1 + tz}{t - z}\, d\,\sigma(t),$$

which corresponds to the assumption that the limits

$$\sigma(-\infty) = \lim_{A \to -\infty} \sigma(A), \quad \sigma(\infty) = \lim_{B \to \infty} \sigma(B)$$

are finite. If, in addition, one applies the normalization conditions

$$\sigma(t - 0) = \sigma(t), \quad \sigma(-\infty) = 0,$$

then the function $\sigma(t)$ *is uniquely determined.*

Proof: Let

$$z = i\,\frac{1 + \zeta}{1 - \zeta},$$

$$\varphi(z) = i f(\zeta).$$

[3] The boundedness of the variation of $\sigma(t)$ follows from the finiteness of the value $\varphi(i)$.

These transformations map the half-plane $\mathfrak{I}\, z > 0$ into the circular region $|\zeta| < 1$, and the function $\varphi(z)$, regular in the half-plane $\mathfrak{I}\, z > 0$, into a function $f(\zeta)$, regular and having non-negative real part in the circle. The integral representation of such a function was treated above. It has the form

$$f(\zeta) = i\beta + \int_0^{2\pi} \frac{e^{is} + \zeta}{e^{is} - \zeta}\, d\rho\,(s) = i\beta + \int_{+0}^{2\pi - 0} \frac{e^{is} + \zeta}{e^{is} - \zeta}\, d\rho\,(s) +$$

$$+ \frac{1 + \zeta}{1 - \zeta} \{ \rho(2\pi) - \rho\,(2\pi - 0) + \rho\,(+0) - \rho\,(0) \} =$$

$$= i\beta + \mu \frac{1 + \zeta}{1 - \zeta} + \int_{+0}^{2\pi - 0} \frac{e^{is} + \zeta}{e^{is} - \zeta}\, d\rho\,(s),$$

where we denote the distribution by $\rho(s)$ instead of $\sigma(s)$. With the aid of the transformation defined above we obtain

$$\varphi\,(z) = -\beta + i\mu \frac{1 + \zeta}{1 - \zeta} + i \int_{+0}^{2\pi - 0} \frac{e^{is} + \zeta}{e^{is} - \zeta}\, d\rho\,(s) \tag{4}$$

or

$$\varphi\,(z) = -\beta + \mu z + \int_{+0}^{2\pi - 0} \frac{z\cot\frac{s}{2} - 1}{\cot\frac{s}{2} + z}\, d\rho\,(s).$$

Letting

$$-\beta = \alpha, \quad -\cot\frac{s}{2} = t, \quad \rho\,(s) = \sigma\,(t),$$

we obtain for $\varphi\,(z)$ the representation

$$\varphi\,(z) = \alpha + \mu z + \int_{-\infty}^{\infty} \frac{1 + tz}{t - z}\, d\sigma\,(t). \tag{5}$$

The transition from (4) to (5) is also correct in the reverse direction. Therefore, Theorem 2 is proved.

THEOREM 3: *In order that a function* $\varphi\,(z)$ *in the half-plane* $\mathfrak{I}\, z > 0$ *admit the representation*

$$\varphi\,(z) = \int_{-\infty}^{\infty} \frac{d\,\omega\,(t)}{t - z}$$

with a nondecreasing function of bounded variation $\omega\,(t)$, *it is necessary and sufficient that* $\varphi(z)$ *be regular and have non-negative imaginary part in the half-plane* $\Im z > 0$ *and*

$$\sup_{y \to \infty} |y \varphi(iy)| < \infty.$$

The function $\omega(t)$ *is uniquely determined by* $\varphi(z)$ *if we require that*

$$\omega(-\infty) = \lim_{A \to -\infty} \omega(A) = 0, \quad \omega(t-0) = \omega(t) \quad (-\infty < t \leqq \infty).$$

Proof: The necessity of the condition is easily verified. We concern ourselves only with the proof of the sufficiency. If the function $\varphi(z)$ is regular and has non-negative imaginary part in the half-plane $\Im z > 0$, then it admits in every case the representation

$$\varphi(z) = \alpha + \mu z + \int_{-\infty}^{\infty} \frac{1 + tz}{t - z} \, d\sigma(t),$$

as we saw in the previous theorem. From this representation it follows that

$$y\varphi(iy) = \alpha y + i\mu y^2 + \int_{-\infty}^{\infty} \frac{y(1 + ity)}{t - iy} \, d\sigma(t).$$

By hypothesis, there exists a constant M such that

$$\left| \alpha y + i\mu y^2 + \int_{-\infty}^{\infty} \frac{y(1 + ity)}{t - iy} \, d\sigma(t) \right| \leqq M \qquad (y > 0).$$

Hence, a fortiori,

$$(6) \qquad \begin{cases} \left| \alpha y + \displaystyle\int_{-\infty}^{\infty} \frac{y(1 - y^2)t}{t^2 + y^2} \, d\sigma(t) \right| \leqq M, \\ \\ \left| \mu y^2 + y^2 \displaystyle\int_{-\infty}^{\infty} \frac{1 + t^2}{t^2 + y^2} \, d\sigma(t) \right| \leqq M. \end{cases}$$

The second inequality shows that $\mu = 0$ and also that

$$\int_{-\infty}^{\infty} \frac{y^2}{t^2 + y^2} (1 + t^2) \, d\sigma(t) \leqq M.$$

Hence, for any $N > 0$,

$$\int_{-N}^{N} \frac{y^2}{t^2 + y^2} (1 + t^2) \, d\sigma(t) \leqq M.$$

Let $y \to \infty$ to obtain

$$\int_{-N}^{N} (1 + t^2)\, d\sigma(t) \leqq \mathrm{M},$$

and, consequently,

$$\int_{-\infty}^{\infty} (1 + t^2)\, d\sigma(t) \leqq \mathrm{M}.$$

We can, therefore, introduce the nondecreasing function

$$\omega(t) = \int_{-\infty}^{t} (1 + t^2)\, d\sigma(t),$$

for which it is obvious that

$$\lim_{t \to -\infty} \omega(t) = 0,$$

and which, together with $\sigma(t)$, is continuous from the left.

From the first inequality in (6) it follows that

$$\alpha = \lim_{y \to \infty} \int_{-\infty}^{\infty} \frac{(y^2 - 1)\, t}{t^2 + y^2}\, d\sigma(t),$$

From this one easily gets

$$\alpha = \int_{-\infty}^{\infty} t\, d\sigma(t).$$

Since, as was shown, $\mu = 0$, (5) takes the form

$$\varphi(z) = \int_{-\infty}^{\infty} t d\sigma(t) + \int_{-\infty}^{\infty} \frac{1 + tz}{t - z}\, d\sigma(t)$$

or

$$\varphi(z) = \int_{-\infty}^{\infty} \frac{d\omega(t)}{t - z},$$

which was to be established.

The uniqueness of this representation (under our normalization conditions) is a consequence of the uniqueness of the function $\sigma(t)$ in Theorem 2, since this follows from the equation

$$\sigma(t) = \int_{-\infty}^{t} \frac{d\omega(t)}{1 + t^2},$$

which relates the functions $\sigma(t)$ and $\omega(t)$. However, it is possible to establish the uniqueness directly by using the following Stieltjes inversion formula:

$$\frac{\omega(t-0)+\omega(t+0)}{2} = \text{const} + \lim_{y\to 0}\frac{1}{\pi}\int_0^t \Im\,\varphi(x+iy)\,dx.$$

60. The Theorem of Bochner

In the present section we consider the continuous analogue of the trigonometric moment problem. The problem consists of finding conditions, which a given function $F(t)$, $-\infty < t < \infty$, must satisfy in order that it have a representation

$$F(t) = \int_{-\infty}^{\infty} e^{ist}\,d\omega(s), \tag{1}$$

where $\omega(s)$ is a nondecreasing function.[4]

Bochner was the first to prove that such a representation exists and is unique under suitable normalization[5] if and only if the function $F(t)$ is continuous and

$$\sum_{\alpha,\beta=1}^{n} F(t_\alpha - t_\beta)\,\rho_\alpha\bar{\rho}_\beta \geqq 0, \tag{2}$$

for any integer n, any real $t_1, t_2, \ldots, t_n$, and any complex $\rho_1, \rho_2, \ldots, \rho_n$. Functions which satisfy (2) are called positive definite.

The necessity of these conditions is easily verified. In fact, the right side of (1) is a continuous function of t, and, on the other hand, by (1),

$$\sum_{\alpha,\beta=1}^{n} F(t_\alpha - t_\beta)\,\rho_\alpha\bar{\rho}_\beta = \int_{-\infty}^{\infty} \Big|\sum_{k=1}^{n} \rho_k e^{ist_k}\Big|^2 d\omega(s),$$

the right side of which is finite and non-negative.

The proof of the sufficiency, to which we now turn, is not so simple. Let $F(t)$ be a continuous function which satisfies (2). We remark that, by (2), for any t $(-\infty < t < \infty)$ the equation

$$\overline{F(t)} = F(-t)$$

and the inequality

$$|F(t)| \leqq F(0) \tag{3}$$

both hold. Let

[4] The boundedness of its variation follows from the finiteness of the value $F(0)$.
[5] cf. Theorem 3 of the preceding section.

$$\Phi(z) = \int_0^\infty e^{itz} F(t)\, dt.$$

By (3), the function $\Phi(z)$ is regular in the region $\Im z > 0$. In addition, for each $y > 0$,

$$| y \Phi(iy) | \leqq \int_0^\infty y e^{-ty} | F(t) | \, dt \leqq F(0) \int_0^\infty y e^{-ty}\, dt = F(0).$$

Now we show that

$$\Re\, \Phi(z) \geqq 0 \qquad (z = x + iy)$$

for $y > 0$. With this aim let us introduce the identity

$$\frac{1}{2y} = \int_0^\infty e^{izv} e^{-i\bar{z}v}\, dv.$$

We have further

$$\Phi(z) + \overline{\Phi(z)} = \int_0^\infty e^{izu} F(u)\, du + \int_0^\infty e^{-i\bar{z}u} F(-u)\, du.$$

Therefore,

$$\frac{\Phi(z) + \overline{\Phi(z)}}{2y} = \int_0^\infty\int_0^\infty e^{iz(u+v)} e^{-i\bar{z}v} F(u)\, du\, dv + \int_0^\infty\int_0^\infty e^{-i\bar{z}(u+v)} e^{izv} F(-u)\, du\, dv =$$

$$= \int_0^\infty d\beta \int_\beta^\infty e^{i\alpha z} e^{-i\beta\bar{z}} F(\alpha - \beta)\, d\alpha + \int_0^\infty d\alpha \int_\alpha^\infty e^{i\alpha z} e^{-i\beta\bar{z}} F(\alpha - \beta)\, d\beta =$$

$$= \int_0^\infty\int_0^\infty F(\alpha - \beta)\, e^{ix(\alpha-\beta)} e^{-y(\alpha+\beta)}\, d\alpha\, d\beta =$$

$$= \lim_{A\to\infty} \int_0^A\int_0^A F(\alpha - \beta)\, e^{ix(\alpha-\beta)} e^{-y(\alpha+\beta)}\, d\alpha\, d\beta =$$

$$= \lim_{A\to\infty} \lim_{n\to\infty} \frac{A^2}{n^2} \sum_{r,s=1}^{n} F\left(\frac{r-s}{n} A\right) e^{ix(r-s)A/n} e^{-y(r+s)A/n},$$

and our assertion follows from the fact that the sum on the right side has the form (2) (it is sufficient to let $t_k = kA/n$, $\rho_k = e^{-y(kA/n)} e^{ix(kA/n)}$).

The function $\Phi(z)$ differs only by the factor i from a function considered in the previous section (cf. Theorem 3). Hence, under our normalization conditions, there exists a unique nondecreasing function $\omega(s)$ such that

$$\Phi(z) = i \int_{-\infty}^{\infty} \frac{d\omega(s)}{s+z} \qquad (\Im z > 0).$$

On the other hand,

$$\frac{i}{s+z} = \int_{0}^{\infty} e^{i(s+z)t}\, dt$$

and, hence,

$$\int_{0}^{\infty} e^{itz} F(t)\, dt = \int_{-\infty}^{\infty} d\omega(s) \int_{0}^{\infty} e^{i(s+z)t}\, dt = \int_{0}^{\infty} e^{itz}\, dt \int_{-\infty}^{\infty} e^{ist}\, d\omega(s).$$

This means that two piece-wise continuous absolutely integrable functions of t, which equal zero for $t < 0$ and equal

$$e^{-ty} F(t),\ e^{-ty} \int_{-\infty}^{\infty} e^{ist}\, d\omega(s)$$

for $t \geqq 0$, have identical Fourier transforms. By a well-known uniqueness theorem these functions are identical. Hence, the representation (1) is obtained and its uniqueness is proved.

From the uniqueness of this representation we have the following more general proposition: if $\omega_1(s)$ and $\omega_2(s)$ are two complex functions of bounded variation which are normalized in the usual manner and if, for $-\infty < t < \infty$,

$$\int_{-\infty}^{\infty} e^{ist}\, d\omega_1(s) = \int_{-\infty}^{\infty} e^{ist}\, d\omega_2(s), \tag{4}$$

then

$$\omega_1(s) = \omega_2(s).$$

Indeed, if (4) holds and

$$\tilde{\omega}(s) = \omega_1(s) - \omega_2(s),$$

then

$$\int_{-\infty}^{\infty} e^{ist}\, d\tilde{\omega}(s) = 0 \qquad (-\infty < t < \infty).$$

If we let

$$\tilde{\omega}(s) = \varphi(s) + i\,\psi(s),$$

where $\varphi(s)$ and $\psi(s)$ are real, then we get

$$\int_{-\infty}^{\infty} \cos st \, d\varphi(s) = \int_{-\infty}^{\infty} \sin st \, d\psi(s),$$

$$\int_{-\infty}^{\infty} \sin st \, d\varphi(s) = -\int_{-\infty}^{\infty} \cos st \, d\psi(s).$$

Since one member of each of the equations is an even function, and the other members are odd functions,

$$\int_{-\infty}^{\infty} \cos st \, d\varphi(s) = \int_{-\infty}^{\infty} \sin st \, d\varphi(s) = \int_{-\infty}^{\infty} \cos st \, d\psi(s) = \int_{-\infty}^{\infty} \sin st \, d\psi(s) = 0.$$

In other words, for all real t,

$$\int_{-\infty}^{\infty} e^{ist} \, d\varphi(s) = 0, \quad \int_{-\infty}^{\infty} e^{ist} \, d\psi(s) = 0.$$

Since each of the functions $\varphi(s)$ and $\psi(s)$ is the difference of two non-decreasing functions, the equations

$$\varphi(s) = 0, \quad \psi(s) = 0 \qquad (-\infty < s < \infty)$$

are direct consequences of Bochner's theorem.

61. The Resolution of the Identity

We recall the results of Section 55 concerning completely continuous self-adjoint operators. Here we assume that the operator—we denote it by A—operates in a Hilbert space H (rather than in an arbitrary inner product space R). As we have seen, the operator A has a finite or countable set of pairwise orthogonal and normalized eigenvectors $e_1, e_2, e_3, \ldots$, which correspond to nonzero eigenvalues. The set H_0 of all vectors from H which are orthogonal to each vector e_k is a subspace. From the theorem of section 55 it follows that for $h \in H_0$, $Ah = 0$. Therefore, it is possible to consider H_0 as the eigenmanifold of the operator A which belongs to the eigenvalue $\lambda = 0$. This space is separable if H is separable, and not separable otherwise. As far as the subspace $H \ominus H_0$ is concerned, it is always separable and $\{e_k\}$ is an orthonormal basis for it. We can represent the subspace $H \ominus H_0$ in the form of an orthogonal sum of eigenmanifolds H_k, which belong to distinct eigenvalues. Then

$$H = H_0 \oplus H_1 \oplus H_2 \oplus \ldots,$$

where

$$Af = \lambda_k f$$

for $f \in H_k$ $(\lambda_0 = 0)$.

Thus, we have decomposed the space H into subspaces, in each of which the operation of A is merely the multiplication of an element by a particular number. Designating the operator of projection on H_k by P_k we can write

$$E = P_0 + P_1 + P_2 + \dots . \tag{1}$$

For each $f \in H$,

$$Af = \lambda_1 P_1 f + \lambda_2 P_2 f + \dots . \tag{2}$$

The numbers λ_k appearing here are real, they lie in a finite interval, and can have as a limit point only the single point 0. To enumerate the numbers λ_k in the order of increasing or decreasing magnitude is possible only if they are finite in number or if all are of one sign. Therefore, the terms of the series (2) follow not in their natural order, but in the order determined by the enumeration of the set of eigenvalues. This defect of the representations (1) and (2) can be removed by using the Stieltjes integral.

With this aim let us introduce, for each real t, the subspace G_t spanned by the eigenvectors which belong to eigenvalues less than t. Here we regard zero as the eigenvalue which belongs to the eigenmanifold H_0. Let E_t be the operator of projection on G_t. The operator E_t has a limit both for increasing and for decreasing t. Therefore, E_{t-0} and E_{t+0} exist. It is easy to see that

$$E_{t-0} = E_t.$$

Thus, E_t is a function of t which is continuous from the left. If λ_k is an eigenvalue, then the difference

$$E_{\lambda_k+0} - E_{\lambda_k} = P_k$$

is the operator of projection on the eigenmanifold H_k. Now formulas (1) and (2) can be represented in the form

$$f = Ef = \int_\alpha^\beta d\,E_t f, \tag{1'}$$

$$Af = \int_\alpha^\beta t\,d\,E_t f, \tag{2'}$$

where the integrals are taken over an interval $[\alpha, \beta]$ which contains all the eigenvalues of the operator A.

These integrals are nothing other than sums of particular series. We have introduced the integral representations because, as we shall see later, they generalize to arbitrary (not necessarily completely continuous) self-adjoint operators in H. Anticipating this generalization we make the following definition.

DEFINITION: *A resolution of the identity is a one-parameter family of projection operators E_t, where t runs through a finite or infinite interval $[\alpha, \beta]$, which satisfy the following conditions.*[6]

$$\text{(a)} \qquad E_\alpha = 0, \; E_\beta = E,$$

$$\text{(b)} \qquad E_{t-0} = E_t \quad (\alpha < t < \beta),$$

$$\text{(c)} \qquad E_u E_v = E_s \quad (s = \min \{u, v\}).$$

From the definition it follows that, for each fixed $f \in$ H, the function of t given by

$$(E_t f, f) = \sigma (t)$$

is nondecreasing, has bounded variation, is continuous from the left, and

$$\sigma (\alpha) = 0, \; \sigma (\beta) = (f, f).$$

Indeed, for $s < t$,

$$(E_s f, f) = \| E_s f \|^2 = \| E_s E_t f \|^2 \leqslant \| E_t f \|^2 = (E_t f, f).$$

For each interval $\Delta = [t', t''] \subset [\alpha, \beta]$, we denote the difference $E_{t''} - E_{t'}$ by $E(\Delta)$. If Δ_1 and Δ_2 are intervals of this type, it follows from (c) that

$$E(\Delta_1) E(\Delta_2) = E(\Delta),$$

where Δ is the intersection of Δ_1 and Δ_2. In particular, if the intervals Δ_1 and Δ_2 do not have common interior points, then

$$E(\Delta_1) E(\Delta_2) = 0,$$

i.e., the subspaces on which $E(\Delta_1)$ and $E(\Delta_2)$ project are orthogonal. On the basis of what has been said, property (c) is called the *orthogonality property* of the decomposition of the identity.

The considerations which preceded our general definition shows that every completely continuous self-adjoint operator generates a resolution of the identity. In the next section we show that arbitrary unitary operators and arbitrary self-adjoint operators in H have well-defined resolutions of the identity. It is true that these resolutions of the identity do not have those simple properties which the resolution of the identity of completely continuous operators possess. This, as we will see, is connected with the fact that the spectra of these operators have more complicated structures than the spectra of completely continuous operators.

62. The Integral Representation of a Unitary Operator

Let U be a unitary operator in H. We choose an arbitrary element $f \in$ H and let

---

[6] If the interval $[\alpha, \beta]$ is infinite then we define $E_{-\infty} = \lim_{t \to -\infty} E_t$, $E_\infty = \lim_{t \to \infty} E_t$.

$$(U^k f, f) = c_k(f) = c_k \quad (\pm k = 0, 1, 2, \ldots), \tag{1}$$

so that

$$c_{-k} = (U^{-k}f, f) = (f, U^k f) = \bar{c}_k.$$

We shall prove that

$$\Phi_n(t) = \sum_{k=-n}^{n} \left(1 - \frac{|k|}{n}\right) c_k e^{-ikt} \geqq 0 \tag{2}$$

for each integer n and each real t. By (1),

$$\Phi_n(t) = \sum_{k=-n}^{n} \left(1 - \frac{|k|}{n}\right) e^{-ikt} (U^k f, f).$$

On the other hand, if

$$T = E + e^{-it}U + e^{-2it}U^2 + \ldots + e^{-(n-1)it}U^{n-1},$$

then

$$(Tf, Tf) = \sum_{r,s=0}^{n-1} e^{i(s-r)t}(U^r f, U^s f) =$$

$$= \sum_{r,s=0}^{n-1} e^{i(s-r)t}(U^{r-s}f, f) = \sum_{k=-n}^{n} (n - |k|) e^{-ikt} (U^k f, f).$$

Therefore,

$$\Phi_n(t) = \frac{1}{n} (Tf, Tf),$$

which implies (2).

The sequence $\{c_k\}_{-\infty}^{\infty}$ satisfies the conditions of the theorem of Section 58. Therefore, there is a uniquely defined normalized nondecreasing function $\sigma(t)$ for which

$$c_k = \int_0^{2\pi} e^{ikt} d\sigma(t) \qquad (\pm k = 0, 1, 2, \ldots).$$

For each fixed t, the function $\sigma(t)$ is some functional of f:

$$\sigma(t) = \sigma(t; f).$$

With the aid of this functional, we now define another functional dependent on a pair of vectors $f, g \in \mathrm{H}$:

$$\sigma(t; f, g) = \frac{1}{4} \sigma(t; f + g) - \frac{1}{4} \sigma(t; f - g) +$$

$$+ \frac{i}{4} \sigma(t; f + ig) - \frac{i}{4} \sigma(t, f - ig).$$

As above, t is any real number in the interval $[0, 2\pi]$. In the equation

$$\int_0^{2\pi} e^{ikt} d\sigma(t; h) = (U^k h, h) \qquad (\pm k = 0, 1, 2, \ldots)$$

substitute, in turn,

$$h = f + g,\ f - g,\ f + ig,\ f - ig.$$

to obtain

$$(3) \qquad \int_0^{2\pi} e^{ikt} d\sigma(t; f, g) = (U^k f, g) \quad (\pm k = 0, 1, 2, \ldots).$$

Thus, we have derived representations in the form of trigonometric Stieltjes integrals not only for the quantity $(U^k f, f)$ but, more generally, for $(U^k f, g)$. This representation is unique (under the normalization conditions on the function $\sigma(t; f, g)$) according to the general theorem at the end of Section 58.

Starting from the uniqueness of the representation (3) we prove that $\sigma(t; f, g)$ is a bilinear functional of f, g, whose norm does not exceed 1. Let

$$f = \alpha_1 f_1 + \alpha_2 f_2.$$

Then

$$(U^k f, g) = \alpha_1 (U^k f_1, g) + \alpha_2 (U^k f_2, g),$$

and, hence,

$$\int_0^{2\pi} e^{ikt} d\sigma(t; f, g) = \alpha_1 \int_0^{2\pi} e^{ikt} d\sigma(t; f_1, g) + \alpha_2 \int_0^{2\pi} e^{ikt} d\sigma(t; f_2, g) =$$

$$= \int_0^{2\pi} e^{ikt} d_t \{\alpha_1 \sigma(t; f_1, g) + \alpha_2 \sigma(t; f_2, g)\}.$$

Since this relation holds for $\pm k = 0, 1, 2, \ldots$, and the normalization condition is satisfied, we have

$$\sigma(t; f, g) = \alpha_1 \sigma(t; f_1, g) + \alpha_2 \sigma(t; f_2, g).$$

We see that σ, as a function of f, g, is linear in the first of these arguments.

We remark now that on one hand

$$(g, U^k f) = (U^{-k} g, f) = \int_0^{2\pi} e^{-ikt} d\sigma(t; g, f),$$

and on the other

$$(g, U^k f) = \overline{(U^k f, g)} = \int_0^{2\pi} e^{-ikt} d\,\overline{\sigma(t; f, g)}.$$

Therefore, for any integer k,

$$\int_0^{2\pi} e^{-ikt} d\,\overline{\sigma(t; f, g)} = \int_0^{2\pi} e^{-ikt} d\sigma(t; g, f).$$

Hence,

$$\overline{\sigma(t; f, g)} = \sigma(t; g, f).$$

From this relation and the linearity in the argument f, it follows that

$$\sigma(t; f, \beta_1 g_1 + \beta_2 g_2) = \bar{\beta}_1 \sigma(t; f, g_1) + \bar{\beta}_2 \sigma(t; f, g_2).$$

We recall now the theorem of Section 20. Since $\sigma(t; f, g)$ is a non-decreasing function of t and since

$$\sigma(0; f, f) = 0,$$

$$\sigma(t; f, f) \leqq \sigma(2\pi, f, f) = \int_0^{2\pi} d\sigma(t; f, f) = (f, f),$$

it follows from the indicated theorem that $\sigma(t; f, g)$ is a bilinear functional of f, g with norm not exceeding 1.

By the theorem on the general form of a bilinear functional, there exists a family of operators E_t, which depend on the parameter $t (0 \leqq t \leqq 2\pi)$, such that

$$\sigma(t; f, g) = (E_t f, g).$$

Now we prove that E_t is a resolution of the identity.

It follows by means of the equation

$$\overline{\sigma(t; f, g)} = \sigma(t; g, f),$$

that

$$(g, E_t f) = (E_t g, f)$$

In other words, E_t is a bounded self-adjoint operator.

We write (3) in the form

$$(U^k f, g) = \int_0^{2\pi} e^{ikt} d(E_t f, g), \tag{3'}$$

and let

$$g = U^{-r} h \qquad (\pm r = 0, 1, 2, \ldots).$$

We obtain

$$(U^k f, U^{-r} h) = \int_0^{2\pi} e^{ikt} d(E_t f, U^{-r} h)$$

and, hence,

$$(U^{k+r} f, h) = \int_0^{2\pi} e^{ikt} d(U^r E_t f, h).$$

But since

$$(U^r E_t f, h) = \int_0^{2\pi} e^{irs} d_s(E_s E_t f, h),$$

we have

$$(4) \qquad (U^{k+r} f, h) = \int_0^{2\pi} e^{ikt} d_t \left\{ \int_0^{2\pi} e^{irs} d_s(E_s E_t f, h) \right\}.$$

On the other hand,

$$(5) \qquad (U^{k+r} f, h) = \int_0^{2\pi} e^{i(k+r)t} d(E_t f, h) = \int_0^{2\pi} e^{ikt} d_t \left\{ \int_0^{t} e^{irs} d(E_s f, h) \right\}.$$

From the validity of the representations (4), (5) for any integer k and from the normalization properties of the functions of t

$$\int_0^{2\pi} e^{irs} d_s(E_s E_t f, h), \quad \int_0^{t} e^{irs} d(E_s f, h),$$

it follows that

$$\int_0^{2\pi} e^{irs} d_s(E_s E_t f, h) = \int_0^{t} e^{irs} d(E_s f, h).$$

This relation is satisfied for any integer r. Therefore, by the uniqueness of the representation, the equation

$$(E_s E_t f, h) = (E_s f, h)$$

is correct for any $f, h \in \mathrm{H}$, whenever $s \leqq t$. We have also shown that

$$(6) \qquad E_s E_t = E_s \qquad (s \leqq t),$$

which implies that

$$E_t^2 = E_t,$$

i.e., E_t is a projection operator. Instead of (6), it is possible to write the more general relation

$$E_u E_v = E_s \qquad (s = \min\{u, v\}).$$

It remains for us to verify that

$$E_0 = O, \; E_{2\pi} = E, \; E_{t-0} = E_t \qquad (0 < t \leqq 2\pi)$$

in order to complete the proof that E_t is a resolution of the identity. The first and second equations need no proof. In order to prove the third relation, we use the normalization conditions which imply that

$$\lim_{\substack{s \to t \\ s < t}} (E_s f, g) = (E_t f, g).$$

Thus,

$$E_{t-0} = E_t$$

in the sense of weak convergence. But this equation also holds in the sense of strong convergence, since, for $t > s$,

$$\| (E_t - E_s) f \|^2 = ((E_t - E_s) f, f) = (E_t f, f) - (E_s f, f),$$

and this implies that

$$\| (E_t - E_s) f \|$$

tends to zero as $s \to t$ $(s < t)$.

We have obtained a resolution of the identity, with the parameter interval $[0, 2\pi]$, which belongs to the operator U in the sense that for any integer k and any $f, g \in \mathrm{H}$

$$(3') \qquad (U^k f, g) = \int_0^{2\pi} e^{ikt} d(E_t f, g).$$

This equation is expressed often in the form

$$U^k f = \int_0^{2\pi} e^{ikt} dE_t f.$$

We show now *that this equation has a valid interpretation, in its own right, in addition to being a symbolic representation of* (3′). As in elementary integral calculus, we introduce a subdivision of the interval $[0, 2\pi]$:

$$0 = t_0 < t_1 < t_2 < \ldots < t_n = 2\pi.$$

Corresponding to this subdivision we construct an operator[7]

$$T_n = e^{ikt_1} E(\Delta_1) + e^{ikt_2} E(\Delta_2) + \ldots + e^{ikt_n} E(\Delta_n),$$

where $\Delta_j = [t_{j-1}, t_j]$. Now we prove that, for each $f \in \mathrm{H}$,

$$(7) \qquad T_n f \to U^k f \qquad (n \to \infty),$$

if the largest of the lengths of the intervals Δ_j tends to zero. In other words, we prove that the sequence of operators T_n converges strongly to the operator U^k. After that, we use the limit of the operators T_n as the definition of the integral

$$\int_0^{2\pi} e^{ikt} dE_t,$$

In order to prove (7) we recall Section 23. In view of Theorem 1 of that section, we should verify that

[7] Translator's Note: The dependence of T_n on k might have been indicated by writing T_{kn} in place of T_n.

(a) for any $f, g \in \mathrm{H}$

$$(T_n f, g) \to (U^k f, g) \qquad (n \to \infty),$$

(b)
$$\| T_n f \| \to \| U^k f \| = \| f \|.$$

The validity of (a) is obvious from (3′), since

$$(T_n f, g) = \sum_{j=1}^{n} e^{ikt_j}(E(\Delta_j)f, g) \to \int_0^{2\pi} e^{ikt} d(E_t f, g).$$

In order to prove (b) we notice that

$$\| T_n f \|^2 = (T_n f, T_n f) = \sum_{r,s=1}^{n} e^{ik(t_r - t_s)}(E(\Delta_r)f, E(\Delta_s)f) =$$

$$= \sum_{r,s=1}^{n} e^{ik(t_r - t_s)}(E(\Delta_s)\, E(\Delta_r)f, f) = \sum_{r=1}^{n} (E(\Delta_r)f, f) = (f, f).$$

These relations show that not only does the limit equation (b) hold, but, simply,

$$\| T_n f \| = \| f \|.$$

From our considerations, we remark that the resolution of the identity $E_t (0 \leqq t \leqq 2\pi)$ is not only completely defined by the operator U, but in turn completely defines this operator.

Completing the present section we show that the operator E_t, for any fixed t, reduces the operator U, and moreover, reduces each integral power of the operator U. This is equivalent to the assertion that

$$(U^k E_t f, g) = (E_t U^k f, g) \quad (\pm\, k = 0, 1, 2, \ldots).$$

for arbitrary $f, g \in \mathrm{H}$, i.e., the operators U^k and E_t are permutable. The proof follows at once from the formula (3′); in fact

$$(U^k E_t f, g) = \int_0^{2\pi} e^{iks} d_s(E_s E_t f, g) = \int_0^{t} e^{iks} d(E_s f, g),$$

and, on the other hand,

$$(E_t U^k f, g) = (U^k f, E_t g) = \int_0^{2\pi} e^{iks} d_s(E_s f, E_t g) =$$

$$= \int_0^{2\pi} e^{iks} d_s(E_t E_s f, g) = \int_0^{t} e^{iks} d(E_s f, g).$$

63. Operators Represented by Stieltjes Integrals

In the preceding section we defined the integral

$$\int_0^{2\pi} e^{ikt} dE_t$$

as the limit, in the sense of strong convergence, of the sequence of operators

$$e^{ikt_1}E(\Delta_1) + e^{ikt_2}E(\Delta_2) + \ldots + e^{ikt_n}E(\Delta_n),$$

which have the form of Stieltjes integral sums. The present section is devoted to a certain generalization of this construction.

Let us take a particular resolution of the identity E_t in a finite or infinite interval $[\alpha, \beta]$ and a particular real or complex function $\varphi(t)$ which we assume to be continuous at every point of $[\alpha, \beta]$ with the possible exception of a finite number of points. We form the expression

$$\int_\alpha^\beta \overline{\varphi(t)}\, d\,\overline{(E_t f, h)},$$

and we denote by M the set (linear manifold) of all those elements f for which this expression is a linear functional of f. By the theorem of F. Riesz there exists a unique element Tf which depends on $f \in$ M, such that

$$\int_\alpha^\beta \overline{\varphi(t)}\, d\,\overline{(E_t f, h)} = (h, Tf),$$

or

$$\int_\alpha^\beta \varphi(t)\, d(E_t f, h) = (Tf, h). \tag{1}$$

It is easy to see that T is a linear operator with domain $D_T =$ M. The present section is devoted to the study of this operator.

For simplicity we assume that the function $\varphi(t)$ is continuous on the half-closed interval $[\alpha, \beta)$. Thus, we will have only one exceptional point (the right end point of the interval). This assumption, however, does not decrease the generality of our conclusions, since the case of any finite number of exceptional points can be reduced to the simpler one by subdividing the interval into parts.

We proceed to show that the domain M of the operator T is the set of all vectors f for which

$$\int_\alpha^\beta |\varphi(t)|^2\, d(E_t f, f) < \infty. \tag{2}$$

The first step is to prove that the inclusion $f \in$ M implies inequality (2). Fix $f \in$ M and, in (1), let

$$h = Tf.$$

Then

$$\|Tf\|^2 = \int_\alpha^\beta \varphi(t)\, d(E_t f, Tf) = \lim_{\gamma\to\beta} \int_\alpha^\gamma \varphi(t)\, d(E_t f, Tf).$$

But

$$(E_t f, Tf) = \overline{(Tf, E_t f)} = \int_\alpha^\beta \overline{\varphi(s)}\, d_s \overline{(E_s f, E_t f)} = \int_\alpha^t \overline{\varphi(s)}\, d(E_s f, f).$$

Therefore,

$$\|Tf\|^2 = \lim_{\gamma\to\beta} \int_\alpha^\gamma |\varphi(t)|^2 d(E_t f, f) = \int_\alpha^\beta |\varphi(t)|^2 d(E_t f, f),$$

which yields inequality (2).

Let us assume now that inequality (2) holds for a fixed element f, and prove that $f \in \mathrm{M}$. Since

$$\int_{\gamma'}^{\gamma''} \varphi(t)\, d(E_t f, h) = \lim_{n\to\infty} \sum_{k=1}^n \varphi(t_k)(E(\Delta_k)f, h),$$

and

$$\left|\sum_{k=1}^n \varphi(t_k)(E(\Delta_k)f, h)\right| \leqq \sum_{k=1}^n |\varphi(t_k)| \,.\, |(E(\Delta_k)f, E(\Delta_k)h| \leqq$$

$$\leqq \sqrt{\sum_{k=1}^n |\varphi(t_k)|^2 (E(\Delta_k)f, f)}\, \sqrt{\sum_{k=1}^n (E(\Delta_k)h, h)},$$

and

$$\sum_{k=1}^n (E(\Delta_k)h, h) \leqq \|h\|^2,$$

$$\lim_{n\to\infty} \sum_{k=1}^n |\varphi(t_k)|^2 (E(\Delta_k)f, f) = \int_{\gamma'}^{\gamma''} |\varphi(t)|^2 d(E_t f, f),$$

we see that

$$\left|\int_{\gamma'}^{\gamma''} \varphi(t)\, d(E_t f, h)\right| \leqq \|h\| \sqrt{\int_{\gamma'}^{\gamma''} |\varphi(t)|^2 d(E_t f, f)}.$$

By this inequality, in which γ', γ'' are any numbers which satisfy the inequality $\alpha \leqq \gamma' < \gamma'' < \beta$, we conclude that the improper integral

$$\int_\alpha^\beta \varphi(t)\, d(E_t f, h)$$

exsits and also that

$$\left|\int_\alpha^\beta \varphi(t)\, d(E_t f, h)\right| \leqslant \|h\| \sqrt{\int_\alpha^\beta |\varphi(t)|^2 d(E_t f, f)}.$$

This implies that $f \in \mathrm{M}$.

From our proof it follows that if the function $\varphi(t)$ is bounded in the interval (α, β), then T is a bounded operator which is defined everywhere in H. In this case T can be defined also as the Stieltjes integral operator

$$\int_{\alpha}^{\beta} \varphi(t)\, dE_t, \tag{3}$$

i.e., as the strong limit of some operator integral sum, similar to that which was done in the preceding section.

Further, in this case it is immediately verified that the operator T^* adjoint to T has the form

$$\int_{\alpha}^{\beta} \overline{\varphi(t)}\, dE_t.$$

If the function $\varphi(t)$ is not bounded, then the domain of the operator does not coincide with the whole space H, but it is dense in H. If, for example, $\varphi(t)$ is bounded on each interval $[\alpha, \gamma]$ with $\gamma \in (\alpha, \beta)$, but is unbounded in the neighborhood of β, then D_T contains the set of all the elements in some manifold

$$(E_\gamma - E_\alpha)\,\mathrm{H}$$

with $\gamma \in (\alpha, \beta)$. This set is dense in H. From what has been said, it follows that T has an adjoint operator T^* also in case $\varphi(t)$ is not bounded. We show now that T^* is defined by the formula

$$(T^* f, g) = \int_{\alpha}^{\beta} \overline{\varphi(t)}\, d(E_t f, g),$$

where $\mathrm{D}_{T^*} = \mathrm{M} = \mathrm{D}_T$. In a similar manner, we show that in the case of an unbounded function $\varphi(t)$, the operator T can be represented by a Stieltjes integral, this time improper. For simplicity we again restrict our consideration to the case when the function $\varphi(t)$ is continuous in the half-closed interval $[\alpha, \beta)$.

First, we introduce the function

$$\varphi_\gamma(t) = \begin{cases} \varphi(t) & (\alpha \leqq t \leqq \gamma < \beta), \\ 0 & (t > \gamma), \end{cases}$$

in terms of which we define an operator T_γ as a proper Stieltjes integral

$$T_\gamma = \int_{\alpha}^{\beta} \varphi_\gamma(t)\, dE_t = \int_{\alpha}^{\gamma} \varphi(t)\, dE_t.$$

Now, T_γ is a bounded operator which is defined everywhere in H and the adjoint operator T_γ^* has the form

$$T_\gamma^* = \int_\alpha^\beta \overline{\varphi_\gamma(t)}\, dE_t = \int_\alpha^\gamma \overline{\varphi(t)}\, dE_t.$$

For any $f, h \in$ H

$$(T_\gamma f, h) = \int_\alpha^\beta \varphi_\gamma(t)\, d(E_t f, h) = \int_\alpha^\gamma \varphi(t)\, d(E_t f, h).$$

Now we prove that for any $f \in$ M the equation

$$Tf = \lim_{\gamma \to \beta} T_\gamma f$$

holds, so that the operator T can be represented as an improper integral

$$\int_\alpha^\beta \varphi(t)\, dE_t,$$

with the linear manifold M as domain. Indeed, according to what has been proved above, for any $f \in$ H

$$\| T_\gamma f \|^2 = \int_\alpha^\gamma | \varphi(t) |^2 d(E_t f, f).$$

Therefore,

$$\lim_{\gamma \to \beta} \| T_\gamma f \|$$

exists if and only if $f \in$ M, in which case

$$\lim_{\gamma \to \beta} \| T_\gamma f \| = \| Tf \|. \tag{4}$$

But on the other hand, for $f \in$ M, and each $h \in$ H

$$\lim_{\gamma \to \beta} (T_\gamma f, h) = \lim_{\gamma \to \beta} \int_\alpha^\gamma \varphi(t)\, d(E_t f, h) = (Tf, h).$$

In other words, for $\gamma \to \beta$,

$$T_\gamma f \xrightarrow{w} Tf.$$

This result and (4) imply that

$$T_\gamma f \to Tf.$$

Replacing $\varphi(t)$ by $\overline{\varphi(t)}$ in the preceding development, we find that, for each $g \in$ M,

$$\int_\alpha^\beta \overline{\varphi(t)}\, dE_t g = \lim_{\gamma \to \beta} \int_\alpha^\gamma \overline{\varphi(t)}\, d E_t g = \lim_{\gamma \to \beta} T_\gamma^* g = Sg,$$

where the limit is in the sense of strong convergence. Therefore, for $f, g \in \mathrm{M}$,

$$(Tf, g) = (f, Sg),$$

which implies that

$$\mathrm{M} \subset \mathrm{D}_{T^*}. \tag{5}$$

We assume now that for some pair of elements $h, h^* \in \mathrm{H}$

$$(Tf, h) = (f, h^*)$$

for any $f \in \mathrm{M}$. For f, we substitute the element $T_\gamma^* h$, which certainly belongs to M. Noticing that

$$(E_t T_\gamma^* h, h) = \int_\alpha^\gamma \overline{\varphi(s)}\, d_s(E_s h, E_t h),$$

we get

$$(Tf, h) = (TT_\gamma^* h, h) = \int_\alpha^\beta \varphi(t)\, d_t\left\{\int_\alpha^\gamma \overline{\varphi(s)}\, d_s(E_s h, E_t h)\right\} = \int_\alpha^\gamma |\varphi(t)|^2 d(E_t h, h) =$$

$$= \| T_\gamma^* h \|^2.$$

From the equation

$$\| T_\gamma^* h \|^2 = (T_\gamma^* h, h^*),$$

it follows that

$$\| T_\gamma^* h \| \leqq \| h^* \|.$$

This yields the inequality

$$\lim_{\gamma \to \beta} \int_\alpha^\gamma |\varphi(t)|^2 d(E_t h, h) \leqq \| h^* \|^2 < \infty,$$

i.e., $h \in \mathrm{M}$. And so, it is proved that $\mathrm{D}_{T^*} \subset \mathrm{M}$. From inclusion (5) it follows that $\mathrm{D}_{T^*} = \mathrm{M} = \mathrm{D}_T$ and also that

$$T^* = S = \int_\alpha^\beta \overline{\varphi(t)}\, dE_t.$$

Since the transition from T to T^* resulted from the replacement of $\varphi(t)$ by the complex conjugate function $\overline{\varphi(t)}$, we have $(T^*)^* = T$. Among other things, it follows that the operator T is closed.

There is still another way to construct the operator T in the case of an unbounded function $\varphi(t)$. In place of approximating $\varphi(t)$ by the functions $\varphi_\gamma(t)$, it is possible to use the function $\varphi(t)$ directly to define the operator

$$\int_\alpha^\beta \varphi(t)\, dE_t$$

first for the elements, for which the integral exists as the strong limit of integral sums (in the proper sense), and then to close that operator. This closure will be our operator T.

Until now we have been considering the class $\mathbf{K}$ of all functions $\varphi(t)$ which are continuous with the possible exception of a finite number of points. A further generalization is related to the larger class $\tilde{\mathbf{K}}$, which is defined by the condition that $\varphi(t) \in \tilde{\mathbf{K}}$ if and only if there exists some $f \in \mathrm{H}$ such that $\varphi(t) \in L^2_\sigma$, where $\sigma(t) = (E_t f, f)$ $(\alpha \leq t \leq \beta)$.

In order to define the integral

$$\int_\alpha^\beta \varphi(t)\, dE_t f$$

for $\varphi(t) \in \tilde{\mathbf{K}}$, we choose a sequence $\{\varphi_n(t)\}_1^\infty$, of functions in $\mathbf{K}$ which converge in the sense of the metric of L^2_σ to the function $\varphi(t)$. We let

$$g_n = \int_\alpha^\beta \varphi_n(t)\, dE_t f \qquad (n = 1, 2, 3, \ldots).$$

Since, for functions of the class $\mathbf{K}$,

$$g_m - g_n = \int_\alpha^\beta \{\varphi_m(t) - \varphi_n(t)\}\, dE_t f$$

and

$$\| g_m - g_n \|^2 = \int_\alpha^\beta | \varphi_m(t) - \varphi_n(t) |^2 d(E_t f, f)$$

the sequence $\{g_k\}_1^\infty$ converges to some element $g \in \mathrm{H}$. We define

$$\int_\alpha^\beta \varphi(t)\, dE_t f = g.$$

Denoting by D_T the set of elements $f \in \mathrm{H}$ such that $\varphi(t) \in L^2_\sigma$, where $\sigma(t) = (E_t f, f)$, we obtain an operator defined by the Lebesgue-Stieltjes integral

$$Tf = \int_\alpha^\beta \varphi(t)\, dE_t f.$$

If the function $\varphi(t)$ belongs to the class $\mathbf{K}$, then this Lebesgue-Stieltjes integral coincides with the integral considered earlier.

64. The Integral Representation of a Group of Unitary Operators

The method which we applied in Section 62 in the study of unitary operators, we apply now in some other cases. We use it to obtain an integral representation of *a group of unitary operators* (in the present section) and the *resolvent of a self-adjoint operator* (in the following section).

Let a family of unitary operators U_s be given which depend on a parameter s $(-\infty < s < \infty)$ and which satisfy the following conditions

1) $U_s U_t = U_{s+t}$,
2) $U_0 = E$,
3) $(U_t f, g)$ is a continuous function of t for any $f, g \in \mathrm{H}$. From 1) and 2) it follows that $U_t^{-1} = U_{-t}$. But since $U_t^* = U_t^{-1}$, we have $U_t^* = U_{-t}$.

The family of operators under consideration comprises a continuous abelian group.

We select an arbitrary element $f \in \mathrm{H}$ and consider the function

$$F(t) = (U_t f, f).$$

This function is positive definite. In fact, it is continuous and, for arbitrary real t_α and complex $\rho_\alpha (\alpha = 1, 2, 3, \ldots, n)$,

$$\sum_{\alpha, \beta=1}^{n} F(t_\alpha - t_\beta)\rho_\alpha \bar{\rho}_\beta = \sum_{\alpha, \beta=1}^{n} (U_{t_\alpha} U_{-t_\beta} f, f)\rho_\alpha \bar{\rho}_\beta =$$

$$= \sum_{\alpha, \beta=1}^{n} (\rho_\alpha U_{t_\alpha} f, \rho_\beta U_{t_\beta} f) = \left\| \sum_{\alpha=1}^{n} \rho_\alpha U_{t_\alpha} f \right\|^2 \geqq 0.$$

Therefore, we can apply the theorem of Bochner to the function $F(t)$. By this theorem, there exists a unique nondecreasing function $\omega(s) = \omega(s; f)$ such that

$$\omega(-\infty) = \lim_{s \to -\infty} \omega(s) = 0, \qquad \omega(s-0) = \omega(s) \qquad (-\infty < s < \infty)$$

and

$$F(t) = \int_{-\infty}^{\infty} e^{ist} d\omega(s) \qquad (-\infty < t < \infty).$$

For $t = 0$ it follows that

$$(f, f) = F(0) = \int_{-\infty}^{\infty} d\omega(s) = \omega(\infty).$$

Further, the function $\omega(s; f)$ determines a function $\omega(s; f, g)$ such that

$$(U_t f, g) = \int_{-\infty}^{\infty} e^{ist} d\omega(s; f, g).$$

Under the normalization conditions

$$\omega(-\infty; f, g) = \lim_{s \to -\infty} \omega(s; f, g) = 0,$$

$$(-\infty < s \leqq \infty),$$

$$\omega(s - 0; f, g) = \omega(s; f, g);$$

this representation is unique. By this uniqueness, as before, it is proved that $\omega(s; f, g)$ is a bilinear functional of f, g, the norm of which does not exceed unity and which has the following property:

$$\overline{\omega(s; f, g)} = \omega(s; f, g).$$

Therefore, there exists a one parameter family of bounded self-adjoint operators E_s such that

$$\omega(s; f, g) = (E_s f, g) \qquad (-\infty \leqq s \leqq \infty).$$

The proof that $E_s(-\infty \leqq s \leqq \infty)$ is a resolution of this identity does not differ much from the analogous proof in Section 62. Indeed, on one hand,

$$(U_{t+\tau} f, g) = \int_{-\infty}^{\infty} e^{is(t+\tau)} d(E_s f, g) = \int_{-\infty}^{\infty} e^{ist} d_s \left\{ \int_{-\infty}^{s} e^{i\sigma\tau} d(E_\sigma f, g) \right\},$$

and, on the other

$$(U_{t+\tau} f, g) = (U_t U_\tau f, g) = \int_{-\infty}^{\infty} e^{ist} d_s (E_s U_\tau f, g).$$

Therefore, by the uniqueness of the representation,

$$(E_s U_\tau f, g) = \int_{-\infty}^{s} e^{i\sigma\tau} d(E_\sigma f, g).$$

We now make use of the representation,

$$(E_s U_\tau f, g) = (U_\tau f, E_s g) = \int_{-\infty}^{\infty} e^{i\sigma\tau} d_\sigma (E_\sigma f, E_s g).$$

Again by the uniqueness of the representation, we obtain the equation

$$(E_\sigma f, g) = (E_\sigma f, E_s g) = (E_s E_\sigma f, g) \qquad (\sigma \leqq s),$$

from which it follows that

$$E_u E_v = E_s \qquad (s = \min\{u, v\}).$$

Further, the normalization conditions satisfied by ω yield

$$E_{s-0} = E_s, \qquad E_{-\infty} = 0, \qquad E_\infty = E$$

first in the sense of weak and then in the sense of strong convergence.

In complete correspondence with the considerations of the preceding section, the integral representation

$$U_t f = \int_{-\infty}^{\infty} e^{ist} dE_s f$$

has a direct meaning as an operator in addition to being a symbolic notation for the equation

$$(U_t f, g) = \int_{-\infty}^{\infty} e^{ist} d(E_s f, g).$$

In conclusion, we leave to the reader the verification that the operator E_s reduces the operator U_t.

65. The Integral Representation of the Resolvent of a Self-Adjoint Operator

Let A be a self-adjoint operator and $R_z = (A - zE)^{-1}$ its resolvent, which we shall consider only for $\Im z \neq 0$. Fix $f \in \mathrm{H}$ and, for z in the upper half-plane, consider the function

$$(R_z f, f) = \varphi(z).$$

From the Hilbert relation

$$R_{z'} - R_z = (z' - z) R_{z'} R_z,$$

we get

$$\frac{\varphi(z') - \varphi(z)}{z' - z} = (R_{z'} R_z f, f) = (R_z R_z f, f) + (z' - z)(R_{z'} R_z R_z f, f).$$

Since the second term of the right member tends to zero as $z' \to z$,

$$\lim_{z' \to z} \frac{\varphi(z') - \varphi(z)}{z' - z} = (R_z R_z f, f).$$

Therefore, $\varphi(z)$ is a regular analytic function in the upper half-plane. We determine its imaginary part. It equals

$$\frac{\varphi(z) - \overline{\varphi(z)}}{2i} = \frac{(R_z f, f) - (f, R_z f)}{2i} = \frac{(R_z f, f) - (R_z^* f, f)}{2i}.$$

But (cf. Section 44)

$$R_z^* = R_{\bar{z}}.$$

Hence, again by Hilbert's relation,

$$(R_z f, f) - (R_z^* f, f) = (z - \bar{z})(R_{\bar{z}} R_z f, f) = (z - \bar{z})(R_z f, R_z f).$$

But, since $z = x + iy$ where $y > 0$,

$$\Im \varphi(z) = y(R_z f, R_z f) \geqq 0,$$

i.e., the imaginary part of the function $\varphi(z)$ is non-negative in the upper half-plane; moreover, it is positive if $f \neq 0$, since $R_z f = 0$ implies that $f = 0$.

We prove also that

$$\sup_{y>0} y \,|\, \varphi(iy) \,| < \infty.$$

Indeed, we saw in Section 43 that

$$\| R_z f \| \leqq \frac{1}{y} \| f \|.$$

Therefore,

$$y \,|\, \varphi(iy) \,| = y \,|\, (R_{iy} f, f) \,| \leqq (f, f).$$

We now recall Theorem 3 of Section 59. By this theorem there exists a unique nondecreasing function of bounded variation $\omega(t) = \omega(t; f)$, which satisfies the normalization conditions

$$\omega(-\infty) = \lim_{t \to -\infty} \omega(t) = 0, \quad \omega(t - 0) = \omega(t) \quad (-\infty < t \leqq \infty),$$

and the equation

$$\varphi(z) = \int_{-\infty}^{\infty} \frac{d\omega(t)}{t - z} \qquad (\Im z > 0).$$

Thus, it is proved that for $\Im z > 0$

$$(R_z f, f) = \int_{-\infty}^{\infty} \frac{d\omega(t; f)}{t - z}. \tag{1}$$

From this representation it follows that

$$(f, R_{\bar z} f) = \int_{-\infty}^{\infty} \frac{d\omega(t; f)}{t - z},$$

and, hence,

$$(R_{\bar z} f, f) = \int_{-\infty}^{\infty} \frac{d\omega(t; f)}{t - \bar z}. \tag{2}$$

If z lies in the upper half-plane, then $\bar z$ lies in the lower. Therefore, from formulas (1) and (2) it follows that the representation

$$(R_z h, h) = \int_{-\infty}^{\infty} \frac{d\omega(t; h)}{t - z}$$

is valid for each nonreal z and each $h \in \mathrm{H}$. We define

$$\omega(t; f, g) = \frac{1}{4} \omega(t; f + g) - \frac{1}{4} \omega(t; f - g) + \frac{i}{4} \omega(t; f + ig) - \frac{i}{4} \omega(t; f - ig),$$

and we find that, for each nonreal z and every $f, g \in \mathrm{H}$,

$$(3) \qquad (R_z f, g) = \int_{-\infty}^{\infty} \frac{d\omega(t; f, g)}{t - z}.$$

Here $\omega(t; f, g)$ is a complex-valued function of bounded variation equal to zero for $t = -\infty$ and continuous from the left at each point $t > -\infty$. It is not difficult to verify that under these normalization conditions, the integral representation (3) is unique. In fact, otherwise there would exist a complex-valued function of bounded variation

$$\sigma(t) = \alpha(t) + i\beta(t)$$

for which

$$(4) \qquad \int_{-\infty}^{\infty} \frac{d\sigma(t)}{t - \bar{z}} = 0$$

for every nonreal z. Then, by (4),

$$\overline{\int_{-\infty}^{\infty} \frac{d\sigma(t)}{t - \bar{z}}} = 0,$$

or

$$(4') \qquad \int_{-\infty}^{\infty} \frac{d\overline{\sigma(t)}}{t - z} = 0.$$

A comparison of (4) with (4′) shows that for every nonreal z

$$\int_{-\infty}^{\infty} \frac{d\alpha(t)}{t - z} = \int_{-\infty}^{\infty} \frac{d\beta(t)}{t - z} = 0.$$

But from theorem (3) of Section 59 and the normalization conditions it follows that

$$\alpha(t) = \beta(t) = 0.$$

Now that the uniqueness of (3) is established, it is easy to prove that

$$(5) \qquad \overline{\omega(t; f, g)} = \omega(t; g, f),$$

and

$$\omega(t; \alpha_1 f_1 + \alpha_2 f_2, g) = \alpha_1 \omega(t; f_1, g) + \alpha_2 \omega(t; f_2, g).$$

We show next that

$$\omega(t; f, f) \leqq (f, f).$$

Then it will follow from the theorem in Section 20 that $\omega(t; f, g)$ is a bilinear functional of f, g with norm not exceeding unity. We already noted that

$$y \,|(R_{iy} f, f)| \leqq (f, f) \qquad (y > 0),$$

or, equivalently,

$$\left| \int_{-\infty}^{\infty} \frac{y d\omega(t; f, f)}{t - iy} \right| \leqq (f, f).$$

We split the integral into three parts and find that

$$\left| \int_{-A}^{A} \frac{y d\omega(t;f,f)}{t - iy} \right| \leqq (f,f) + \int_{-\infty}^{-A} d\omega(t;f,f) + \int_{A}^{\infty} d\omega(t;f,f).$$

Let $y \to \infty$ to obtain

$$\int_{-A}^{A} d\omega(t;f,f) \leqq (f,f) + \int_{-\infty}^{-A} d\omega(t;f,f) + \int_{A}^{\infty} d\omega(t;f,f).$$

Now let $A \to \infty$ to get

$$\omega(\infty; f, f) \leqq (f, f).$$

Since $\omega(t; f, g)$ is a nondecreasing function of t, the inequality

$$\omega(t; f, f) \leqq (f, f)$$

is proved.

By the theorem on the general form of a bilinear functional, there exists a family of operators $E_t(-\infty \leqq t \leqq \infty)$ such that

$$\omega(t; f, g) = (E_t f, g).$$

From the relation (5) it follows that

$$(E_t f, g) = (f, E_t g),$$

i.e., the operator E_t is self-adjoint. Further, the function $(E_t f, f)$ is a nondecreasing function of t.

These considerations yield the equation

$$(R_z f, g) = \int_{-\infty}^{\infty} \frac{1}{t - z} d(E_t f, g),$$

which is valid for arbitrary $f, g \in \mathrm{H}$ and each nonreal z. From this equation it follows that

$$(R_z f, R_{\bar{z}'} g) = \int_{-\infty}^{\infty} \frac{1}{t - z} d(E_t f, R_{\bar{z}'} g), \tag{6}$$

and, on the other hand,

$$(R_z f, R_{\bar{z}'} g) = (R_{z'} R_z f, g) = \frac{1}{z' - z} \{(R_{z'} f, g) - (R_z f, g)\} =$$

$$= \frac{1}{z' - z} \int_{-\infty}^{\infty} \left(\frac{1}{t - z'} - \frac{1}{t - z} \right) d(E_t f, g) = \int_{-\infty}^{\infty} \frac{1}{(t - z)(t - z')} d(E_t f, g).$$

And so, besides representation (6), we also have the representation

$$(R_z f, R_{\bar{z}'} g) = \int_{-\infty}^{\infty} \frac{1}{t-z} d_t \left\{ \int_{-\infty}^{t} \frac{1}{s-z'} d(E_s f, g) \right\}.$$

These representations must agree. Therefore,

$$\int_{-\infty}^{t} \frac{1}{s-z'} d(E_s f, g) = (E_t f, R_{\bar{z}'} g) = (R_{\bar{z}'} E_t f, g),$$

or

$$\int_{-\infty}^{t} \frac{1}{s-z'} d(E_s f, g) = \int_{-\infty}^{\infty} \frac{1}{s-z'} d_s(E_s E_t f, g).$$

Since these representations are identical,

$$(E_s f, g) = (E_s E_t f, g) \qquad (s \leqq t).$$

It follows that

$$E_u E_v = E_s \qquad (s = \min\{u, v\}),$$

which implies that

$$E_s \leqq E_t \qquad (s \leqq t).$$

From the normalization conditions of the function $\omega(t; f, g)$ it follows immediately that

$$E_{-\infty} = 0, \quad E_{t-0} = E_t$$

first in the sense of weak, and second in the sense of strong convergence.

Furthermore, the strong limit

$$\lim_{t \to \infty} E_t = E_\infty$$

exists. We prove now that

$$E_\infty = E,$$

from which it will follow that E_t is a resolution of the identity. Let

$$E - E_\infty = F.$$

Then

$$FE_t = (E - E_\infty) E_t = E_t - \lim_{s \to \infty} E_s E_t = E_t - E_t = 0.$$

Hence, for every $f, g \in \mathrm{H}$,

$$(R_z F f, g) = \int_{-\infty}^{\infty} \frac{1}{t-z} d(E_t F f, g) = 0.$$

This implies that, for every $f \in \mathrm{H}$,

$$R_z F f = 0.$$

Therefore,

$$F f = 0.$$

Hence, $F = 0$, which was to be shown.

The representation derived for the resolvent R_z of a self-adjoint operator A is also an illustration of the general representation of Section 63, which refers to operator integrals. This representation can be written in the form

$$R_z = \int_{-\infty}^{\infty} \frac{1}{t-z}\, dE_t \qquad (\Im z \neq 0).$$

We propose that the reader verify that each of the operators E_t reduces the resolvent R_z, where z is an arbitrary nonreal number, i.e.,

$$E_t R_z = R_z E_t.$$

66. The Integral Representation of Self-Adjoint Operators

From the general considerations of Section 63, we obtain, as a particular case, the following theorem.

THEOREM 1: *To each resolution of the identity* E_t $(-\infty \leqq t \leqq \infty)$ *there corresponds a uniquely defined self-adjoint operator*

$$B = \int_{-\infty}^{\infty} t dE_t. \tag{1}$$

The domain D_B *of this operator is the set of all vectors* f *for which the inequality*

$$\int_{-\infty}^{\infty} t^2 d(E_t f, f) < \infty$$

is satisfied. The left member of this inequality is equal to $\| Bf \|^2$.

The basic problem of the present section is the proof of the converse proposition. For the purposes of this proof, let A denote a given self-adjoint operator and let R_z denote its resolvent. According to the theorem of the preceding section there is a resolution of the identity which corresponds to A. We shall prove that this resolution of the identity belongs to the operator A in the sense that it generates A in the way indicated by Theorem 1. The first step of this proof is the following lemma.

LEMMA: *If* E_t $(-\infty \leqq t \leqq \infty)$ *is a resolution of the identity which belongs to the resolvent* R_z *of a self-adjoint operator* A, *then the set of all vectors* $f \in H$, *for which inequality* (2) *holds coincides with the domain* D_A *of the operator* A.

Proof: We know that the vector $f = R_z h$ runs through D_A, when the vector h runs through H and z is any fixed nonreal number. We take $z=i$ and, for brevity, write

$$R_i = R, \ R_{-i} = R^*.$$

Our proposition will be proved if we verify the equivalence of the following two assertions:

α) The element f has the property

$$\int_{-\infty}^{\infty} t^2 d(E_t f, f) < \infty.$$

β) There exists a vector h such that

$$f = Rh.$$

We prove first that β implies α. Therefore, let

$$f = Rh.$$

Then

$$(E_t f, f) = (E_t Rh, Rh) = (RE_t h, Rh) = (R^* RE_t h, h) =$$

$$= \frac{1}{2i}(\{R - R^*\} E_t h, h) = \frac{1}{2i} \int_{-\infty}^{\infty} \left(\frac{1}{s-i} - \frac{1}{s+i}\right) d_s(E_s E_t h, h) =$$

$$= \int_{-\infty}^{t} \frac{1}{1+s^2} d(E_s h, h),$$

and, hence,

$$\int_{-M}^{M} t^2 d(E_t f, f) = \int_{-M}^{M} \frac{t^2}{1+t^2} d(E_t h, h) \leqq \int_{-M}^{M} d(E_t h, h) \leqq (h, h).$$

This inequality shows that

$$\int_{-\infty}^{\infty} t^2 d(E_t f, f) \leqq (h, h) < \infty,$$

so that assertion α is proved.

Now we prove that α implies β. Assume α. This implies that the vector f belongs to the domain of the operator B which was defined in Theorem 1. Let

$$h = (B - iE) f.$$

Assertion β) will be proved if we verify that

$$R(B - iE) f = f.$$

For an arbitrary vector $g \in$ H, we have the equations

$$(R(B - iE)f, g) = \int_{-\infty}^{\infty} \frac{1}{t-i} d(E_t(B - iE)f, g),$$

and

$$((B - iE)f, E_t g) = \int_{-\infty}^{\infty} (s - i) d_s(E_s f, E_t g) = \int_{-\infty}^{t} (s - i) d(E_s f, g).$$

Therefore,

$$(R(B - iE)f, g) = \int_{-\infty}^{\infty} \frac{t - i}{t - i} d(E_t f, g) = (f, g).$$

Since g is arbitrary,

(3) $$R(B - iE)f = f.$$

THEOREM 2: *The resolution of the identity $E_t(-\infty \leqq t \leqq \infty)$ determined by the resolvent R_z of a self-adjoint operator A is the resolution of the identity which belongs to the operator A in the sense that, first, D_A is the set of all vectors f for which*

$$\int_{-\infty}^{\infty} t^2 d(E_t f, f) < \infty,$$

and, second, for any $f \in D_A$

$$Af = \int_{-\infty}^{\infty} t dE_t f.$$

Proof: We construct the operator B, which corresponds, in the sense of Theorem 1, to the resolution of the identity determined by the resolvent R_z of the operator A. By the lemma, $D_B = D_A$. Hence, it remains to prove that the equation

(4) $$Bf = Af$$

holds for each $f \in D_A$. But, in Section 44, it was shown that the equation,

$$R_z g = 0,$$

with z nonreal, is satisfied only for $g = 0$. Therefore, instead of (4), it is sufficient to prove that

$$RBf = RAf,$$

or

$$R(B - iE)f = R(A - iE)f.$$

This equation holds since the left side equals f by (3) and the right side equals f by the definition of the resolvent.

THEOREM 3: *The resolution of the identity $E_t(-\infty \leqq t \leqq \infty)$ belongs to the self-adjoint operator A if and only if*

1° $E(\Delta)$ *reduces the operator* A *for any interval* $\Delta \subset [-\infty, \infty]$,

2° *the relation*

$$f \in (E_t - E_s)H \qquad (-\infty \leqq s < t \leqq \infty)$$

implies the inequality

$$s\|f\|^2 \leqq (Af, f) \leqq t\|f\|^2.$$

Proof: The necessity of condition 1° was noted earlier. The necessity of condition 2° follows from the representation

$$(Af, f) = \int_s^t \tau d(E_\tau f, f),$$

which is valid for $f \in (E_t - E_s)H$.

We turn now to the proof of the sufficiency of the conditions. Therefore, assume conditions 1° and 2° and choose an arbitrary element $f \in D_A$. By condition 1°, $(E_\beta - E_\alpha)f$ belongs to the domain D_A, for any α and β. Supposing that α and β are finite, we form the partition

$$\alpha = \alpha_0 < \alpha_1 < \alpha_2 < \ldots < \alpha_{n-1} < \alpha_n = \beta \tag{5}$$

of the interval $[\alpha, \beta]$, and we represent $(E_\beta - E_\alpha)f$ in the form

$$(E_\beta - E_\alpha)f = \sum_{k=0}^{n-1}(E_{\alpha_{k+1}} - E_{\alpha_k})f.$$

By 1°,

$$(E_\beta - E_\alpha)Af = \sum_{k=0}^{n-1} A(E_{\alpha_{k+1}} - E_{\alpha_k})f. \tag{6}$$

It follows from condition 2° that, for $f \in (E_t - E_s)H$,

$$-\frac{t-s}{2}(f, f) \leqq \left(Af - \frac{s+t}{2}f, f\right) \leqq \frac{t-s}{2}(f, f).$$

In other words, the restriction of the self-adjoint operator

$$A - \frac{s+t}{2}E$$

to the subspace $(E_t - E_s)H$ has norm $\leqq \frac{t-s}{2}$. Therefore, we write (6) in the form

$$(E_\beta - E_\alpha)Af = \sum_{k=0}^{n-1} \frac{\alpha_k + \alpha_{k+1}}{2}(E_{\alpha_{k+1}} - E_{\alpha_k})f + \tag{7}$$

$$+ \sum_{k=0}^{n-1}\left(A - \frac{\alpha_k + \alpha_{k+1}}{2}E\right)(E_{\alpha_{k+1}} - E_{\alpha_k})f,$$

and note that

$$\left\| \sum_{k=0}^{n-1} \left(A - \frac{\alpha_k + \alpha_{k+1}}{2} E \right) (E_{\alpha_{k+1}} - E_{\alpha_k}) f \right\|^2 \leqq$$

$$\leqq \sum_{k=0}^{n-1} \left(\frac{\alpha_{k+1} - \alpha_k}{2} \right)^2 \| (E_{\alpha_{k+1}} - E_{\alpha_k}) f \|^2 \leqq \varepsilon^2 \| f \|^2,$$

where

$$\varepsilon = \max_k \frac{\alpha_{k+1} - \alpha_k}{2}.$$

In (7), let the diameter 2ε of the subdivision (5), tend to zero. This passage to the limit yields

$$(E_\beta - E_\alpha) A f = \int_\alpha^\beta t dE_t f,$$

and

$$\| (E_\beta - E_\alpha) A f \|^2 = \int_\alpha^\beta t^2 d (E_t f, f).$$

The second of these relations shows that

$$\int_{-\infty}^{\infty} t^2 d (E_t f, f) = \| A f \|^2 < \infty,$$

and the first gives

$$A f = \int_{-\infty}^{\infty} t dE_t f.$$

Therefore, the theorem is proved.

In the derivation of the integral representation of the resolvent of a self-adjoint operator (Section 65) we used only the fact that $R_z(\Im z \neq 0)$ is a family of operators, defined everywhere in H, which satisfy the following conditions:

1° $\| R_z f \| \leqq \frac{1}{|y|} \| f \| \quad (y = \Im z).$

2° $R_z^* = R_{\bar{z}}.$

3° $R_{z'} - R_z = (z' - z) R_{z'} R_z.$

4° If $R_z f = 0$ for any z, then $f = 0$.

Now we can assert that every family of operators, which satisfy these conditions is the resolvent of some self-adjoint operator.[8] Indeed, having such a family of operators, we can find its integral representation by some resolution of the identity E_t, and then construct a self-adjoint operator A with this resolution of the identity. The resolvent of this operator A is the family R_z.

Completing the present section we mention several simple facts about self-adjoint operators which are immediate consequences of the integral representation of these operators.

1° If A is a self-adjoint operator for which

$$\inf_{f \in D_A} \frac{(Af,f)}{(f,f)} = \alpha,$$

$$\sup_{f \in D_A} \frac{(Af,f)}{(f,f)} = \beta,$$

then the integral representation of A has the form

$$A = \int_{\alpha}^{\beta} t dE_t,$$

i.e.,

$$E_t = 0 \qquad (\text{for } t \leqq \alpha),$$
$$E_t = E \qquad (\text{for } t \geqq \beta).$$

2° In order that a vector $f \in H$ admit the n-fold application of the self-adjoint operator A, i.e., in order that

$$A^k f = A(A^{k-1} f) \qquad (k = 1, 2, \ldots, n)$$

have meaning, it is necessary and sufficient that the inequality

$$\int_{-\infty}^{\infty} t^{2n} d(E_t f, f) < \infty$$

be satisfied. If this inequality is satisfied, then for $k = 1, 2, \ldots, n$

$$A^k f = \int_{-\infty}^{\infty} t^k dE_t f, \quad \| A^k f \|^2 = \int_{-\infty}^{\infty} t^{2k} d(E_t f, f).$$

3° There exists a linear manifold dense in H which is in the domain of every positive integral power of the self-adjoint operator A. One such manifold is the set of vectors of the form $E(\Delta) h$, where h runs through H, and Δ runs through the set of all finite intervals on the real axis.

The proof of these assertions we leave to the reader.

---

[8] This fact can be established without the use of the integral representation of the family of operators R_z. If one uses only properties 2°, 3°, 4°, then it is easy to show that the operator A defined by the equation $A = zE + R_z^{-1}$ does not depend on z, is self-adjoint, and its resolvent is R_z. Thus, it is revealed that property 1° is a corollary which need not be assumed.

67. The Cayley Transform

In the present section we give another construction of the integral representation of a self-adjoint operator—a construction which is based on the so-called *Cayley transform.*

Let A be a closed symmetric operator and let z be a nonreal number. Let h run through D_A and let

$$(A - \bar{z}E)\,h = f, \tag{1}$$

$$(A - zE)\,h = g. \tag{2}$$

The vector f will run through some linear manifold F, and the vector g through some linear manifold G. We prove that F and G are closed and, hence, are subspaces of H. With this aim, we assume that

$$f_n = (A - \bar{z}E)\,h_n \qquad (n = 1, 2, 3, \ldots),$$

and

$$f_n \to f. \tag{3}$$

If $z = x + iy$, where $y \neq 0$, then

$$f_n = (A - xE)\,h_n - iyh_n,$$

which yields

$$\|f_n - f_m\|^2 = \|(A - xE)(h_n - h_m)\|^2 + y^2\|h_n - h_m\|^2.$$

Therefore, from (3), it follows that

$$\lim_{n\to\infty} (A - xE)\,h_n \text{ and } \lim_{n\to\infty} h_n$$

exist. We denote the second of these limits by h and use the fact that A is a closed operator to conclude that

$$f = (A - xE)\,h + iyh = (A - \bar{z}E)\,h.$$

Hence, $f \in$ F, so that the manifold F is closed. Analogously, it is proved that G is closed.

By means of relations (1) and (2), the manifold D_A is mapped in one-to-one manners onto F and G, respectively. Therefore, for each $f \in$ F there is one and only one element $h \in D_A$ which satisfies relation (1). Having found this element h, we determine the element g by formula (2). Thus, to each element $f \in$ F is associated a unique element $g \in$ G, i.e., we have an operator V with domain $D_V =$ F and range $\Delta_V =$ G:

$$g = Vf.$$

It is easy to see that V is an isometric operator. Indeed, it is linear and

$$\|f\|^2 = \|(A - xE)\,h\|^2 + y^2\|h\|^2,$$

$$\|g\|^2 = \|(A - xE)\,h\|^2 + y^2\|h\|^2.$$

The isometric operator V is called the Cayley transform of the symmetric operator A.

Let $z = i$. Then equations (1) and (2) have the form

$$(A + iE)\,h = f, \tag{1'}$$

$$(A - iE)\,h = g = Vf. \tag{2'}$$

From these relations it follows that

$$h = \frac{1}{2i}(f - g) = \frac{1}{2i}(E - V)f,$$

$$Ah = \frac{1}{2}(f + g) = \frac{1}{2}(E + V)f.$$

Therefore,

$$Ah = i(E + V)(E - V)^{-1}h. \tag{4}$$

Thus, the symmetric operator A is expressed in terms of its Cayley transform V.

Especially important is the case when V is a unitary operator.

THEOREM: *Let the Cayley transform of a symmetric operator A be a unitary operator U with a resolution of the identity $F_s (0 \leqq s \leqq 2\pi)$. If*

$$E_t = F_s \qquad \left(t = -\cot\frac{s}{2}\right),$$

then the domain D_A of A is the set of all vectors h for which

$$\int_{-\infty}^{\infty} t^2 d\,(E_t h, h) < \infty.$$

The operator A has the integral representation

$$Ah = \int_{-\infty}^{\infty} t dE_t h.$$

Proof: Let $h \in D_A$ and f be the element which corresponds to h by the mapping (1′). Then

$$h = \frac{1}{2i}(E - U)f.$$

In this case,

$$F_s h = \frac{1}{2i}(E - U)F_s f = \frac{1}{2i}\int_0^{2\pi}(1 - e^{i\tau})\,d_\tau(F_\tau F_s f) = \tag{5}$$

$$= \frac{1}{2i}\int_0^{s}(1 - e^{i\tau})d\,F_\tau f,$$

and

$$(F_s h, h) = \frac{1}{4}(\{E - U\} F_s f, \{E - U\} f) =$$

$$= \frac{1}{4}(\{E - U^{-1}\} \{E - U\} F_s f, f) =$$

$$(6) \qquad = \frac{1}{4}(\{2E - U - U^{-1}\} F_s f, f) =$$

$$= \frac{1}{4}\int_0^{2\pi} (2 - e^{i\tau} - e^{-i\tau}) d_\tau (F_\tau F_s f, f) =$$

$$= \int_0^s \sin^2 \frac{\tau}{2} d(F_\tau f, f).$$

On the other hand

$$(7) \qquad Ah = \frac{1}{2}(E + U) f = \frac{1}{2}\int_0^{2\pi} (1 + e^{is})\, dF_s f,$$

and

$$\| Ah \|^2 = \frac{1}{4}(\{E + U\} f, \{E + U\} f) = \frac{1}{4}(\{2E + U + U^{-1}\} f, f) =$$

$$(8) \qquad = \int_0^{2\pi} \cos^2 \frac{s}{2} d(F_s f, f).$$

A comparison of (6) and (8) reveals that

$$\| Ah \|^2 = \int_0^{2\pi} \cot^2 \frac{s}{2} \sin^2 \frac{s}{2} d(F_s f, f) = \int_0^{2\pi} \cot^2 \frac{s}{2} d(F_s h, h).$$

In the same manner it follows from (5) and (7) that, for each $h' \in H$,

$$(Ah, h') = -\int_0^{2\pi} \cot \frac{s}{2}\, d(F_s h, h').$$

We now let

$$t = -\cot \frac{s}{2}, \quad E_t = F_s.$$

Then E_t is a resolution of the identity in the interval $[-\infty, \infty]$. If $h \in D_A$, then

$$(9) \qquad \int_{-\infty}^{\infty} t^2 d(E_t h, h) < \infty$$

and

$$Ah = \int_{-\infty}^{\infty} t dE_t h. \tag{10}$$

In order to complete the proof of the theorem, it remains to prove that if

$$\int_{-\infty}^{\infty} t^2 d(E_t h, h) = \int_0^{2\pi} \cot^2 \frac{s}{2} d(F_s h, h) < \infty, \tag{11}$$

then $h \in D_A$. Therefore, assume that (11) holds with a particular element h. Then the operator

$$-\int_0^{2\pi} e^{-is/2} \frac{1}{\sin \frac{s}{2}} dF_s$$

is defined for the element h. Let

$$-\int_0^{2\pi} e^{-is/2} \frac{1}{\sin \frac{s}{2}} dF_s h = f.$$

Then, for each $h' \in H$,

$$-\int_0^{2\pi} e^{-is/2} \frac{1}{\sin \frac{s}{2}} d(F_s h, h') = (f, h').$$

Setting

$$h' = -\frac{1}{2i}(E - U^{-1}) f',$$

we find that

$$\left(\frac{1}{2i}\{E - U\} f, f'\right) = (f, h') = -\frac{1}{2i} \int_0^{2\pi} \frac{e^{-is/2}}{\sin \frac{s}{2}} d(F_s h, \{E - U^{-1}\} f') =$$

$$= -\frac{1}{2i} \int_0^{2\pi} \frac{e^{-is/2}}{\sin \frac{s}{2}} (1 - e^{is}) d(F_s h, f') = \int_0^{2\pi} d(F_s h, f') = (h, f').$$

Since f' is arbitrary this implies that

$$h = \frac{1}{2i}(E - U) f.$$

In other words, the element h has a representation, which implies that it belongs to D_A.

Since the Cayley transform of a self-adjoint operator is a unitary operator, it follows from the theorem that every self-adjoint operator has an integral representation, which fact we also obtained in the preceding section.

But the proof of the theorem yields another fact: If the Cayley transform of a symmetric operator A is a unitary operator, then A is a self-adjoint operator. Indeed, from the theorem it follows that the operator A has the representation (10) and is defined for all the elements h for which inequality (9) holds. By Theorem 1 of the preceding section, such an operator is necessarily self-adjoint.

68. The Spectra of Self-Adjoint and Unitary Operators

Let E_t be a resolution of the identity. Without loss of generality we suppose that E_t is defined for $-\infty \leqq t \leqq \infty$. Indeed, if E_t is defined originally for $t \in [\alpha, \beta]$, then we can extend the definition by letting $E_t = E$ for $t \geqq \beta$ and $E_t = 0$ for $t \leqq \alpha$.

We call the point t a *point of constancy* of E_t if there exists an $\epsilon > 0$ such that

$$E_{t+\varepsilon} - E_{t-\varepsilon} = 0,$$

and a *point of growth* otherwise. Furthermore, we call the point t a *jump point* if

$$E_{t+0} - E_t \neq 0,$$

and a *continuity point* if

$$E_{t+0} - E_t = 0.$$

Continuity points which are also points of growth we call *points of continuous growth.*

Let E_t be the resolution of the identity of a self-adjoint operator A. We prove that

(a) A real number λ is a regular point of A if and only if λ is a point of constancy of E_t.

(b) A real number λ is an eigenvalue of A if and only if λ is a jump point of E_t.

In order to prove assertion (a) we use the equation

$$\|(A - \lambda E)f\|^2 = \int_{-\infty}^{\infty} (t - \lambda)^2 \, d(E_t f, f), \tag{1}$$

which is valid for $f \in D_A$. If λ is a point of constancy of E_t, then the function $(E_t f, f)$ is constant in some neighborhood of the point λ. Let ϵ be the radius of this neighborhood. Equation (1) yields the inequality

$$\|(A-\lambda E)f\|^2 \geqq \int_{-\infty}^{\lambda-\epsilon} + \int_{\lambda+\epsilon}^{\infty} (t-\lambda)^2 d(E_t f, f) \geqq \varepsilon^2 (f, f),$$

from which it follows that λ is a regular point of A.

Conversely, if λ is a regular point, then, for some $\epsilon > 0$ and every $f \in D_A$,

$$\|(A-\lambda E)f\| \geqq \epsilon \|f\|,$$

and this implies that, for $f \in D_A$,

$$\int_{-\infty}^{\infty} (t-\lambda)^2 d(E_t f, f) \geqq \epsilon^2 \int_{-\infty}^{\infty} d(E_t f, f). \tag{2}$$

Let us assume that λ is not a point of constancy of E_t. Then there exists an element g and a positive $\eta < \epsilon$ such that

$$(E_{\lambda+\eta} - E_{\lambda-\eta})g \neq 0.$$

Applying inequality (2) to the element

$$f = (E_{\lambda+\eta} - E_{\lambda-\eta})g,$$

which is known to belong to D_A, we get

$$\int_{\lambda-\eta}^{\lambda+\eta} (t-\lambda)^2 d(E_t g, g) \geqq \epsilon^2 \int_{\lambda-\eta}^{\lambda+\eta} d(E_t g, g).$$

Since this is obviously false, assertion a) is proved.

We proceed now to the proof of assertion b). Let λ be an eigenvalue of the operator A and let f be an eigenvector associated with it. Then

$$0 = \|(A-\lambda E)f\|^2 = \int_{-\infty}^{\infty} (t-\lambda)^2 d(E_t f, f).$$

This equation shows that the only point of growth of the function $(E_t f, f)$ can be the point $t = \lambda$. Since

$$0 \neq (f, f) = \int_{-\infty}^{\infty} d(E_t f, f),$$

the point $t = \lambda$ is definitely a point of growth. But an isolated point of growth is a jump point. Therefore

$$(E_{\lambda+0} f, f) \neq (E_\lambda f, f),$$

and, consequently,

$$E_{\lambda+0} \neq E_\lambda.$$

Conversely, let λ be a point such that

$$E_{\lambda+0} \neq E_\lambda.$$

This implies that for some vector g

$$(E_{\lambda+0} - E_\lambda)g = f \neq 0.$$

The vector f belongs to D_A and

$$\| (A - \lambda E) f \|^2 = \int_{-\infty}^{\infty} (t - \lambda)^2 d\,(E_t f, f).$$

But the function

$$(E_t f, f) = (E_t \{E_{\lambda+0} - E_\lambda\} g, f)$$

is equal to zero for $t < \lambda$ and does not depend on t for $t > \lambda$. Hence,

$$\| (A - \lambda E) f \| = 0,$$

i.e., λ is an eigenvalue of the operator A.

From the propositions (a) and (b) it follows that each point of continuous growth of the resolution of the identity of a self-adjoint operator belongs to its continuous spectrum.

A trivial consequence of our considerations is that the spectrum of each self-adjoint operator is nonvoid.

Turning now from self-adjoint operators to unitary operators we recall that every eigenvalue of a unitary operator has absolute value unity, i.e., each has the form $e^{i\lambda}$ where λ is called the *eigenfrequency* of the unitary operator. The equation

$$U^k f = e^{ik\lambda} f$$

holds for each integer k if f is an eigenvector associated with the eigenfrequency λ. Let U_s $(-\infty < s < \infty)$ be a continuous group of unitary operators. In this case λ is called an eigenfrequency of the group of operators if an element $f \neq 0$ (an eigenvector) exists such that, for every real s,

$$U_s f = e^{is\lambda} f$$

We now prove that: *a real number λ is an eigenfrequency of a unitary operator U (respectively, a continuous group of unitary operators U_s) if and only if λ is a jump point of the corresponding resolution of the identity.*

For definiteness, we consider the case of a group of unitary operators. Let λ be an eigenfrequency of this group and let f be an associated eigenvector. Then, for each real s,

$$U_s f - e^{is\lambda} f = 0.$$

But since

$$\|(U_s - e^{is\lambda}E)f\|^2 = (\{U_{-s} - e^{-is\lambda}E\}\{U_s - e^{is\lambda}E\}f,f) =$$
$$= (\{2E - e^{is\lambda}U_{-s} - e^{-is\lambda}U_s\}f,f) =$$
$$= \int_{-\infty}^{\infty}(2 - e^{is(\lambda-t)} - e^{-is(\lambda-t)})\,d(E_t f,f) =$$
$$= \int_{-\infty}^{\infty} 4\sin^2\frac{s(\lambda-t)}{2}\,d(E_t f,f),$$

the only possible point of growth of the function $(E_t f, f)$ is $t = \lambda$. This must be a point of growth, since otherwise the relation

$$0 \neq (f,f) = \int_{-\infty}^{\infty} d(E_t f,f)$$

would not be possible. Hence,

$$(E_{\lambda+0}f,f) \neq (E_\lambda f,f)$$

and

$$E_{\lambda+0} \neq E_\lambda.$$

We remark that the role of the non-negative factor $(\lambda - t)^2$ in the case of the self-adjoint operator is played now by $4\sin^2\frac{s(\lambda-t)}{2}.$ Other than this, there is no difference in the proof. We refer to the second part of the assertion and also to the following proposition: the point λ is a point of constancy of the resolution of the identity of a unitary operator U (respectively, a group of unitary operators U_s) if and only if $e^{ik\lambda}$ for any integer k (respectively, $e^{is\lambda}$ for any real s) is a regular point of the operator U^k (respectively, of the operator U_s).

Since the resolution of the identity E_t of a self-adjoint or unitary operator completely determines the spectrum of the operator, one often calls E_t the *spectral function.*

Whether or not a given point λ belongs to the spectrum of a self-adjoint operator A is determined by the behavior of its resolvent R_z for z in the neighborhood of the point λ. The circumstances are contained in the proposition: *a real point λ does not belong to the spectrum of a self-adjoint operator A, and hence, is a regular point of this operator if and only if, for each fixed element $f \in$ H, the function $(R_z f, f)$*

1) *is regular for* z *in some neighborhood of the point* λ, *and*

2) *takes on real values in some interval* $\lambda - \delta < t < \lambda + \delta$.

The necessity of these conditions follows immediately from the integral representation of the resolvent and assertion a) of the present section. The sufficiency follows from the Stieltjes inversion formula (cf. Section 59) and assertion a) of the present section.

In the following section we shall become acquainted with a class of operators for which the above conditions need to be verified only for one (specially chosen) element f.

69. The Simple Spectrum

In linear algebra and the elementary theory of integral equations, the spectrum of an operator is called *simple*, if the multiplicity of each eigenvalue of this operator equals one. This definition does not apply to arbitrary operators in Hilbert space, since in general the set of eigenvalues of an operator does not exhaust its spectrum. We make the following definition.

DEFINITION: *The spectrum of a self-adjoint (respectively, unitary) operator is called simple if there exists a vector* $g \in \mathrm{H}$ *(generating vector) such that the linear envelope of the set of vectors* $E(\Delta)g$, *where* Δ *runs through the set of all intervals on the real axis (respectively, the segment* $[0, 2\pi]$*) is dense in* H.

By this definition the spectra of a self-adjoint operator and of its Cayley transform are simultaneously simple or not simple. In addition, from the given definition it follows that in a nonseparable space a self-adjoint (unitary) operator cannot have a simple spectrum.

For operators, which occur in linear algebra and the elementary theory of integral equations, it is characteristic that the linear envelope of the set of all eigenvectors of these operators coincides with the space, or in any case is dense in it. It is not difficult to show that for an arbitrary self-adjoint (or unitary) operator in H, which possesses these properties, the new definition adopted above for the simplicity of the spectrum is in complete accord with the elementary definition mentioned at the beginning of the present section. This fact can be regarded as a justification of the general definition.

We consider as an example the multiplication operator[9] Q in L^2_σ $(-\infty, \infty)$ (cf. Section 48). The spectral function E_t of this operator is defined by the equation

$$E(\Delta)f(t) = \chi_\Delta(t)f(t) \qquad (f(t) \in L^2_\sigma),$$

[9] For simplicity we write Q instead of Q_σ.

where $\chi_\Delta(t)$ is the characteristic function of the interval Δ. In order to ascertain this it is sufficient to recall Theorem 3 of Section 66. The operator Q has a simple spectrum. Indeed, let us take a step function $g(t)$ which satisfies the conditions

$$g(t) = a_k > 0 \qquad (k - 1 < t \leqq k),$$

$$\sum_{k=-\infty}^{\infty} a_k^2 \{\sigma(k) - \sigma(k - 1 + 0)\} < \infty.$$

This is a generating function for the operator Q in the sense of our definition. Hence, the linear envelope of the set of functions $E(\Delta)\, g(t)$ coincides with the set of all step functions which are equal to zero outside of some interval, and this set is dense in $L_\sigma^2(-\infty, \infty)$ (cf. Section 42).

It is evident that a function $g(t) \in L_\sigma^2(-\infty, \infty)$ which is equal to zero on a set of positive σ-measure, cannot be a generator for the operator Q.

For a generating element one may take any function $g(t) \in L_\sigma^2(-\infty, \infty)$, which is distinct from zero everywhere with the possible exception of a set of zero σ-measure. For the proof of this assertion it is sufficient to establish that any function $f(t) \in L_\sigma^2(a, b)$ can be approximated with any accuracy by products of the form $g(t)\, q(t)$, where $q(t)$ is a step function:

$$\int_a^b |f(t) - g(t)\, q(t)|^2 d\sigma(t) < \epsilon^2. \tag{1}$$

We introduce the space L_ρ^2 (a, b), where

$$\rho(t) = \int_a^t |g(s)|^2\, d\sigma(s). \tag{2}$$

Since the function $\dfrac{f(t)}{g(t)}$ belongs to L_ρ^2 (a, b), there exists a step function $q(t)$ for which

$$\int_a^b \left| \frac{f(t)}{g(t)} - q(t) \right|^2 d\rho(t) < \epsilon^2.$$

After the replacement of $\rho(t)$ according to formula (2), we obtain (1).

THEOREM 1: *If the spectrum of a unitary operator U is simple and g is a generating element, then the linear envelope of the vectors $U^k g$ ($\pm k =$ 0, 1, 2, . . .) is dense in* H. *Conversely, if there exists a vector g such that the linear envelope of the vectors $U^k g$ ($\pm k = 0, 1, 2, \ldots$) is dense in* H, *then the spectrum of the operator U is simple and the vector g is a generating element.*

Proof: Let the spectrum of the operator U be simple and let g be a generating element. Assume that the linear envelope of the set of the vectors $U^k g$ $(\pm k = 0, 1, 2, \ldots)$ is not dense in H. Then there exists a vector $h \neq 0$ for which

$$(U^k g, h) = 0 \qquad (\pm k = 0, 1, 2, \ldots).$$

From the integral representation of the unitary operator U it follows that

$$\int_0^{2\pi} e^{ikt} d\,(E_t g, h) = 0 \qquad (\pm k = 0, 1, 2, \ldots).$$

These equations and the uniqueness theorem yield

$$(E_t g, h) = 0 \qquad (0 \leqq t \leqq 2\pi).$$

This implies that the vector h is orthogonal to each vector of the form $E(\Delta)\,g$. But this is impossible, since g is a generating element. Thus, the first part of the theorem is proved.

The second part is proved very easily in the same way.

THEOREM 2: (*The canonical form of a self-adjoint operator with a simple spectrum*). *Let A be a self-adjoint operator with a simple spectrum, g be a generating element, and $\sigma(t) = (E_t g, g)$. Then the formula*

$$f = \int_{-\infty}^{\infty} f(t)\, dE_t g$$

associates each function $f(t) \in L^2_\sigma(-\infty, \infty)$ with a vector $f \in$ H, and this correspondence is an isometric mapping of $L^2_\sigma(-\infty, \infty)$ into H. It transforms the domain $D_Q \subset L^2_\sigma(-\infty, \infty)$ of the multiplication operator Q in $L^2_\sigma(-\infty, \infty)$ into the domain $D_A \subset$ H of the operator A. If the element $f \in D_A$ corresponds to the function $f(t) \in L^2_\sigma(-\infty, \infty)$, then the element Af corresponds to the function $tf(t)$.

Proof: We denote by G the set of all vectors of H which can be represented in the form (3). Since G is a linear manifold which contains all the vectors of the form

$$\int_{-\infty}^{\infty} \chi_\Delta(t)\, dE_t g = E(\Delta)\, g,$$

G is dense in H. If $f(t)$ belong to the set **K** which was introduced in Section 63, then

$$(E_t g, f) = \int_{-\infty}^{\infty} \overline{f(s)}\, d_s\, (E_t g, E_s g) = \int_{-\infty}^{t} \overline{f(s)}\, d(E_s g, g)$$

and

$$(f,f) = \int_{-\infty}^{\infty} f(t)\, d(E_t g, f) = \int_{-\infty}^{\infty} |f(t)|^2 d(E_t g, g) = \int_{-\infty}^{\infty} |f(t)|^2 d\,\sigma(t).$$

Thus, a linear manifold which is dense in $L^2_\sigma(-\infty, \infty)$ is transformed isometrically by (3) into a linear manifold which is dense in G. Therefore, the completeness of $L^2_\sigma(-\infty, \infty)$ implies that G is closed and, hence, G coincides with H, so that the first part of the theorem is proved.

In order to prove the second part of the theorem, we allow $f(t)$ to run through the set of all continuous functions which are equal to zero outside of a finite interval. If $f(t)$ is such a function and f is the vector in H which corresponds to it, then $f \in D_A$ and by (3), for each $h \in H$,

$$(Af, h) = \int_{-\infty}^{\infty} t d(E_t f, h) = \int_{-\infty}^{\infty} t d(f, E_t h) = \int_{-\infty}^{\infty} t d\left\{\int_{-\infty}^{\infty} f(s)\, d_s(E_s g, E_t h)\right\} =$$

$$= \int_{-\infty}^{\infty} t d\left\{\int_{-\infty}^{t} f(s)\, d(E_s g, h)\right\} = \int_{-\infty}^{\infty} t f(t)\, d(E_t g, h). \tag{4}$$

It follows that

$$Af = \int_{-\infty}^{\infty} t f(t)\, dE_t g. \tag{5}$$

We let in (4)

$$h = Af,$$

and get

$$\| Af \|^2 = \int_{-\infty}^{\infty} | t f(t) |^2 d\sigma(t). \tag{6}$$

Recalling the definition of an operator represented by a Lebesgue-Stieltjes integral, we conclude from formulas (5) and (6) that an arbitrary function $f(t) \in L^2_\sigma(-\infty, \infty)$ belongs to the domain D_Q if and only if $f \in D_A$. We see also that the application of the operator A to the vector f corresponds to the multiplication of the function $f(t)$ by t.

By the theorem just proved, each self-adjoint operator with a simple spectrum in a separable space is isomorphic to the operator of multiplication by the independent variable in $L^2_\sigma(-\infty, \infty)$. The latter will be referred to as the *canonical form* of the operator A.

This isomorphism determined by

$$f = \int_{-\infty}^{\infty} f(t)\, dE_t g \tag{3}$$

implies that

$$E(\Delta) f = \int_{\Delta} f(t)\, dE_t g \tag{3'}$$

first for continuous functions $f(t)$ which are equal to zero outside of a finite interval, and then for arbitrary $f(t) \in L^2_\sigma(-\infty, \infty)$. Choosing in (3) distinct generating elements g, we obtain an entire class of multiplication operators which are isomorphic to A. In the following section we characterize all distribution functions $\sigma(t) = (E_t g, g)$.

Here we note one general property of all generating elements g: every point of growth of the spectral function E_t is a point of growth of the function $(E_t g, g)$. The validity of this assertion follows from the fact that if Δ_0 is an interval of constancy of the function $(E_t g, g)$, then the linear manifold consisting of all the vectors $E(\Delta) g$, where Δ runs through the set of all real intervals, is orthogonal to the subspace $E(\Delta_0)$ H.

Let us recall the criterion given at the end of the previous section for a point λ to belong to the spectrum of a self-adjoint operator A which has a simple spectrum. In view of our present considerations, it is sufficient to verify this criterion for any one generating element.

We complete the present section with a proposition concerning self-adjoint operators which is the analogue of Theorem 1.

THEOREM 3: *If a self-adjoint operator A has a simple spectrum then there exists a vector h such that $A^k h$ is defined for $k = 0, 1, 2, \ldots$ and such that the linear envelope of this set of vectors is dense in* H. *Conversely, if there exists such a vector h, then the spectrum of the operator A is simple and the vector h is a generating element.*

Proof: The proof of the second part is very simple. Indeed, let the linear envelope of the set $A^k h$ $(k = 0, 1, 2, \ldots)$ be dense in H. The integral representation

$$(A^k h, f) = \int_{-\infty}^{\infty} t^k d(E_t h, f)$$

shows that there does not exist a vector $f \neq 0$ which is orthogonal to $E_t h$

for all t. Hence the linear envelope of the set $E(\Delta)h$ is dense in H, so that the spectrum of the operator is simple and h is a generating element.

Turning to the first part of the theorem, we prove that the vector h can be taken as

$$h = \int_{-\infty}^{\infty} e^{-t^2} dE_t g,$$

where g is some generating element of the operator. In this case,

$$A^k h = \int_{-\infty}^{\infty} t^k e^{-t^2} dE_t g \qquad (k = 0, 1, 2, \ldots).$$

If there exists a vector $f \neq 0$ which is orthogonal to all the vectors $A^k h$, then

$$\int_{-\infty}^{\infty} t^k e^{-t^2} f(t)\, d\sigma(t) = 0 \qquad (k = 0, 1, 2, \ldots),$$

where $\sigma(t) = (E_t g, g)$ and

$$0 \neq \int_{-\infty}^{\infty} |f(t)|^2 d\sigma(t) < \infty. \tag{7}$$

Letting

$$\omega(t) = \int_{-\infty}^{t} f(s)\, d\sigma(s) - C, \tag{8}$$

with a suitable choice of the constant C, we find that

$$\int_{-\infty}^{\infty} t^k e^{-t^2} \omega'(t)\, dt = 0 \quad (k = 0, 1, 2, \ldots).$$

Since the system of Tchebysheff-Hermite functions is complete, it follows that $\omega(t) = 0$, which by (7) and (8) reduces to the contradiction

$$0 \neq \int_{-\infty}^{\infty} |f(t)|^2 d\,\sigma(t) = \int_{-\infty}^{\infty} \overline{f(t)}\, d\omega(t) = 0.$$

Orthogonalizing the sequence of vectors $\{A^k h\}_0^\infty$, we get an orthonormal basis $\{e_k\}_0^\infty$ all elements of which belong to D_A. Obviously, for $i > k + 1$,

$$(Ae_k, e_i) = 0,$$

and, since the operator A is symmetric, this equation holds also for $i < k - 1$.

Therefore, the matrix $((Ae_k, e_i))_{i,k=0}^{\infty}$ of the operator A relative to the basis $\{e_k\}_0^{\infty}$ is a Jacobi matrix. It is necessary to note that, in general, the constructed basis is not the basis of the matrix representative of A in the sense of Section 47.

70. Spectral Types

A *distribution function* is any left continuous nondecreasing function of bounded variation which is defined on the whole real axis. If $\sigma(t)$ is such a function, then $\sigma(\Delta) = \sigma(t'') - \sigma(t')$ (t' and t'' are the end points of Δ) is an *additive interval function*. (We also use the name distribution function for $\sigma(\Delta)$.)

Let us agree to say that a distribution function $\sigma(t)$ is *inferior*[10] to the distribution function $\rho(t)$ and write

$$\sigma(t) \prec \rho(t) \tag{1}$$

if $\sigma(t)$ is absolutely continuous with respect to $\rho(t)$, i.e., if for any $\Delta \subset [-\infty, \infty]$

$$\sigma(\Delta) = \int_{\Delta} \varphi(t)\, d\rho(t),$$

where $\varphi(t)$ is a ρ-measurable non-negative function.

If simultaneously with (1)

$$\rho(t) \prec \sigma(t) \tag{2}$$

holds, then the distribution functions $\sigma(t)$ and $\rho(t)$ are said to be of the same *spectral type*. If (1) holds but (2) does not, then we say that the spectral type of $\sigma(t)$ is less than the spectral type of $\rho(t)$.

Now let A be an arbitrary self-adjoint operator and E_t be its spectral function. The function $(E_t f, f)$, it is evident, is a distribution function in t for each fixed $f \in \mathrm{H}$. We call the spectral type of the function $(E_t f, f)$ the *spectral type of the element* f (with respect to the operator A). We say that the spectral types of the elements f belong to A. If among the elements $f \in \mathrm{H}$ there exists an element g of maximal spectral type with respect to A (i.e., an element g such that $(E_t f, f) \prec (E_t g, g)$ for all $f \in \mathrm{H}$), then we ascribe this spectral type to the operator A.

If A is a self-adjoint operator with a simple spectrum, then there exist elements of *maximal spectral type* with respect to A. This follows from the next theorem which, at the same time, gives the answer to the question of

[10] Translator's Note: Inferior has the meaning of "less than" in the partial ordering defined by the concept of absolute continuity.

the preceding section about the characterization of the set of generating elements of an operator with a simple spectrum.

THEOREM 1: *Let A be a self-adjoint operator with a simple spectrum. In order that an element g be a generating element it is necessary and sufficient that g is of maximal spectral type with respect to A.*

Proof: Let g be a generating element and f an arbitrary vector of H. In this case

$$f = \int_{-\infty}^{\infty} f(t)\, dE_t g,$$

where $f(t)$ is some function of $L^2_\sigma(-\infty, \infty)$ and $\sigma(t) = (E_t g, g)$. Furthermore, for each interval $\Delta \subset [-\infty, \infty]$,

$$E(\Delta) f = \int_{\Delta} f(t)\, dE_t g = \int_{-\infty}^{\infty} \chi_\Delta(t) f(t)\, dE_t g.$$

Thus, under the mapping of H into $L^2_\sigma(-\infty, \infty)$ determined by g, the vectors f and $E(\Delta)f$ correspond to the functions $f(t)$ and $\chi_\Delta(t) f(t)$. Since the mapping is an isometry,

$$(E(\Delta)f, f) = \int_{-\infty}^{\infty} |f(t)|^2 \chi_\Delta(t)\, d\sigma(t) = \int_{\Delta} |f(t)|^2 d\sigma(t).$$

Therefore,

$$(E_t f, f) \prec (E_t g, g),$$

so that the necessity is proved.

Let us assume now that g is an element of maximal spectral type, so that, for each $f \in$ H,

$$(E_t f, f) \prec (E_t g, g).$$

In particular, this relation holds for the generating element g_0, (which the operator A possesses since its spectrum is simple), i.e.,

$$(E_t g_0, g_0) \prec (E_t g, g).$$

On the other hand, by the first part of the theorem,

$$(E_t g, g) \prec (E_t g_0, g_0).$$

Thus, the elements g and g_0 are of the same spectral type, which implies that

$$\sigma(\Delta) \equiv (E(\Delta)\, g, g) = \int_{\Delta} p^2(t)\, d\sigma_0(t),$$

$$\sigma_0(\Delta) \equiv (E(\Delta)\, g_0, g_0) = \int_{\Delta} p_0^2(t)\, d\sigma(t),$$

where the functions $p(t) \geqq 0$ and $p_0(t) \geqq 0$ belong respectively to $L^2_{\sigma_0}(-\infty, \infty)$ and $L^2_\sigma(-\infty, \infty)$. These equations imply that the sets of zero σ-measure and zero σ_0-measure coincide. In addition, it is not difficult to see that

$$p(t)\, p_0(t) = 1$$

everywhere except a set of zero σ-measure. Therefore, each of the functions $p(t)$ and $p_0(t)$ can equal zero only on a set of zero σ-measure.

We must prove that g is a generating element for the operator A. In other words, we must show that the equation

$$(E(\Delta)\, g, h) = 0$$

is satisfied for every $\Delta \subset [-\infty, \infty]$ only when $h = 0$. But, by the isomorphism between H and $L^2_{\sigma_0}(-\infty, \infty)$,

$$(E(\Delta)\, g, h) = \int_\Delta g(t)\, \overline{h(t)}\, d\sigma_0(t),$$

where $g(t), h(t) \in L^2_{\sigma_0}(-\infty, \infty)$ and $|\, g(t)\, | = p(t)$. Hence, if g is not a generating element then, for some function $h(t) \in L^2_{\sigma_0}(-\infty, \infty)$ and for arbitrary $\Delta \subset [-\infty, \infty]$

$$\int_\Delta g(t)\, \overline{h(t)}\, d\sigma_0(t) = 0,$$

or, equivalently,

$$\int_\Delta p(t)\, \overline{h_0(t)}\, d\sigma_0(t) = 0.$$

This is impossible, since the linear envelope of the set of functions $\chi_\Delta(t)\, p(t)$, as we showed in the previous section, is dense in $L^2_{\sigma_0}(-\infty, \infty)$, since the function $p(t) \geqq 0$ can be zero only on a set of zero σ_0-measure.

In connection with the theorem just proved the question arises: is it possible to represent, in the form $(E_t f, f)$, every distribution function of a type which does not exceed the spectral type of A. An affirmative answer is given to this question in the following theorem.

THEOREM 2: *Let A be a self-adjoint operator with a simple spectrum and $\sigma(t)$ a given distribution function of a type which does not exceed the spectral type of A. Then there exists a vector $f \in$ H which this distribution function generates in the sense that*

$$\sigma(t) = (E_t f, f).$$

Proof: From the conditions of the theorem it follows that the distribution function $\sigma(t)$ can be represented in the form

$$\sigma(\Delta) = \int_{\Delta} \varphi(t)\, d(E_t g, g),$$

where g is a generating element and $\varphi(t) \geqq 0$. Let

$$f = \int_{-\infty}^{\infty} f(t)\, dE_t g = \int_{-\infty}^{\infty} \sqrt{\varphi(t)}\, dE_t g.$$

The element f generates the distribution function $\sigma(t)$, since

$$(E(\Delta)f, f) = \int_{\Delta} |f(t)|^2 d(E_t g, g) = \sigma(\Delta).$$

71. The Multiple Spectrum

If the linear envelope of the set of all eigenvectors of a self-adjoint operator is dense in the space, then it is natural to define the multiplicity (total multiplicity) of the spectrum of this operator as the maximal multiplicity of its eigenvalues.

Before we give the definition of the multiplicity of the spectrum of an arbitrary self-adjoint operator in H, we introduce the concept of the generating subspace. A subspace G is called a *generating subspace* of the self-adjoint operator A with the spectral function E_t if the linear envelope of the union of the sets $E(\Delta)$ G, where Δ runs through the set of all intervals, is dense in H.

DEFINITION 1: *The multiplicity* (*or total multiplicity*) *of the spectrum of a self-adjoint*[11] *operator A is the minimum dimension of the generating subspaces of this operator if A has a finite-dimensional generating subspace; otherwise, the multiplicity of the spectrum of A is infinite. The multiplicity of the operator A in the interval $\Delta_0 = [t', t'']$ is defined as the total multiplicity of the spectrum of the operator $E(\Delta_0) A$.*

It is easy to see that if G is a generating subspace of the operator $E(\Delta_0) A$, then for $\Delta_1 \subset \Delta_0$, the subspace $E(\Delta_1)$ G is a generating subspace of the operator $E(\Delta_1) A$. Therefore, for $\Delta_1 \subset \Delta_0$, the multiplicity of the operator A in the interval Δ_1 does not exceed its multiplicity in the interval Δ_0.

Hence, we are justified in making the next definition.

DEFINITION 2: *The multiplicity of the spectrum of a self-adjoint operator*

[11] We omit the analogous definition for unitary operators. It is evident that the multiplicities of the spectra of a self-adjoint operator and of its Cayley transform coincide.

A at a point λ is the limit as $n \to \infty$ of the (monotone decreasing) sequence of the multiplicities of the spectrum of A in the intervals $\left[\lambda-\frac{1}{n}, \lambda+\frac{1}{n}\right]$.

It is not difficult to verify that, if λ is an isolated point of the spectrum, then Definition 2 coincides with the definition given earlier for the multiplicity of an eigenvalue. If the linear envelope of the set of all eigenvectors is dense in the space, then Definition 1 gives as the total multiplicity of the spectrum the maximal multiplicity of the eigenvalues.

It follows from Definition 1 that in a nonseparable space the total multiplicity of the spectrum of a self-adjoint operator is necessarily infinite.

In conclusion, we prove a lemma which will be used in the following section.

LEMMA: *Let a self-adjoint operator A have an n-fold spectrum and a generating subspace* G. *If the linear operator $A_0 \subset A$ is defined only on the linear envelope of the sets $E(\varDelta)$* G, *where $\varDelta$ runs through the set of all finite intervals, then the operator A is the closure of the operator A_0.*

Proof: Since the relation $\bar{A}_0 \subset A$ is evident, we must show only that $A \subset \bar{A}_0$. With this aim we choose an arbitrary vector $f \in \mathrm{D}_A$ and let

$$f_k = E(\varDelta_k)f \qquad (k = 1, 2, 3, \ldots),$$

where $\varDelta_k = [-k, k]$. Since G is a generating subspace, the linear envelope of the sets $E(\varDelta')$ G, where $\varDelta'$ runs through all subintervals of the finite interval $\varDelta$, coincides with $E(\varDelta)$ H. Therefore, the vectors f_k belong to the domain of A_0 and

$$A_0 f_k = AE(\varDelta_k)f = E(\varDelta_k)\, Af.$$

Let $k \to \infty$ to obtain

$$E(\varDelta_k)f \to f, \; E(\varDelta_k)\, Af \to Af.$$

Therefore, the vector f belongs to the domain of $\bar{A}_0$ and

$$A_0 f = Af,$$

so that the relation

$$A \subset \bar{A}_0$$

is proved.

72. The Canonical Form of a Self-Adjoint Operator with a Spectrum of Finite Multiplicity

In the present section we sketch a generalization of Theorem 2 of Section 69 for the case of an operator with a spectrum of finite multiplicity. With this aim we proceed to define the space $L^2_S(-\infty, \infty)$ of vector-valued functions, which is a generalization of the function space $L^2_\sigma(-\infty, \infty)$.

Let

$$S(t) = (\sigma_{ik}(t))_{i,k=1}^{m} \quad (m < \infty; \; -\infty < t < \infty)$$

be a Hermitean matrix function which satisfies the conditions:

1° *for arbitrary complex* ξ_k,

$$\sum_{i,k=1}^{m} \{\sigma_{ik}(t'') - \sigma_{ik}(t')\} \xi_i \bar{\xi}_k \geqq 0 \qquad (t' \leqq t''),$$

2° $$S(-\infty) = 0, \; S(t-0) = S(t).$$

A matrix function $S(t)$ *defined by conditions* 1° *and* 2° *we call a matrix distribution function and we write* $S(\Delta)$ *in place of* $S(t'') - S(t')$ *if* $\Delta = [t', t'']$.

From condition 1° it is easy to conclude that $\sigma_{ii}(t)$ $(i=1, 2, \ldots m)$ are nondecreasing functions of t and that the functions $\sigma_{ik}(t)$ $(i, k=1, 2, \ldots, m)$ are of bounded variation in every finite interval.

In the construction of the space $L_S^2(-\infty, \infty)$ we start with the set of all vector functions of the form

$$\overrightarrow{\chi(t)} = \{\chi_1(t), \chi_2(t), \ldots, \chi_m(t)\},$$

where $\chi_k(t)$ is the characteristic function of the interval $\Delta_k (k=1, 2, \ldots m)$. Such vector functions will be called *characteristic vector functions.* We form the linear envelope R of the set of all characteristic vector functions, letting

$$\alpha_1\overrightarrow{\chi^{(1)}(t)} + \alpha_2\overrightarrow{\chi^{(2)}(t)} = \{\alpha_1\chi_1^{(1)}(t) + \alpha_2\chi_1^{(2)}(t), \alpha_1\chi_2^{(1)}(t) + \alpha_2\chi_2^{(2)}(t), \ldots\}.$$

In terms of the matrix distribution function $S(t)$, we define a scalar product. First, let

$$(\overrightarrow{\chi^{(1)}(t)}, \overrightarrow{\chi^{(2)}(t)}) = \sigma_{ik}(\Delta_i^{(1)} \cap \Delta_k^{(2)}),$$

for vector functions of the form

$$\overrightarrow{\chi^{(1)}(t)} = \{0, \ldots, 0, \chi_i^{(1)}(t), 0, \ldots, 0\},$$

$$\overrightarrow{\chi^{(2)}(t)} = \{0, \ldots, 0, \chi_k^{(2)}(t), 0, \ldots, 0\},$$

where $\chi_i^{(1)}(t)$ and $\chi_k^{(2)}(t)$ are the characteristic functions of the intervals $\Delta_i^{(1)}$ and $\Delta_k^{(2)}$ and then, by linearity, extend the definition of the scalar product to the entire system R.[12]

---

[12] Translator's Note: The closure of R in the norm introduced is $L_S^2(-\infty, \infty)$.

The construction described does not give an analytic characterization of the space $L_S^2(-\infty, \infty)$. Moreover, this construction does not give us a justification for calling $L_S^2(-\infty, \infty)$ a space of vector functions, since those of its elements which do not belong to R are defined only as limits in norm and cannot be interpreted as vector functions on the basis of their definitions.

A functional analytic characterization of the elements of the space $L_S^2(-\infty, \infty)$ can be given,[13] in which the distribution function

$$\nu(t) = \sum_{i=1}^{m} \sigma_{ii}(t)$$

appears. There it is shown that the space $L_S^2(-\infty, \infty)$ consists of all vector functions

$$\overrightarrow{f(t)} = \{f_1(t), \ldots, f_m(t)\}$$

with components which are measurable and finite almost everywhere with respect to the measure induced by $\nu(t)$, and such that

$$\lim_{N\to\infty} (\overrightarrow{f_N(t)}, \overrightarrow{f_N(t)}) < \infty,$$

where

$$f_N(t) = \begin{cases} \overrightarrow{f(t)} & \text{if } \max\limits_{1 \leq i \leq m} |f_i(t)| < N, \\ 0 & \text{otherwise.} \end{cases}$$

However, for our purpose the functional analytic characterization of the space $L_S^2(-\infty, \infty)$ is not necessary. We remark only that our construction of $L_S^2(-\infty, \infty)$ reveals that this space contains the vector functions with continuous components which are zero outside of a finite interval. Therefore, we can define on the manifold of these vector functions

$$\overrightarrow{f(t)} = \{f_1(t), f_2(t), \ldots, f_m(t)\}$$

an operator Q_0 by the equation

$$Q_0\overrightarrow{f(t)} = \{tf_1(t), tf_2(t), \ldots, tf_m(t)\}.$$

The closure Q of the symmetric operator Q_0 we call the operator of multiplication by the variable t in the space $L_S^2(-\infty, \infty)$. With the aid of the remark at the end of Section 67 it is easy to establish that Q is a self-adjoint operator. Furthermore, Theorem 3 of Section 66 enables us

---

[13] cf., I. Kaz [1], Vol. I.

to obtain the spectral function E_t of the operator Q. On the vector functions

$$\overrightarrow{f(t)} = \{f_1(t), f_2(t), \ldots, f_m(t)\}$$

with continuous components which are equal to zero outside of a finite interval, the spectral function E_t is defined by the equation

$$E(\Delta)\overrightarrow{f(t)} = \{\chi_\Delta(t)f_1(t), \chi_\Delta(t)f_2(t), \ldots, \chi_\Delta(t)f_m(t)\},$$

where $\chi_\Delta(t)$ is the characteristic function of the interval Δ.

Repeating the argument of Section 69, we find that the multiplicity of the spectrum of the operator Q in $L_S^2(-\infty, \infty)$ does not exceed the order m of the matrix $S(t)$.

Now let A be a self-adjoint operator with a spectrum of multiplicity n ($n < \infty$) and the spectral function E_t. Let us agree to call every basis of any generating subspace of the operator A a *generating basis* of the operator. If $g_1, g_2, \ldots, g_m (m \geqq n)$ is a generating basis of the operator A then we denote the closure of the linear envelope of the set of all vectors $E(\Delta)g_i$, for fixed i, by G_i.

A basis of an n-dimensional generating subspace we call a *minimal generating basis*. It is possible to select a minimal generating basis so that the corresponding subspaces $G_i (i = 1, 2, \ldots, n)$ are pairwise orthogonal. To obtain such a minimal basis $g_1, g_2, \ldots, g_n$ we start from an arbitrary basis $h_1, h_2, \ldots, h_n$ and, letting $g_1 = h_1$, we generate the subspace G_1. It is evident that h_2 is not in G_1 (since otherwise the multiplicity would be less than n) and, hence, $(E - P_{G_1})h_2 \neq 0$. Letting $g_2 = (E - P_{G_1})h_2$, we generate the subspace G_2. It is not difficult to see that $G_2 \perp G_1$. We continue this process in order to obtain the desired basis.

We do not presuppose that the generating basis $g_1, g_2, \ldots, g_m$ $(m \geqq n)$ is chosen such that the corresponding subspaces $G_1, G_2, \ldots, G_m$ are pairwise orthogonal, and we restrict ourselves only to the requirement that these subspaces be linearly independent. In what follows, when speaking of a generating basis, we shall assume always that this last condition is satisfied.

It is easy to see that for any generating basis $g_1, g_2, \ldots, g_m$ $(n \leqq m < \infty)$, the matrix function

$$S(t) = ((E_t g_i, g_k))_{i,k=1}^m$$

is a matrix distribution function. Thus, each generating basis defines some matrix distribution function.

Now it is possible to formulate the theorem about the canonical form of an operator with a spectrum of finite multiplicity.

THEOREM: *Let A be a self-adjoint operator in* H *with a multiple spectrum of multiplicity n. Let $g_1, g_2, \ldots, g_m$ $(n \leqq m < \infty)$ be an arbitrary generating basis and $S(t) = ((E_t g_i, g_k))_{i,k=1}^m$. Then there exists an isometric mapping of the space* H *into $L_S^2(-\infty, \infty)$ with the following properties. The domain $D_A \subset$ H of A is mapped into the domain $D_Q \subset L_S^2(-\infty, \infty)$ of the multiplication operator Q. If the element $f \in$ H maps into the vector function $\overrightarrow{f(t)} \in L_S^2(-\infty, \infty)$, then the element Af maps into the vector function $Q\overrightarrow{f(t)}$.*

Proof: We associate with the vector

$$f = E(\Delta_1) g_1 + E(\Delta_2) g_2 + \ldots + E(\Delta_m) g_m \tag{1}$$

of H the characteristic vector function

$$\overrightarrow{f(t)} = \{\chi_{\Delta_1}(t), \chi_{\Delta_2}(t), \ldots, \chi_{\Delta_m}(t)\} \tag{2}$$

of $L_S^2(-\infty, \infty)$. Then

$$\|f\| = (f, f) = \left(\sum_{i=1}^m E(\Delta_i) g_i, \sum_{k=1}^m E(\Delta_k) g_k \right) =$$

$$= \sum_{i,k=1}^m (E(\Delta_i \cap \Delta_k) g_i, g_k) = \sum_{i,k=1}^m \sigma_{ik}(\Delta_i \cap \Delta_k) = \|\overrightarrow{f(t)}\|^2.$$

It is easy also to verify that to orthogonal vectors of the form (1) correspond orthogonal vector functions of the form (2). The extension of this correspondence by linearity and continuity to the whole space H is an isometric mapping of the space H into $L_S^2(-\infty, \infty)$. It is not difficult to verify also that for continuous vector functions $\overrightarrow{f(t)} = \{f_1(t), \ldots, f_m(t)\}$, with components equal to zero outside of a finite interval, formula (3) of Section 69 now takes the form

$$f = \sum_{i=1}^m \int_{-\infty}^{\infty} f_i(t)\, dE_t g_i.$$

We find next the vector function which corresponds to the vector Af, where f is a vector of the form (1). We have

$$Af = \int_{-\infty}^{\infty} t dE_t f = \int_{-\infty}^{\infty} t dE_t \left(\sum_{i=1}^m E(\Delta_i) g_i \right) = \sum_{i=1}^m \int_{\Delta_i} t dE_t g_i = \sum_{i=1}^m \int_{-\infty}^{\infty} t \chi_i(t)\, dE_t g_i.$$

If one takes into account the method of construction of the multiplication operator Q and the lemma of the preceding section, then the theorem on the canonical form is proved.

In Section 69 we showed that for every self-adjoint operator with a simple spectrum there exists an orthogonal basis with respect to which the matrix of the operator is a Jacobi matrix.

Completing the present section we show that this result generalizes to self-adjoint operators with spectra of finite multiplicity. Let A be such an operator. We select a generating basis $g_1, g_2, \dots g_n$ such that

$$\mathrm{H} = \mathrm{G}_1 \oplus \mathrm{G}_2 \oplus \mathrm{G}_3 \oplus \dots \oplus \mathrm{G}_n \quad (\mathrm{G}_i \perp \mathrm{G}_k,\ i \neq k).$$

Repeating the argument of Section 69, we get in each G_r a corresponding Jacobi matrix

$$((Ae_{ri}, e_{rk})).$$

If we select in H a particular orthonormal basis $e_s (s=1, 2, 3, \dots)$, numbering the unit-vectors $e_{rk}(r=1, 2, 3, \dots n;\ k=1, 2, 3, \dots)$ first in the order of increasing first index and then in the order of increasing second index, i.e.,

$$e_1 = e_{11},\ e_2 = e_{21},\ \dots,\ n_n = e_{n1}\ ;\ e_{n+1} = e_{12},\ e_{n+2} = e_{22},\ \dots$$

then, with respect to the basis $\{e_s\}_{s=1}^{\infty}$ we have a generalized Jacobi matrix $((Ae_i, e_k))$. It will be of a special type since its only elements different from zero lie only on the principal diagonal and on the pair of nth neighboring diagonals.

73. Some Remarks about Unitary Invariants of Self-Adjoint Operators

Two operators A_1 and A_2 which act respectively in the Hilbert spaces H_1 *and* H_2 *are called*[14] *isomorphic (or unitarily equivalent) if there exists an isometric mapping V of the space* H_1 *into* H_2 *such that*

$$\mathrm{D}_{A_2} = V\mathrm{D}_{A_1}, \tag{1}$$

and

$$A_2 = VA_1V^{-1}. \tag{2}$$

If the operator A_1 is symmetric, then the operator A_2 isomorphic to it is also symmetric. This follows immediately from (1) and (2).

The spectra of a pair of isometric operators coincide, since (1) and (2) yield

$$\begin{aligned}\Delta_{A_2 - \lambda E_2} &= (A_2 - \lambda E_2)\mathrm{D}_{A_2} = (VA_1V^{-1} - \lambda VV^{-1})\mathrm{D}_{A_2} = \\ &= V(A_1 - \lambda E_1)\mathrm{D}_{A_1} = V\Delta_{A_1 - \lambda E_1}.\end{aligned} \tag{3}$$

Moreover, it follows from the last equation that the total spectrum of A_1 and each of its parts (the point spectrum and the continuous spectrum) are *unitary invariants*, i.e., they do not change under the isomorphic transition

[14] We repeat here the definition which was given in Section 36.

from A_1 to A_2. Finally, it follows from the fact that the operator A_1 is self-adjoint that the operator A_2, isomorphic to A_1, is self-adjoint.

We limit ourselves here to some remarks concerning unitarily equivalent self-adjoint operators. Let E_{1t} be the resolution of the identity of the operator A_1. Let

$$E_{2t} = VE_{1t}V^{-1}. \tag{4}$$

This formula defines a family of bounded self-adjoint operators in H_2 and it is easy to see that E_{2t} is a resolution of the identity in the space H_2. We verify, for example, that

$$E_{2u}E_{2v} = E_{2s},$$

where $s = \min\{u, v\}$. Hence,

$$E_{2u}E_{2v} = VE_{1u}V^{-1}VE_{1v}V^{-1} = VE_{1u}E_{1v}V^{-1} = VE_{1s}V^{-1} = E_{2s}.$$

We shall prove that E_{2t} is the decomposition of the identity of the operator A_2. With this aim, we consider the integral representation

$$(A_1f_1, g_1) = \int_{-\infty}^{\infty} td(E_{1t}f_1, g_1).$$

Letting

$$f_1 = V^{-1}f_2, \; g_1 = V^{-1}g_2,$$

we get

$$(VA_1V^{-1}f_2, g_2) = \int_{-\infty}^{\infty} td(VE_{1t}V^{-1}f_2, g_2)$$

or

$$(A_2f_2, g_2) = \int_{-\infty}^{\infty} td(E_{2t}f_2, g_2).$$

This is essentially what we set out to prove.

From relation (4) it follows that the multiplicity of the spectrum (the total multiplicity or multiplicity at a point) also is a unitary invariant of the self-adjoint operator.

If the linear envelope of the sequence of all eigenvectors of a self-adjoint operator A is dense in H, then the spectrum of A and the multiplicity of the spectrum at each point represent a complete system of unitary invariants of A, i.e., every other operator with identical spectrum and identical multiplicity of the spectrum at each point is isomorphic to A.

For self-adjoint operators with arbitrary spectra, the problem of finding a complete system of unitary invariants is extremely complicated, even in a separable space.

We show next how to solve this problem in the case of an arbitrary self-adjoint operator with a simple spectrum. Let A_1 and A_2 be two unitarily equivalent self-adjoint operators with simple spectra acting in the spaces H_1 and H_2, respectively. From the equation

$$(E_{2t}f_2, f_2) = (E_{1t}V^{-1}f_2, V^{-1}f_2) = (E_{1t}f_1, f_1),$$

it follows that the spectral types of the element f_1 with respect to A_1 and of the element $f_2 = Vf_1$ with respect to A_2 coincide. Therefore, the operators A_1 and A_2 are of the same spectral type.

The spectrum and multiplicity of the spectrum at each point do not represent complete systems of unitary invariants if the operator A has a continuous spectrum, as the following example indicates. The multiplication operator Q_{σ_1} in $L^2_{\sigma_1}(0, 1)$ and Q_{σ_2} in $L^2_{\sigma_2}(0, 1)$, where $\sigma_1(t) = t$ and $\sigma_2(t) = t^2$, have total spectra each of multiplicity one. But they are not isomorphic, since the spectral types of the distribution functions $\sigma_1(t)$ and $\sigma_2(t)$ do not coincide.

On the other hand, two operators with simple spectra and identical spectral types are isomorphic since, by Sections 69 and 70, they are both isomorphic to the same multiplication operator. Thus, the following theorem holds.

THEOREM: *In order that two operators with simple spectra be isomorphic it is necessary and sufficient that their spectral types coincide.*

For the extension from the case of simple spectra to the general case of multiple spectra, the determination of a complete system of unitary invariants is extremely complicated. The question of a complete system of unitary invariants of a self-adjoint operator with a multiple spectrum cannot be decided easily by a decomposition of such operators into orthogonal sums of operators with simple spectra, because such a decomposition is not defined uniquely. For the solution of the problem in the general case, one must choose the decomposition of the operator into an orthogonal sum in a special manner. This requires the introduction of the notion of the so-called *independent spectral types.*

The theory of unitary invariants of self-adjoint operators was developed by Hellinger for the case of a separable space and was developed later by A. I. Plesner for the case of a nonseparable space. The examination of these topics is beyond the scope of the present book. We recommend to the reader who desires to become acquainted with the theory of invariants the paper of A. I. Plesner and V. A. Rochlin in Vol. II of the journal *Uspekhi Math. Nauk* where this theory is presented with exhaustive completeness.

74. Some Remarks about Functions of Self-Adjoint Operators

In Section 63 we constructed the integral

$$\int_{-\infty}^{\infty} \varphi(t)\, dE_t f = Tf \tag{1}$$

by starting from a given resolution of the identity $E_t(-\infty \leqq t \leqq \infty)$, an element $f \in \mathrm{H}$, and a function $\varphi(t)$ which belongs to $L^2_\sigma(-\infty, \infty)$, where $\sigma(t) = (E_t f, f)$. For a fixed function $\varphi(t)$, this integral defines an operator with domain D_T which consists of all vectors $f \in \mathrm{H}$ such that $\varphi(t)$ is measurable with respect to the distribution function $\sigma(t) = (E_t f, f)$ and

$$\int_{-\infty}^{\infty} |\varphi(t)|^2 d\sigma(t) < \infty.$$

We now consider the self-adjoint operator A to which the resolution of the identity E_t corresponds. If we take as the function $\varphi(t)$ some positive integer power of the variable t, then the operator T defined by formula (1) coincides[15] with the same power of the operator A. Therefore, if for $\varphi(t)$ we take the polynomial

$$p(t) = a_0 + a_1 t + \dots + a_n t^n,$$

then the operator integral

$$\int_{-\infty}^{\infty} p(t)\, dE_t f$$

becomes

$$(a_0 E + a_1 A + \dots + a_n A^n) f = p(A) f.$$

We see that a polynomial in a self-adjoint operator A, which has a direct definition, can also be defined as an operator integral with respect to the spectral function E_t of A. The latter method, i.e., the definition in terms of an integral, we can apply also in cases when an immediate definition of the function of an operator is not possible.

DEFINITION: *If $\varphi(t)$ is an arbitrary complex function defined on the whole real axis, then $\varphi(A)$ is the operator defined by the formula*

$$\varphi(A) f = \int_{-\infty}^{\infty} \varphi(t)\, dE_t f$$

---

[15] Cf. the end of Section 66.

for all vectors $f \in \mathrm{H}$ for which $\varphi(t)$ is measurable with respect to the distribution function $(E_t f, f)$ and for which

$$\int_{-\infty}^{\infty} |\varphi(t)|^2 d(E_t f, f) < \infty.$$

We remark that the values of the function $\varphi(t)$ at regular points of the operator A do not affect the definition of the operator $\varphi(A)$. Therefore, we can assume that the function $\varphi(t)$ is defined only on the spectrum of the operator A.

For the construction of a productive theory of functions of operators it is expedient to introduce a somewhat restricted class of admissible functions $\varphi(t)$. We limit our considerations to functions $\varphi(t)$ which are measurable with respect to the distribution function $(E_t f, f)$ for any $f \in \mathrm{H}$. We denote this class of functions by $\mathbf{K}_0$.

Thus, let $\varphi(t) \in \mathbf{K}_0$. We prove first that the operator $\varphi(A)$ is linear. Evidently, it is sufficient to prove that if $f \in \mathrm{D}_{\varphi(A)}$ and $g \in \mathrm{D}_{\varphi(A)}$, then $f + g \in \mathrm{D}_{\varphi(A)}$. But, for arbitrary $h_1, h_2 \in \mathrm{H}$,

$$\| h_1 + h_2 \|^2 + \| h_1 - h_2 \|^2 = 2 \| h_1 \|^2 + 2 \| h_2 \|^2.$$

Therefore, for any $f, g \in \mathrm{H}$ and any $\Delta \subset [-\infty, \infty]$

$$(E(\Delta)\{f+g\}, f+g) \leqq 2(E(\Delta)f, f) + 2(E(\Delta)g, g),$$

and this implies that

$$\int_{-\infty}^{\infty} |\varphi(t)|^2 d(E_t\{f+g\}, f+g) \leqq 2\int_{-\infty}^{\infty} |\varphi(t)|^2 d(E_t f, f) + 2\int_{-\infty}^{\infty} |\varphi(t)|^2 d(E_t g, g).$$

This proves our assertion.

THEOREM: *Let $\varphi(t) \in \mathbf{K}_0$ and*

$$\varphi(A) = \int_{-\infty}^{\infty} \varphi(t)\, dE_t.$$

Then, for each $f \in \mathrm{D}_{\varphi(A)}$ and each $h \in \mathrm{H}$,

$$(\varphi(A) f, h) = \int_{-\infty}^{\infty} \varphi(t)\, d(E_t f, h).$$

Proof: First, we notice that for the functions $\varphi(t)$ in the class $\mathbf{K}$, introduced in Section 63, the theorem was proved in Section 63. Using this, it is easy to establish the theorem for any bounded function in the class $\mathbf{K}_0$. Indeed, let $\varphi(t)$ be such a function. We fix f and h, and let

$$\rho(t) = \tfrac{1}{4}(E_t\{f+h\}, f+h) + \tfrac{1}{4}(E_t\{f-h\}, f-h) +$$
$$+ \tfrac{1}{4}(E_t\{f+ih\}, f+ih) + \tfrac{1}{4}(E_t\{f-ih\}, f-ih) = (E_t f, f) + (E_t h, h),$$

and introduce the space $L^2_\rho(-\infty, \infty)$ which is generated by the distribution function $\rho(t)$. Further, we construct a sequence $\{\psi_n(t)\}_1^\infty$ of functions in **K** which converge to $\varphi(t)$ in the metric of $L^2_\rho(-\infty, \infty)$. For each of the operators $\psi_n(A)$ the theorem is valid. Therefore,

$$(\psi_n(A) f, h) = \int_{-\infty}^{\infty} \psi_n(t)\, d(E_t f, h).$$

On the other hand,

$$\int_{-\infty}^{\infty} |\varphi(t) - \psi_n(t)|^2 d(E_t f, f) \leqq \int_{-\infty}^{\infty} |\varphi(t) - \psi_n(t)|^2 d\rho(t).$$

Hence, in view of the definition (cf. Section 63) of the integral

$$\int_{-\infty}^{\infty} \varphi(t)\, dE_t f$$

as the limit in the sense of strong convergence of

$$\int_{-\infty}^{\infty} \psi_n(t)\, dE_t f,$$

we conclude that

$$\lim_{n\to\infty} (\psi_n(A) f, h) = (\varphi(A) f, h).$$

It remains to prove that

$$\lim_{n\to\infty} \int_{-\infty}^{\infty} \psi_n(t)\, d(E_t f, h) = \int_{-\infty}^{\infty} \varphi(t)\, d(E_t f, h). \tag{2}$$

But

$$\int_{-\infty}^{\infty} \{\varphi(t) - \psi_n(t)\}\, d(E_t f, h) =$$
$$= \tfrac{1}{4} \int_{-\infty}^{\infty} \{\varphi(t) - \psi_n(t)\}\, d\{\sigma_1(t) - \sigma_2(t) + i\sigma_3(t) - i\sigma_4(t)\},$$

where

$$\sigma_1(t) = (E_t\{f+h\}, f+h), \quad \sigma_2(t) = (E_t\{f-h\}, f-h),$$
$$\sigma_3(t) = (E_t\{f+ih\}, f+ih), \quad \sigma_4(t) = (E_t\{f-ih\}, f-ih).$$

Hence,

$$\left| \int_{-\infty}^{\infty} \{\varphi(t) - \psi_n(t)\} d(E_t f, h) \right| \leqq \int_{-\infty}^{\infty} | \varphi(t) - \psi_n(t) | \, d\rho(t) \leqq$$

$$\leqq \sqrt{\int_{-\infty}^{\infty} d\rho(t)} \sqrt{\int_{-\infty}^{\infty} | \varphi(t) - \psi_n(t) |^2 d\rho(t)},$$

and the limit relation (2) is proved. Thus, the theorem is proved for bounded functions in $\mathbf{K}_0$.

We assume now that $\varphi(t)$ is an unbounded function in $\mathbf{K}_0$. Let $\sigma(t) = (E_t f, f)$, where f is an arbitrary element in $D_{\varphi(A)}$ such that

$$\int_{-\infty}^{\infty} | \varphi(t) |^2 d\sigma(t) < \infty.$$

Letting

$$\varphi_n(t) = \begin{cases} \varphi(t) & \text{for} \quad | \varphi(t) | \leqq n, \\ 0 & \text{for} \quad | \varphi(t) | > n, \end{cases}$$

we obtain

$$(\varphi_n(A) f, h) = \int_{-\infty}^{\infty} \varphi_n(t) \, d(E_t f, h). \tag{3}$$

Now, since $\varphi_n(t)$ converges to $\varphi(t)$ in the metric of $L^2_\sigma(-\infty, \infty)$ as $n \to \infty$,

$$\varphi_n(A) f \to \varphi(A) f.$$

Hence, the right side of formula (3) has a limit as $n \to \infty$. According to the definition of the integral of an unbounded function, this limit is the integral

$$\int_{-\infty}^{\infty} \varphi(t) \, d(E_t f, h).$$

Thus,

$$(\varphi(A) f, h) = \int_{-\infty}^{\infty} \varphi(t) \, d(E_t f, h),$$

and the proof of the theorem is complete.

For bounded functions which belong to the class $\mathbf{K}_0$, it is easy to see that if

$$\varphi(t) = \alpha_1 \varphi_1(t) + \alpha_2 \varphi_2(t),$$

then

$$\varphi(A) = \alpha_1\varphi_1(A) + \alpha_2\varphi_2(A),$$

where α_1 and α_2 are arbitrary numbers.

Furthermore, if $\varphi(t)$ is a bounded function in class $\mathbf{K}_0$, then $[\varphi(A)]^* = \bar{\varphi}(A)$, where $\bar{\varphi}$ denotes the function defined by $\bar{\varphi}(t) = \overline{\varphi(t)}$. Hence, for any $f, g \in \mathrm{H}$,

$$(4) \qquad (\varphi(A)f, g) = \int_{-\infty}^{\infty} \varphi(t)\, d(E_t f, g) = \overline{\int_{-\infty}^{\infty} \bar{\varphi}(t)\, d(E_t g, f)} =$$

$$= \overline{(\bar{\varphi}(A) g, f)} = (f, \bar{\varphi}(A) g).$$

It is not difficult to verify that if $\varphi_1(t)$ and $\varphi_2(t)$ are bounded functions in class $\mathbf{K}_0$ and

$$\varphi(t) = \varphi_1(t)\, \varphi_2(t),$$

then

$$\varphi(A) = \varphi_1(A)\, \varphi_2(A).$$

In particular, it follows that each pair of bounded functions of a self-adjoint operator commute.

Another consequence of the facts established above is formulated: if $\varphi(t) \in \mathbf{K}_0$, then $\varphi(A)$ is a projection operator if and only if $\varphi(t)$ assumes only the two values zero and one, i.e., it is the characteristic function of some set $e \subset [-\infty, \infty]$ which is measurable with respect to the distribution function $(E_t f, f)$ for every $f \in \mathrm{H}$.

If $\chi_\Delta(t)$ is the characteristic function of some interval $\Delta \subset [-\infty, \infty]$, then

$$\chi_\Delta(A) = \int_{-\infty}^{\infty} \chi_\Delta(t)\, dE_t = \int_{\Delta} dE_t = E(\Delta).$$

Therefore, it seems natural to define $E(e)$, for an arbitrary set $e \subset [-\infty, \infty]$, which is measurable with respect to all distribution functions $(E_t f, f)$, by the formula

$$E(e) = \chi_e(t),$$

where $\chi_e(t)$ is the characteristic function of the set e.

We prove now that the relation

$$[\varphi(A)]^* = \bar{\varphi}(A).$$

holds not only for bounded functions in $\mathbf{K}_0$ but also for unbounded functions $\varphi(t) \in \mathbf{K}_0$ if the domain of the operator $\varphi(A)$ is dense in H (otherwise the operator $\varphi(A)$ would not have an adjoint). Let $g \in \mathrm{D}_{[\varphi(A)]^*}$, so that, for every $f \in \mathrm{D}_{\varphi(A)}$,

$$(5) \qquad (\varphi(A)f, g) = (f, g^*).$$

Letting e_n denote the set of points on the real axis for which

$$|\varphi(t)| \leqq n,$$

we define

$$\varphi_n(t) = \begin{cases} \varphi(t) & (t \in e_n), \\ 0 & (t \notin e_n). \end{cases}$$

Since $\varphi_n(A)$ is bounded, we have, for every $h \in \mathrm{H}$,

$$(6) \qquad (\varphi_n(A)h, g) = (h, \overline{\varphi}_n(A)g).$$

We define the vector f by

$$f = E(e_n)h.$$

It is not difficult to convince oneself that this vector belongs to $\mathrm{D}_{\varphi(A)}$. Indeed

$$E_t f = E_t E(e_n) h = \int_{-\infty}^{t} \chi_{e_n}(s)\, dE_s h,$$

and, therefore,

$$\int_{-\infty}^{\infty} |\varphi(t)|^2 d(E_t f, f) = \int_{-\infty}^{\infty} |\varphi(t)|^2 \chi_{e_n}(t)\, d(E_t h, h) =$$

$$= \int_{-\infty}^{\infty} |\varphi_n(t)|^2 d(E_t h, h) < \infty.$$

By means of a similar calculation we find that, for each $h' \in \mathrm{H}$,

$$(\varphi(A)f, h') = \int_{-\infty}^{\infty} \varphi(t)\, d(E_t f, h') = \int_{-\infty}^{\infty} \varphi(t)\, \chi_{e_n}(t)\, d(E_t h, h') =$$

$$= \int_{-\infty}^{\infty} \varphi_n(t)\, d(E_t h, h') = (\varphi_n(A)h, h'),$$

so that

$$\varphi(A)f = \varphi_n(A)h.$$

On the basis of (5) and (6), we conclude that

$$(h, \overline{\varphi}_n(A)g) = (E(e_n)h, g^*),$$

and, consequently,

$$(h, \overline{\varphi}_n(A)g) = (h, E(e_n)g^*).$$

Since h is an arbitrary vector,

$$\bar{\varphi}_n(A)\, g = E\,(e_n)\, g^*.$$

Therefore,

$$\|\, \bar{\varphi}_n(A)\, g\,\| = \|\, E\,(e_n)\, g^*\| \leqq \|\, g^*\|,$$

and

$$\int_{-\infty}^{\infty} |\, \bar{\varphi}_n\,(t)\,|^2 d\,(E_t g, g) \leqq \|\, g^*\|^2.$$

As n increases, the left side does not decrease. This implies that

$$\int_{-\infty}^{\infty} |\, \bar{\varphi}\,(t)\,|^2 d\,(E_t g, g) \leqq \|\, g^*\|^2.$$

In other words, it is proved that if $g \in D_{[\varphi(A)]^*}$ then $g \in D_{\bar{\varphi}(A)}$. But if $g \in D_{\bar{\varphi}(A)}$ then, for every $f \in D_{\varphi(A)}$,

$$(\varphi\,(A) f, g) = (f, \bar{\varphi}\,(A)\, g), \tag{7}$$

which may be proved in exactly the same way as equation (4). Comparing (5) and (7) we obtain the equation

$$g^* = \bar{\varphi}\,(A)\, g,$$

and our assertion is completely proved.

The product

$$\varphi_1(A)\, \varphi_2(A) \tag{8}$$

can be examined also in case the functions $\varphi_1(t)$, $\varphi_2(t) \in \mathbf{K}_0$ are not bounded. It is defined only for those elements $f \in D_{\varphi_2(A)}$ for which $\varphi_2(A) f \in D_{\varphi_1(A)}$. Therefore, generally speaking, it does not coincide with $\varphi\,(A)$, where $\varphi\,(t) = \varphi_1(t)\, \varphi_2(t)$. But, if an element f is in the domain of the operator (8), then, for every $h \in H$,

$$(\varphi_1(A)\, \varphi_2(A) f, h) = \int_{-\infty}^{\infty} \varphi_1(t)\, d\,(E_t \varphi_2(A) f, h)$$

and

$$(\varphi_2(A) f, E_t h) = \int_{-\infty}^{t} \varphi_2(s)\, d\,(E_s f, h),$$

so that

$$(\varphi_1(A)\, \varphi_2(A) f, h) = \int_{-\infty}^{\infty} \varphi_1(t)\, \varphi_2(t)\, d\,(E_t f, h) = \int_{-\infty}^{\infty} \varphi\,(t)\, d\,(E_t f, h) = (\varphi\,(A) f, h),$$

and, hence,

$$\varphi_1(A)\, \varphi_2(A) \subset \varphi\,(A).$$

We consider now the question of the inverse operators. The operator $\varphi(A)$, where $\varphi(t) \in \mathbf{K}_0$, has an inverse if and only if the relation

$$\varphi(A) f = 0,$$

or, equivalently,

$$\int_{-\infty}^{\infty} |\varphi(t)|^2 d(E_t f, f) = 0$$

is satisfied only for $f = 0$. This condition is satisfied if and only if the point set on which $\varphi(t) = 0$ has zero σ-measure for each distribution function $\sigma(t) = (E_t f, f), f \in \mathrm{H}$. Assume this condition and let

$$\psi(t) = \frac{1}{\varphi(t)}.$$

It is evident that the operator $\psi(A)$ is the inverse of the operator $\varphi(A)$. This justifies the equation

$$[\varphi(A)]^{-1} = \frac{1}{\varphi(A)}.$$

In light of the considerations of the present section, the Cayley transform of a self-adjoint operator A attains a new aspect. It is a linear fractional function of the operator A:

$$V = (A - iE)(A + iE)^{-1} = \frac{A - iE}{A + iE}.$$

The Cayley transform of a self-adjoint operator is a unitary operator. In general an operator $\varphi(A)$ will be unitary if and only if $|\varphi(t)| \equiv 1$. As an example, consider the operators

$$e^{isA} = \int_{-\infty}^{\infty} e^{ist} dE_t \qquad (-\infty < s < \infty),$$

which form the abelian group studied in Section 64.

Completing the present section, we consider what simplifications the assumption that the spectrum of the operator A is simple brings to the theory. Making this assumption, we let g denote some generating element of the operator and we let $\sigma(t)$ denote the distribution function $(E_t g, g)$. It is easy to see that the class $\mathbf{K}_0$ consists now of all functions which are measurable with respect to this particular distribution function. Indeed, let e be a set with finite σ-measure. This σ-measure equals

$$\int_{-\infty}^{\infty} \chi_e(t)\, d\sigma(t), \tag{9}$$

where $\chi_e(t)$ is the characteristic function of the set e. If f is an arbitrary vector, then for any $\Delta \subset [-\infty, \infty]$

$$(E(\Delta)f, f) = \int_{\Delta} p(t)\, d(E_t g, g), \tag{10}$$

where $p(t)$ is some non-negative σ-integrable function. Comparing (9) and (10), we see that the measure of the set e with respect to the distribution function $(E_t f, f)$ exists and equals

$$(E(e)f, f) = \int_{-\infty}^{\infty} \chi_e(t)\, d(E_t f, f) = \int_{-\infty}^{\infty} p(t)\, \chi_e(t)\, d(E_t g, g).$$

Let us take a particular function $\varphi(t) \in \mathbf{K}_0$ and construct the operator $\varphi(A)$. Each vector $f \in \mathrm{D}_{\varphi(A)}$ can be represented in the form

$$f = \int_{-\infty}^{\infty} f(t)\, dE_t g, \tag{11}$$

where $f(t)$ belongs to $L^2_\sigma(-\infty, \infty)$. This representation shows that

$$f = f(A)g$$

so that

$$\varphi(A)f = \varphi(A)f(A)g = \int_{-\infty}^{\infty} \varphi(t)f(t)\, dE_t g.$$

The relation (11) maps H isomorphically into $L^2_\sigma(-\infty, \infty)$. Hence, the operator $\varphi(A)$ in H corresponds to the operator of multiplication by $\varphi(t)$ in $L^2_\sigma(-\infty, \infty)$.

75. Commutative Operators

In Section 14 we defined the commutativity of a pair of operators S and T, of which at least one (say, S) is bounded. According to this definition, S and T commute provided that if $f \in \mathrm{D}_T$ then $Sf \in \mathrm{D}_T$ and

$$STf = TSf.$$

Now we replace the operator T by an arbitrary self-adjoint operator A. We continue to let S be an arbitrary bounded operator. We shall prove that

S and A commute if and only if S and E_t commute for each real t, where E_t is the resolution of the identity for the operator A.

Thus, let the operators S and A commute. Then for each nonreal z and each $f \in D_A$,

$$S(A - zE)f = (A - zE)Sf.$$

Consider the resolvent operator $R_z = (A - zE)^{-1}$ of A. Since the range of R_z is D_A, we have, for each $h \in H$,

$$S(A - zE)R_z h = (A - zE)SR_z h$$

and

$$R_z Sh = SR_z h\,.$$

By the integral representation of the resolvent, we have, for each $g \in H$,

$$\int_{-\infty}^{\infty} \frac{1}{t-z}\, d(E_t Sh, g) = \int_{-\infty}^{\infty} \frac{1}{t-z}\, d(SE_t h, g).$$

Since this relation holds for every nonreal z, it follows by means of an argument applied earlier that

$$E_t S = SE_t$$

for each $t \in [-\infty, \infty]$. Thus, one of the assertions is proved. The same formulas in the reverse order yield the other assertion.

From the considerations of the preceding section it follows that each bounded function of a self-adjoint operator of class $\mathbf{K}_0$ commutes with this operator. The converse proposition is not valid, i.e., the commutativity of a bounded operator S with a self-adjoint operator A, does not imply that S is a function of A. However, the following theorem holds.

THEOREM 1: *If a bounded operator S commutes with a self-adjoint operator A which has a simple spectrum, then S is a function of A.*

We shall prove this theorem below, but first we pause to consider a general question: what are necessary and sufficient conditions for a bounded operator S to be a function of a given self-adjoint operator A. One answer to this question is given in the following theorem.

THEOREM 2: (Riesz-Neumann) *A bounded operator S is a function of a self-adjoint operator A if and only if S commutes with every bounded operator T which commutes with A.*

The proof of the necessity of the criterion given in the theorem is very simple. Indeed, let

$$S = \varphi(A).$$

Let T be a bounded operator which commutes with A. Then

$$E_t T = TE_t$$

for each t. It follows that

$$(\varphi(A)Tf, g) = \int_{-\infty}^{\infty} \varphi(t)\, d(E_t Tf, g) = \int_{-\infty}^{\infty} \varphi(t)\, d(TE_t f, g) =$$

$$= \int_{-\infty}^{\infty} \varphi(t)\, d(E_t f, T^* g) = (\varphi(A)f, T^* g) = (T\varphi(A)f, g)$$

for all $f, g \in \mathrm{H}$. Therefore,

$$\varphi(A)T = T\varphi(A).$$

However, the proof of the sufficiency is not so simple. It will not be given in complete detail in this book. We restrict ourselves to the case for which the eigenvectors $e_k (k=1, 2, 3, \ldots)$ of the operator A form a complete orthonormal system in H (this condition is satisfied, for example, for a completely continuous operator in a separable space).

Let $P_k (k=1, 2, 3, \ldots)$ denote the operator of projection on the eigenvector e_k. Since the equations

$$Ae_k = \lambda_k e_k \qquad (k=1, 2, 3, \ldots)$$

imply that

$$AP_k = P_k A \qquad (k=1, 2, 3, \ldots),$$

we have, by the assumption of the theorem,

$$SP_k = P_k S.$$

The last equation implies that each eigenvector of A is an eigenvector of S, and that eigenvectors which belong to equal eigenvalues of A also belong to equal eigenvalues of S.

Let $\mu_k (k=1, 2, 3, \ldots)$ be the eigenvalues of S, so that

$$Se_k = \mu_k e_k \qquad (k = 1, 2, 3, \ldots).$$

The operators A and S are represented by the formulas

$$Af = \sum_{k=1}^{\infty} \lambda_k (f, e_k) e_k \qquad (f \in \mathrm{D}_A),$$

$$Sh = \sum_{k=1}^{\infty} \mu_k (h, e_k)\, e_k \qquad (h \in \mathrm{H}).$$

Since to equal values λ_k correspond equal values μ_k, we can define a function $\varphi(t)$ on the spectrum $\{\lambda_k\}_1^{\infty}$ by the equations

$$\varphi(\lambda_k) = \mu_k \qquad (k = 1, 2, 3, \ldots).$$

Thus, we obtain a representation of S in the form

$$Sh = \sum_{k=1}^{\infty} \varphi(\lambda_k)(h, e_k) e_k.$$

Thus,

$$S = \varphi(A).$$

Now we shall prove Theorem 1. Let g be a generating element of A. Then each vector $h \in \mathrm{H}$ is represented in the form

$$h = \int_{-\infty}^{\infty} h(t)\, d\, E_t g,$$

where $h(t)$ belongs to $L^2_\sigma(-\infty, \infty)$ with $\sigma(t) = (E_t g, g)$. In particular, we have the representation

$$Sg = \int_{-\infty}^{\infty} \varphi(t)\, dE_t g.$$

We prove that the function $\varphi(t)$, which appears here, is bounded. With this aim, we assume the contrary. Thus, we assume that there exists an infinite sequence of sets $e_1, e_2, e_3, \ldots$ of positive σ-measure such that

$$|\varphi(t)| > n$$

for $t \in e_n$. Consider the elements

$$f_n = E(e_n)\, g.$$

These elements are different from zero since

$$\|f_n\|^2 = (E(e_n)\, g, g) = m_\sigma\{e_n\} \neq 0.$$

Since the operator S commutes with A, it follows from the necessity of the criterion[16] of Theorem 2 that S commutes with every function of A and, in particular, with the operator

$$E(e_n) = \chi_{e_n}(A).$$

Thus

$$E(e_n)\, S = SE(e_n),$$

and, therefore,

$$\|Sf_n\|^2 = \|SE(e_n) g\|^2 = \|E(e_n) Sg\|^2 = (Sg, E(e_n) Sg) =$$

$$= \int_{-\infty}^{\infty} \varphi(t)\, d(E_t g, E(e_n) Sg) = \int_{-\infty}^{\infty} \chi_{e_n}(t)\, \varphi(t)\, d(E_t g, Sg) =$$

$$= \int_{-\infty}^{\infty} \chi_{e_n}(t)\, |\varphi(t)|^2 d(E_t g, g) > n^2 \int_{-\infty}^{\infty} \chi_{e_n}(t)\, d(E_t g, g) = n^2 \|f_n\|^2.$$

---

[16] Recall that the necessity was proved in complete detail.

This implies that the operator S is not bounded, which contradicts the original assumption. Therefore, the function $\varphi(t)$ is bounded. Now choose $f \in \mathrm{H}$ arbitrarily and let

$$f = \int_{-\infty}^{\infty} f(t)\, d\, E_t g,$$

where $f(t) \in L^2_\sigma(-\infty, \infty)$. Then, for each $h \in \mathrm{H}$,

$$(Sf, h) = (f, S^*h) = \int_{-\infty}^{\infty} f(t)\, d(E_t g, S^*h) =$$

$$= \int_{-\infty}^{\infty} f(t)\, d(SE_t g, h) = \int_{-\infty}^{\infty} f(t)\, d(E_t Sg, h) = \int_{-\infty}^{\infty} f(t)\, d(Sg, E_t h).$$

But since

$$(Sg, E_t h) = \int_{-\infty}^{\infty} \varphi(s)\, d_s(E_s g, E_t h) = \int_{-\infty}^{t} \varphi(s)\, d(E_s g, h),$$

we have

$$(Sf, h) = \int_{-\infty}^{\infty} f(t)\, \varphi(t)\, d(E_t g, h),$$

and, hence,

$$Sf = \int_{-\infty}^{\infty} \varphi(t) f(t)\, d\, E_t g.$$

By the concluding remarks of the preceding section, this implies that

$$S = \varphi(A),$$

so that Theorem 1 is proved.

Remark: Since each bounded operator T has the representation

$$T = B_1 + iB_2,$$

where B_1 and B_2 are bounded self-adjoint operators, it is sufficient in Theorem 2 to require that S commute with every bounded self-adjoint operator B which commutes with A.

76. Rings of Bounded Self-Adjoint Operators

A set $\mathfrak{M}$ of bounded self-adjoint operators $A, B, C, \ldots,$ is called a *ring*, if $A, B \in \mathfrak{M}$ implies that $AB \in \mathfrak{M}$ and $\alpha A + \beta B \in \mathfrak{M}$ for arbitrary real α and β.

A ring is called *weakly closed* if $A_n \in \mathfrak{M}$ $(n=1, 2, 3, \ldots)$ and $A_n \overset{w}{\to} A$, imply that $A \in \mathfrak{M}$.

Since the product of two bounded self-adjoint operators is self-adjoint if and only if these operators commute, the operators in such a ring commute in pairs. Conversely, every set $\mathfrak{N}$ of pairwise commutative bounded self-adjoint operators determine a weakly closed ring, namely the *smallest weakly closed ring* which contains $\mathfrak{N}$. We donate this ring by $\mathfrak{R}(\mathfrak{N})$. It can be defined as the intersection of all weakly closed rings which contain $\mathfrak{N}$.

The following important theorem is due to von Neumann.

THEOREM 1: *For each weakly closed ring* $\mathfrak{M}$ *in a separable Hilbert space* H, *there exists a bounded self-adjoint operator A such that* $\mathfrak{M} = \mathfrak{R}(A)$.

Since the presentation of the proof of this theorem of von Neumann does not fall into the plan of the present book, we restrict ourselves to the statement of the theorem and to some of its implications.

Consider a finite or infinite set $\mathfrak{N}$ of pairwise commutative bounded self-adjoint operators in a separable space. We form the ring $\mathfrak{R}(\mathfrak{N})$. By the theorem of von Neumann, there exists a bounded self-adjoint operator A such that $\mathfrak{R}(\mathfrak{N}) = \mathfrak{R}(A)$. Let $\mathfrak{P}(A)$ denote the set of all bounded self-adjoint operators which commute with A. Let $\mathfrak{P}\{\mathfrak{P}(A)\}$ denote the set of all bounded self-adjoint operators which commute with each operator in $\mathfrak{P}(A)$. Since this set contains the operator A, every operator in $\mathfrak{P}\{\mathfrak{P}(A)\}$ commutes with A. Hence,

$$\mathfrak{P}\{\mathfrak{P}(A)\} \subset \mathfrak{P}(A).$$

We show next that $\mathfrak{P}\{\mathfrak{P}(A)\}$ is a weakly closed ring. Let $C', C'' \in \mathfrak{P}\{\mathfrak{P}(A)\}$. It is evident that the operator $\alpha'C' + \alpha''C''$ belongs to $\mathfrak{P}\{\mathfrak{P}(A)\}$ for arbitrary real α' and α''. Since C' and C'' belong to $\mathfrak{P}\{\mathfrak{P}(A)\}$, they commute with each bounded self-adjoint operator which commutes with A. But, by Theorem 2 of the preceding section and the remark at the end of that section, C' and C'' are functions of the operator A, so that $C'C''$ is contained in $\mathfrak{P}\{\mathfrak{P}(A)\}$. Hence, $\mathfrak{P}\{\mathfrak{P}(A)\}$ is a ring.

It remains to show that the ring $\mathfrak{P}\{\mathfrak{P}(A)\}$ is weakly closed. Let

$$C_n \in \mathfrak{P}\{\mathfrak{P}(A)\}$$

for $n=1, 2, \ldots$ and, for arbitrary $f, g \in$ H, let

$$\lim_{n\to\infty} (C_n f, g) = (Cf, g).$$

If C' is an arbitrary operator in $\mathfrak{P}(A)$, then

$$(C_n C'f, g) = (C'C_n f, g) = (C_n f, C'g),$$

which yields

$$(CC'f, g) = (Cf, C'g).$$

It follows that

$$CC' = C'C,$$

so that the operator C commutes with every operator of $\mathfrak{P}(A)$. Therefore, $C \in \mathfrak{P}\{\mathfrak{P}(A)\}$ and $\mathfrak{P}\{\mathfrak{P}(A)\}$ is a weakly closed ring.

This result and the relation

$$A \in \mathfrak{P}\{\mathfrak{P}(A)\}$$

imply that

$$\mathfrak{R}(A) \subset \mathfrak{P}\{\mathfrak{P}(A)\}.$$

Since

$$\mathfrak{N} \subset \mathfrak{R}(\mathfrak{N}) = \mathfrak{R}(A),$$

we have

$$\mathfrak{N} \subset \mathfrak{P}\{\mathfrak{P}(A)\}.$$

Every operator in $\mathfrak{P}\{\mathfrak{P}(A)\}$ is, as we know, a function of the operator A. Hence, the same property is possessed by all the operators of $\mathfrak{N}$. Thus, we obtain the following theorem.

THEOREM 2: *If the bounded self-adjoint operators*

$$C', C'', C''', \ldots \tag{1}$$

in a separable space are pairwise commutative, then there exists a bounded self-adjoint operator A such that the operators (1) *are functions of A.*

There is a case for which this theorem can be proved simply without the application of Theorem 1, viz., if the number of operators (1) is finite and if the linear envelope of the eigenvectors of each of these operators is dense in H. We simplify the proof, without loss of generality, by assuming that the number of operators (1) is two. Thus, we have two commutative bounded self-adjoint operators L and M such that the linear envelope of the eigenvectors of each of these operators is dense in H.

Let E_s and F_t be the spectral functions for L and M, respectively. Further, let $\mathfrak{L}_\lambda$ and $\mathfrak{M}_\mu$ be the eigenmanifolds of these operators which correspond to the eigenvalues λ and μ, i.e.,

$$\mathfrak{L}_\lambda = (E_{\lambda+0} - E_\lambda)\,\mathrm{H},$$

$$\mathfrak{M}_\mu = (F_{\mu+0} - F_\mu)\,\mathrm{H}.$$

Since L and M commute,

$$(E_{\lambda+0} - E_\lambda)(F_{\mu+0} - F_\mu) = (F_{\mu+0} - F_\mu)(E_{\lambda+0} - E_\lambda).$$

Therefore,

$$P_{\lambda,\mu} = (E_{\lambda+0} - E_\lambda)(F_{\mu+0} - F_\mu)$$

is a projection operator for which

$$G_{\lambda,\mu} = P_{\lambda,\mu} H = \mathfrak{L}_\lambda \cap \mathfrak{M}_\mu.$$

It is easy to verify that if (λ', μ') and (λ'', μ'') are different pairs, then the subspaces $G_{\lambda',\mu'}$, $G_{\lambda'',\mu''}$ are orthogonal.

We prove that there does not exist a nonzero vector f which is orthogonal to all the subspaces $G_{\lambda,\mu}$. Indeed, if $f \neq 0$, there exists at least one λ for which

$$(E_{\lambda+0} - E_\lambda) f \neq 0.$$

We denote this value λ by λ_0 and let

$$(E_{\lambda_0+0} - E_{\lambda_0}) f = f'.$$

There exists μ_0 such that

$$(F_{\mu_0+0} - F_{\mu_0}) f' \neq 0.$$

Hence,

$$P_{\lambda_0,\mu_0} f \neq 0,$$

which implies that the vector f is not orthogonal to the subspace G_{λ_0,μ_0}. Thus, our assertion is proved.

In each of the subspaces $G_{\lambda,\mu}$ we construct a complete orthonormal system of vectors

$$e_{\lambda\mu}^{(1)},\ e_{\lambda\mu}^{(2)},\ e_{\lambda\mu}^{(3)},\ \ldots .$$

The set of all these vectors is countable, since the space H is separable. We enumerate this set of vectors $e_{\lambda\mu}^{(k)}$ and obtain orthonormal basis

$$e_1,\ e_2,\ e_3,\ \ldots$$

for H. Each vector e_k is an eigenvector of both of the operators L and M. Let

$$\begin{aligned} Le_k &= \lambda_k e_k, \\ Me_k &= \mu_k e_k, \end{aligned} \qquad (k = 1, 2, 3, \ldots).$$

Now we choose an arbitrary bounded sequence of real numbers $\{\alpha_k\}_1^\infty$ and define an operator A by the formula

$$Af = \sum_{k=1}^{\infty} \alpha_k (f, e_k)\, e_k.$$

Note that A is a bounded self-adjoint operator. Moreover, A is completely continuous if the sequence $\{\alpha_k\}_1^\infty$ converges to zero. The only restriction which we impose on the numbers α_k is that they all be distinct.

Let $\varphi(t)$ and $\psi(t)$ be real functions which satisfy the conditions

$$\varphi(\alpha_k) = \lambda_k, \quad \psi(\alpha_k) = \mu_k \qquad (k = 1, 2, 3, \ldots)$$

and otherwise are arbitrary. We find without difficulty that

$$L = \varphi(A), \qquad M = \psi(A).$$

77. Examples

We illustrate some of the general results of the present chapter in terms of the differential operator

$$P = i\frac{d}{dt}$$

and the operator Q of multiplication by an independent variable, both defined in $L^2(-\infty, \infty)$. These operators are self-adjoint. Furthermore, they are unitarily equivalent, i.e.,

$$P = \mathfrak{F}\, Q\, \mathfrak{F}^*,$$

where $\mathfrak{F}$ is the Fourier-Plancherel operator. Since the spectrum of the operator Q is simple and is the set of all real numbers (cf. Section 69), the spectrum of the operator P has the same properties.

A. Let $E_t^{(Q)}$ be the spectral function of the operator Q and $E_t^{(P)}$ that of the operator P. Then

$$E^{(Q)}(\Delta)\, h = \chi_\Delta(t)\, h(t),$$

where $\chi_\Delta(t)$ is the characteristic function of the interval Δ and $h = h(t)$ is an arbitrary element of the space $L^2(-\infty, \infty)$. Since (cf. Section 73)

$$E_t^{(P)} = \mathfrak{F}\, E_t^{(Q)} \mathfrak{F}^*,$$

we have, for each finite interval $\Delta = [\alpha, \beta]$,

$$E^{(P)}(\Delta) f = \frac{1}{2\pi} \int_{-\infty}^{\infty} \frac{e^{i\beta(u-t)} - e^{i\alpha(u-t)}}{i(u-t)} f(u)\, du,$$

where $f = f(t)$ is an arbitrary element of $L^2(-\infty, \infty)$.

B. We recall Theorem 2 of Section 69. By this theorem, if g is any generating element of the operator Q, then the formula

$$f = \int_{-\infty}^{\infty} \varphi(t)\, d\, E_t^{(Q)} g,$$

which associates the element $f = f(t) \in L^2(-\infty, \infty)$ with the element $\varphi(t) \in L_\sigma^2(-\infty, \infty)$, where $\sigma(t) = (E_t g, g)$, establishes an isometric mapping of $L^2(-\infty, \infty)$ into $L_\sigma^2(-\infty, \infty)$. Since Q is the multiplication operator in $L^2(-\infty, \infty)$, the question naturally arises as to the choice of a generating element g_0 for which this isometric mapping is the identity transformation:

$$f(s) = \int_{-\infty}^{\infty} f(t)\, d\, E_t^{(Q)} g_0$$

This equation holds if and only if

$$E^{(Q)}(\Delta)\, g_0 = \chi_\Delta(t),$$

where Δ is an arbitrary finite interval and $\chi_\Delta(t)$ is the characteristic function of Δ. But, as $\Delta \to [-\infty, \infty]$, the functions $\chi_\Delta(t)$ do not tend to a limit in $L^2(-\infty, \infty)$. Hence, the desired vector g_0 does not exist. Nevertheless, if we want an affirmative answer to our question, then we must augment the space $L^2(-\infty, \infty)$ by adding to it an *improper element* g_0 such that the projection of g_0 on each subspace $L^2(a, b)$, where $-\infty < a < b < \infty$, is the characteristic function of $[a, b]$. Thus, the element g_0 is a *unit function* which does not belong to the space $L^2(-\infty, \infty)$.

C. We consider now the operator P. Let $h = h(t)$ be an arbitrary generating element of this operator. Then, for each $f \in L^2(-\infty, \infty)$, there exists an element $\psi(t) \in L^2(-\infty, \infty)$, where $\sigma(t) = (E_t^{(P)}h, h)$, such that

$$f(s) = \int_{-\infty}^{\infty} \psi(t)\, d\, E_t^{(P)} h.$$

The question naturally arises as to the choice of a generating element h_0 for which $\psi(t)$ is the Fourier-Plancherel transform (or the inverse transform) of the function $f(t)$. We desire that

$$\psi(t) = \mathfrak{F}^* f(t).$$

The representations

$$f(s) = \int_{-\infty}^{\infty} \psi(t)\, d\, E_t^{(P)} h_0$$

and

$$f(s) = \text{l. i. m.} \frac{1}{\sqrt{2\pi}} \int_{-\infty}^{\infty} \psi(t)\, e^{-ist} dt$$

yield

$$E^{(P)}(\Delta)\, h_0 = \frac{1}{\sqrt{2\pi}} \frac{e^{-i\beta t} - e^{-i\alpha t}}{-it}$$

for each finite interval $\Delta = [\alpha, \beta]$. Thus, the desired vector h_0 should satisfy the condition

$$E^{(P)}(\Delta)\, h_0 = \mathfrak{F}\, E^{(Q)}(\Delta)\, g_0, \tag{1}$$

where g_0 is the identity function. Such a vector h_0 does not exist in $L^2(-\infty, \infty)$. We introduce h_0 as another improper element[17] of the space $L^2(-\infty, \infty)$. Equation (1) shows that the element h_0 should be considered as the Fourier-Plancherel transformation of the improper element g_0.

D. Turning to functions of the operator P, we should, first of all, define the class $\mathbf{K}_0$. This class is the set of all functions $\varphi(t)$ which are measurable on each finite real interval. We obtain the same class for the investigation of functions of the operator Q. Bounded functions are of greatest interest since they generate bounded operators.

Let $\varphi(t)$ be such a function. If $f = f(t)$ is an arbitrary function of $L^2(-\infty, \infty)$ and $g = g(t)$ is its inverse Fourier-Plancherel transform, then

$$\varphi(P) f = \mathfrak{F}\{\varphi(t)\, g(t)\}. \tag{2}$$

The situation is particularly simple when the function $\varphi(t)$ not only is bounded but also belongs to $L^2(-\infty, \infty)$. Indeed, then

$$\psi(t) = \mathfrak{F}\, \varphi(t)$$

exists and the right member of the equation is equal to

$$\frac{1}{\sqrt{2\pi}} \int_{-\infty}^{\infty} \psi(t-s) f(s)\, ds.$$

Thus, in this case,

$$\varphi(P) f = \frac{1}{\sqrt{2\pi}} \int_{-\infty}^{\infty} \psi(t-s) f(s)\, ds. \tag{3}$$

Since the right side of formula (2) becomes $\psi(t)$ when $g(t)$ is replaced by the identity function $g_0(t)$, $\psi(t)$ should be considered as the image of the improper element h_0 by the operator $\varphi(P)$:

$$\psi(t) = \varphi(P)\, h_0.$$

Thus, if $\varphi(t)$ is bounded and belongs to $L^2(-\infty, \infty)$ then it is sufficient to find $\varphi(P)\, h_0$, and the determination of $\varphi(P) f$ is reduced to finding some convolution according to (3).

Let us assume now that the function $\varphi(t)$ is bounded, but does not belong to $L^2(-\infty, \infty)$. Then it is natural to let

$$\varphi(t) = (t + i)\, \varphi_1(t),$$

---

[17] We remark that the introduction of improper elements is often useful not only in $L^2(-\infty, \infty)$ but also in other spaces.

where

$$\varphi_1(t) = \frac{\varphi(t)}{t+i}$$

and φ_1 belongs to $L^2(-\infty, \infty)$. Let

$$\varphi_1(P)\, h_0 = \psi_1(t).$$

Then

$$\varphi_1(P) f = \frac{1}{\sqrt{2\pi}} \int_{-\infty}^{\infty} \psi_1(t-s) f(s)\, ds$$

and

$$\varphi(P) f = (P+i)\,\varphi_1(P) f = \frac{i}{\sqrt{2\pi}} \left(\frac{d}{dt} + 1\right) \int_{-\infty}^{\infty} \psi_1(t-s) f(s)\, ds.$$

E. Integral operators with kernels as functions of the difference of two arguments often occur in problems of analysis. We see that such operators are functions of the operator of differentiation. As an example, we introduce the operator defined by the formula

$$D_\lambda f = \frac{1}{\pi} \int_{-\infty}^{\infty} \frac{\sin \sqrt{\lambda}\,(t-s)}{t-s} f(s)\, ds,$$

where λ is a non-negative parameter. This operator plays an important role in the theory of the Fourier integral. It is of interest that D_λ is a resolution of the identity. This fact follows from the general circumstance: if E_t is the resolution of the identity for a self-adjoint operator A, then

$$E_{\sqrt{\lambda}} - E_{-\sqrt{\lambda}} \qquad (\lambda \geqq 0)$$

is the resolution of the identity for the operator A^2. But, it is easy to see that

$$\mathrm{D}_\lambda = E^{(P)}_{\sqrt{\lambda}} - E^{(P)}_{-\sqrt{\lambda}}.$$

Therefore, D_λ is the resolution of the identity for P^2. Now we show that the operator P^2 coincides with the operator L which is defined by the formula

$$Lf = -\frac{d^2 f}{dt^2}$$

for each function $f(t)$ in $L^2(-\infty, \infty)$ with a second derivative almost everywhere which also belongs to $L^2(-\infty, \infty)$.

To begin with, it is clear that

$$P^2 \subset L.$$

Therefore, it remains only to prove that if

$$\int_{-\infty}^{\infty} |f(t)|^2\, dt < \infty, \quad \int_{-\infty}^{\infty} |f''(t)|^2\, dt < \infty,$$

then

$$\int_{-\infty}^{\infty} |f'(t)|^2\, dt < \infty.$$

It is sufficient to establish this fact for the case when the function $f(t)$ is real and the integration is over the positive real semi-axis. Consider the identity

$$\int_0^s [f'(t)]^2 dt = f(s)f'(s) - f(0)f'(0) - \int_0^s f(t)f''(t)\, dt,$$

$$\int_0^t f(s)f'(s)\, ds = \tfrac{1}{2}[f(t)]^2 - \tfrac{1}{2}[f(0)]^2.$$

Assume that $f'(t)$ does not belong to $L^2(-\infty, \infty)$. Then

$$\lim_{s\to\infty} \int_0^s [f'(t)]^2 dt = \infty.$$

From the first identity, of which the last term of the right member is bounded, we conclude that

$$\lim_{s\to\infty} f(s)f'(s) = \infty.$$

Therefore, from the second identity, we find that

$$\lim_{t\to\infty} [f(t)]^2 = \infty,$$

which contradicts the assumption that

$$\int_0^{\infty} [f(t)]^2 dt < \infty.$$

We proceed to prove that the multiplicity of the spectrum of the operator L equals two. First we show that this spectrum is not simple. Assuming the contrary we choose a generating element g of the operator L such that the linear envelope of the set of vectors $D_\lambda g (\lambda \geqq 0)$ is dense in $L^2(-\infty, \infty)$. Hence, the element g is *a fortiori* a generating element of P. But since the spectrum of P is the entire real line, the vector $E^{(P)}(\Delta)g$ can equal zero for no finite interval Δ. Let us take a finite interval $\Delta = [\alpha, \beta]$, where $\beta > \alpha > 0$, and represent the vector $E^{(P)}(\Delta)g$ in the form

$$E^{(P)}(\Delta)\, g = \int_0^\infty \varphi\,(t)\, d\mathrm{D}_t g,$$

which is possible since g is a generating element of L. This representation can be written in the form

$$\int_\alpha^\beta dE_t^{(P)} g = \int_0^\infty \varphi\,(s)\, d\,\{E_{\sqrt{s}}^{(P)} - E_{-\sqrt{s}}^{(P)}\}\, g = \int_{-\infty}^\infty \varphi\,(t^2)\, dE_t^{(P)} g$$

from which it follows that

$$\varphi\,(t^2) = \begin{cases} 1 & (\alpha < t < \beta), \\ 0 & (-\beta < t < -\alpha), \end{cases}$$

which is impossible.

Thus, the spectrum of the operator L is not simple. In order to prove that the multiplicity is two, it is necessary to show that the operator L has a generating basis which consists of two vectors. Let us choose two generating elements $h_1 = h_1(t)$ and $h_2 = h_2(t)$ of P, such that the first is an odd function and the second is an even function. We represent an arbitrary function $f(t)$ of $L^2(-\infty, \infty)$ in the form of the sum

$$f(t) = f_1(t) + f_2(t)$$

of an odd function $f_1(t)$ and an even function $f_2(t)$. Since h_1 is a generating element of P,

$$f_1 = \int_{-\infty}^\infty \varphi_1(s)\, d(E_s^{(P)} - E_0^{(P)})h_1. \tag{4}$$

But

$$(E_s^{(P)} - E_0^{(P)})\, h_1 \equiv g_1(t; s) = \frac{1}{2\pi} \int_{-\infty}^\infty \frac{e^{is(u-t)} - 1}{i\,(u - t)}\, h_1(u)\, du,$$

and, hence,

$$g_1(-t, -s) = \frac{1}{2\pi} \int_{-\infty}^\infty \frac{e^{is(-u-t)} - 1}{i(u + t)}\, h_1(u)\, du =$$

$$= \frac{1}{2\pi} \int_{-\infty}^\infty \frac{e^{is(u-t)} - 1}{i(-u + t)}\, h_1(-u)\, du =$$

$$= \frac{1}{2\pi} \int_{-\infty}^\infty \frac{e^{is(u-t)} - 1}{i\,(u - t)}\, h_1(u)\, du = g_1(t; s).$$

Since $f_1(t)$ is odd, it follows from (4) that

$$f_1 = \int_{-\infty}^{\infty} \varphi_1(-s)\, d\,(E_s^{(P)} - E_0^{(P)})\, h_1.$$

This implies that $\varphi_1(t)$ is also an even function and the representation (4) takes the form

$$f_1 = \int_0^{\infty} \varphi_1(s)\, d\,(E_s^{(P)} - E_{-s}^{(P)})\, h_1 = \int_0^{\infty} \varphi_1(\sqrt{t})\, d\mathrm{D}_t h_1.$$

Analogously, it is proved that

$$f_2 = \int_0^{\infty} \varphi_2(\sqrt{t})\, d\mathrm{D}_t h_2.$$

Hence, the vectors h_1 and h_2 form a generating basis for the operator L. It may be observed that this basis is orthogonal.

Chapter VII

THEORY OF EXTENSIONS OF SYMMETRIC OPERATORS

78. Deficiency Indices

Let T be an arbitrary linear operator. For the present we require only that the domain of T be dense in H.

We call a complex number λ *a point of regular type of the operator* T *if there exists* $k = k(\lambda) > 0$ *such that for all* $f \in D_T$

$$\|(T - \lambda E)f\| \geqq k\|f\|.$$

From this definition it follows that λ is a point of regular type of the operator T if and only if $(T - \lambda E)^{-1}$ exists and is bounded. In particular, the eigenvalues of the operator T are not points of regular type of T.

If λ_0 is a point of regular type and

$$|\lambda - \lambda_0| \leqq \delta \leqq \tfrac{1}{2} k(\lambda_0),$$

then, for each $f \in D_T$,

$$\|(T - \lambda E)f\| \geqq \|(T - \lambda_0 E)f\| - |\lambda - \lambda_0| \cdot \|f\| \geqq$$
$$\geqq \{k(\lambda_0) - \delta\} \|f\| \geqq \tfrac{1}{2} k(\lambda_0) \|f\|.$$

This fact, used once before (cf. the proof of Theorem 4, Section 43), shows that the set of points of regular type is open. We call this set of points the field of regularity of the operator T.

If A is a symmetric operator and $z = x + iy$ $(y \neq 0)$, then (cf. the proof of Theorem 3, Section 43)

$$\|(A - zE)f\|^2 = \|(A - xE)f\|^2 + y^2\|f\|^2 \geqq y^2\|f\|^2$$

for each $f \in D_A$. Hence, the upper and lower z-half-planes are connected subsets of the field of regularity of an arbitrary symmetric operator.

The field of regularity of an isometric operator V also contains two particular connected subsets: the region inside and the region outside of the unit circle. Indeed, for $|\zeta| < 1$

$$\|(V - \zeta E)f\| \geqq \|Vf\| - |\zeta| \cdot \|f\| = (1 - |\zeta|)\|f\|;$$

and, analogously, for $|\zeta| > 1$

$$\|(V - \zeta E)f\| \geqq |\zeta| \cdot \|f\| - \|Vf\| = (|\zeta| - 1)\|f\|.$$

THEOREM: *If Γ is a connected subset of the field of regularity of a linear operator T, then the dimension of the subspace* $H \ominus \Delta_T(\lambda)$ *is the same for each* $\lambda \in \Gamma$.

Proof: We denote by P_λ the projection operator on the subspace

$$\mathfrak{N}_\lambda = H \ominus \Delta_T(\lambda).$$

If we show that for arbitrary $\lambda_1, \lambda_2 \in \Gamma$,

$$\| P_{\lambda_1} - P_{\lambda_2} \| < 1, \tag{1}$$

then it follows from the theorem of Section 34 that the dimensions of the subspaces $\mathfrak{N}_{\lambda_1}$ and $\mathfrak{N}_{\lambda_2}$ are equal. In order to prove the inequality (1) for arbitrary $\lambda_1, \lambda_2 \in \Gamma$, it is sufficient, by using the Heine-Borel theorem, to show that for each $\lambda_0 \in \Gamma$ there exists $\delta = \delta(\lambda_0) > 0$ such that

$$\| P_\lambda - P_{\lambda_0} \| < 1$$

for $|\lambda - \lambda_0| < \delta$. Thus, let λ_0 be an arbitrary point of the region Γ and let

$$\delta = \delta(\lambda_0) \leqq \tfrac{1}{3} k(\lambda_0).$$

Since

$$k(\lambda_0) \|f\| \leqq \|(T - \lambda_0 E)f\| \leqq \|(T - \lambda E)f\| + |\lambda - \lambda_0| \cdot \|f\|,$$

we have, for $|\lambda - \lambda_0| \leqq \delta$,

$$\|(T - \lambda E)f\| \geqq \tfrac{2}{3} k(\lambda_0) \|f\|.$$

Let $|\lambda - \lambda_0| \leqq \delta$. Then, for each $h \in \mathfrak{N}_\lambda$ ($\|h\| = 1$),

$$\begin{aligned} \|(E - P_{\lambda_0})h\| &= \sup_{f \in D_T} \frac{|(h, \{T - \lambda_0 E\} f)|}{\|(T - \lambda_0 E)f\|} = \\ &= \sup_{f \in D_T} \frac{|(h, \{T - \lambda E\} f + (\lambda - \lambda_0) f)|}{\|(T - \lambda_0 E)f\|} = \\ &= \sup_{f \in D_T} \frac{|\lambda - \lambda_0| \cdot |(h, f)|}{\|(T - \lambda_0 E)f\|} \leqq \frac{1}{2} \end{aligned} \tag{2}$$

and, for each $h \in \mathfrak{N}_{\lambda_0}$ ($\|h\| = 1$),

$$\|(E - P_\lambda)h\| \leqq \tfrac{1}{2}. \tag{3}$$

By the second definition of the aperture of two linear manifolds (cf. Section 34) it follows from inequalities (2) and (3) that

$$\| P_\lambda - P_{\lambda_0} \| \leqq \tfrac{1}{2}.$$

Thus, the theorem is proved.

From our considerations, it follows that

$$\lim_{\lambda \to \lambda_0} \| P_\lambda - P_{\lambda_0} \| = 0 \qquad (\lambda, \lambda_0 \in \Gamma).$$

This relation expresses that as the point λ varies in the region Γ, the subspace $\mathfrak{N}_\lambda$ rotates continuously around the point $f = 0$.

We define the *deficiency* of a linear manifold as the dimension of its orthogonal complement $\mathfrak{N} = \mathrm{H} \ominus \mathfrak{M}$ and write

$$\operatorname{def} \mathfrak{M} = \dim \mathfrak{N}.$$

The deficiency can be finite or infinite.

The theorem just proved motivates the following definition.

DEFINITION: *The deficiency of the linear manifold* $\mathfrak{M}_\lambda = \Delta_T(\lambda)$ *for an arbitrary point* λ *of a given connected subset of the field of regularity of the operator* T *is called the deficiency number of the operator* T *with respect to this subset of the field of regularity. Furthermore,* $\mathfrak{N}_\lambda = \mathrm{H} \ominus \mathfrak{M}_\lambda$ *is called the deficiency subspace of the operator* T *for the point* λ *and each nonzero element of the deficiency subspace is called a deficiency element.*

Every symmetric operator A has two particularly important deficiency numbers, $\mathfrak{m}$ and $\mathfrak{n}$, corresponding to the lower and upper half-planes:

$$\operatorname{def} \Delta_A(z) = \begin{cases} \mathfrak{m} & (\Im z < 0), \\ \mathfrak{n} & (\Im z > 0). \end{cases}$$

Every isometric operator V has two analogous deficiency numbers, $\mathfrak{m}$ and $\mathfrak{n}$, corresponding to the regions inside and outside the unit circle:

$$\operatorname{def} \Delta_V(\zeta) = \begin{cases} \mathfrak{m} & (|\zeta| < 1), \\ \mathfrak{n} & (|\zeta| > 1). \end{cases}$$

These particular deficiency numbers are called the deficiency indices of the symmetric (respectively, isometric) operator. They are written in the form of an ordered pair $(\mathfrak{m}, \mathfrak{n})$.

From the theorem proved above immediately follow three propositions:

1° *If a symmetric operator has a real point of regular type, then its deficiency indices are equal:* $\mathfrak{m} = \mathfrak{n}$. *The same fact is true for an isometric operator if it has a point of regular type on the unit circle.*

2° *If* A *is a symmetric operator, then each nonreal number* z *is an eigenvalue of multiplicity* $\mathfrak{m}$ *for the self-adjoint operator* A^* *if* $\Im z > 0$, *and an eigenvalue of multiplicity* $\mathfrak{n}$ *if* $\Im z < 0$.

Indeed[1] if $f \in \mathrm{D}_A$ and $g \in \mathfrak{N}_{\bar{z}}$, then

$$(g, \{A - \bar{z}E\} f) = 0.$$

[1] Proposition 2° was first discovered by H. Weyl [2], Vol. I, for differential operators of the second order. Then T. Carleman proved it for integral operators. For arbitrary symmetric operators in H the proposition was established first by J. von Neumann. The proof of the more general theorem given earlier is due to M. G. Krein and M. A. Krasnoselski.

It follows that

$$(A^*g - zg, f) = 0,$$

so that

$$A^*g = zg.$$

Thus, z is an eigenvalue of the operator A^* with multiplicity dim $\mathfrak{N}_{\bar{z}}$.

3° *The deficiency indices of an isometric operator V can be defined by means of the equations*

$$\mathfrak{m} = \operatorname{def} \mathrm{D}_V,$$

$$\mathfrak{n} = \operatorname{def} \Delta_V.$$

It is enough to prove just the first equation. For each $\zeta \neq 0$,

$$\Delta_V(\zeta) = (V - \zeta E)\mathrm{D}_V = \left(\frac{1}{\zeta}V - E\right)\mathrm{D}_V =$$

$$= \left(\frac{1}{\zeta}E - V^{-1}\right)V\mathrm{D}_V = \left(V^{-1} - \frac{1}{\zeta}E\right)\mathrm{D}_{V^{-1}} = \Delta_{V^{-1}}\left(\frac{1}{\zeta}\right).$$

If $|\zeta| > 1$, then

$$\mathfrak{m} = \operatorname{def} \Delta_V(\zeta) = \operatorname{def} \Delta_{V^{-1}}\left(\frac{1}{\zeta}\right) =$$

$$= \operatorname{def} \Delta_{V^{-1}}(0) = \operatorname{def} \Delta_{V^{-1}} = \operatorname{def} \mathrm{D}_V.$$

79. Further Remarks on the Cayley Transform

In Section 67 we introduced the idea of the Cayley transform V of a closed symmetric operator A with the aid of the formulas

$$(1) \qquad \begin{cases} (A - \bar{z}E)h = f. \\ (A - zE)h = Vf, \end{cases}$$

where z is any nonreal number and $h \in \mathrm{D}_A$. We assume in the present section that $\Im z > 0$. The operator V is expressed by the operator A in the form

$$Vf = (A - zE)(A - \bar{z}E)^{-1}f,$$

where the domain D_V of the operator V is $\Delta_A(\bar{z})$. By formula (1),

$$(2) \qquad \begin{cases} h = \dfrac{(V - E)f}{\bar{z} - z}, \\[2ex] Ah = \dfrac{(\bar{z}V - zE)f}{\bar{z} - z}, \end{cases}$$

and, therefore,

$$Ah = (\bar{z}V - zE)(V - E)^{-1}h.$$

For what follows, it is very important that the deficiency indices $(\mathfrak{m}, \mathfrak{n})$ of the operator A are the same as the deficiency indices of the operator V. According to the definition,

$$\mathfrak{m} = \operatorname{def} \Delta_A(\bar{z}).$$

But

$$\Delta_A(\bar{z}) = \mathrm{D}_V,$$

so that

$$\operatorname{def} \mathrm{D}_V = \mathfrak{m}.$$

On the other hand,

$$\mathfrak{n} = \operatorname{def} \Delta_A(z)$$

by definition, and

$$\Delta_A(z) = \Delta_V,$$

so that

$$\operatorname{def} \Delta_V = \mathfrak{n}.$$

It remains to apply proposition 3 of the preceding section.

THEOREM 1: *If V is an isometric operator and if the manifold $\Delta_V(1)$ is dense in* H, *then the operator A which is defined by formula* (2′) *is symmetric and the operator V is its Cayley transform.*

Proof: Since $\Delta_V(1)$ is dense in H, the inverse operator $(V - E)^{-1}$ exists. Indeed, if this operator does not exist, then unity is an eigenvalue of the operator V. But if

$$Vg = g \qquad (g \neq 0)$$

then, for each $f \in \mathrm{D}_V$,

$$(Vf - f, g) = (Vf, g) - (f, g) = (Vf, Vg) - (f, g) = 0,$$

i.e., $g \perp \Delta_V(1)$.

Since the operator $(V - E)^{-1}$ exists, the operator

$$A = (\bar{z}V - zE)(V - E)^{-1}$$

exists and its domain is dense in H. We shall show that this operator is symmetric.

Let f and g be arbitrary elements of $\mathrm{D}_A = \Delta_V(1)$:

$$\begin{aligned} f &= V\varphi - \varphi, \\ g &= V\psi - \psi, \end{aligned} \qquad (\varphi, \psi \in \mathrm{D}_V).$$

Then

$$Af = (\bar{z}V - zE)\,\varphi = \bar{z}V\varphi - z\varphi,$$
$$Ag = (\bar{z}V - zE)\,\psi = \bar{z}V\psi - z\psi.$$

Therefore,

$$(Af, g) = (\bar{z}V\varphi - z\varphi,\, V\psi - \psi) =$$
$$= (z + \bar{z})(\varphi, \psi) - \bar{z}(V\varphi, \psi) - z(\varphi, V\psi)$$

and

$$(f, Ag) = (V\varphi - \varphi,\, \bar{z}V\psi - z\psi) =$$
$$= (z + \bar{z})(\varphi, \psi) - \bar{z}(V\varphi, \psi) - z(\varphi, V\psi),$$

so that

$$(Af, g) = (f, Ag).$$

The proof of the relation

$$V = \frac{A - zE}{A - \bar{z}E}$$

is not difficult. Thus, the operator V is the Cayley transform of the operator A.

In what follows we apply the name "Cayley transform" to each of the two operators V and A which are connected by the relations (1) and (2), i.e., we not only call the operator V the Cayley transform of the operator A but also we call the operator A the Cayley transform of the operator V.

From the propositions proved above immediately follows the next theorem.

THEOREM 2: *Let A_1 and A_2 be symmetric operators and let V_1 and V_2 be their Cayley transforms. In order that A_2 be an extension of A_1 it is necessary and sufficient that V_2 be an extension of V_1.*

By Theorem 2 the question of symmetric extensions of a given operator A is reduced to the question of isometric extensions of its Cayley transform V.

From Section 9 we know that closed linear manifolds F and G can be the domain and range, respectively, of an isometric operator if and only if their dimensions are equal. In the following manner the isometric extensions of an operator V can be obtained.

In the deficiency subspaces $\mathrm{H} \ominus \mathrm{D}_V$ and $\mathrm{H} \ominus \Delta_V$ we choose two subspaces F and G of the same dimensions and we construct an arbitrary isometric operator V' with domain F and range G. We further define a linear operator $\tilde{V}$ with domain $D_{\tilde{V}} = D_V \oplus F$ and range $\Delta_{\tilde{V}} = D_V \oplus G$ by the formulas

$$\tilde{V}f = \begin{cases} [Vf,\ f \in D_V], \\ [V'f, f \in F]. \end{cases}$$

It is evident that $\tilde{V}$ is an isometric extension of V, and for all possible choices of F, G, and V', we get all isometric extensions $\tilde{V}$ of the operator V and each one just once.

In order to find a symmetric extension $\tilde{A}$ of the operator A we pass to the Cayley transform of the operator, find an extension $\tilde{V}$ of the operator V by the method described above and, finally, let $\tilde{A}$ be the Cayley transform of $\tilde{V}$. The formula which corresponds to this procedure will be given in the following section.

From the argument outlined above it follows, in particular, that the operator A is a maximal symmetric (self-adjoint) operator if and only if its Cayley transform V is a maximal isometric (correspondingly, unitary) operator. Therefore, we have

THEOREM 3: *In order that a symmetric operator be maximal, it is necessary and sufficient that one of its deficiency indices equal zero. In order that a symmetric operator be self-adjoint, it is necessary and sufficient that both its deficiency indices equal zero.*

By the procedure described above, we also obtain

THEOREM 4: *Let A be an arbitrary symmetric operator with deficiency indices $(\mathfrak{m}, \mathfrak{n})$. Then A has a maximal symmetric extension. If $\mathfrak{m} \neq \mathfrak{n}$, then none of the extensions is self-adjoint. If $\mathfrak{m} = \mathfrak{n} < \infty$, then every maximal extension of the operator A is self-adjoint. If the deficiency indices $\mathfrak{m}$ and $\mathfrak{n}$ are infinite and equal, then A has maximal self-adjoint extensions and non-self-adjoint extensions.*

80. The Neumann Formulas

We begin with some remarks about linear independence. In the very beginning of the book we introduced the concept of a linearly independent set of vectors. Now we introduce two more general concepts.

Let us agree to call the linear manifolds $M_1, M_2, \ldots, M_n$ $(n < \infty)$ linearly independent if the equation

$$f_1 + f_2 + \ldots + f_n = 0 \qquad (f_k \in M_k;\ k = 1, 2, \ldots, n)$$

implies that

$$f_1 = f_2 = \ldots = f_n = 0.$$

If the manifolds $M_1, M_2, \ldots, M_n$ are linearly independent, then it is possible to form their direct sum

$$M_1 \oplus M_2 \oplus \ldots \oplus M_n$$

(we defined this in Section 7).

Further, let M and $\tilde{M}$ be two linear manifolds such that $M \subset \tilde{M}$. The vectors $f_1, f_2, \ldots, f_k$ of $\tilde{M}$ are called *linearly independent modulo* M if from

$$\alpha_1 f_1 + \alpha_2 f_2 + \ldots + \alpha_k f_k \in M,$$

it follows that

$$\alpha_1 = \alpha_2 = \ldots = \alpha_k = 0.$$

It is evident that a set of vectors in $\tilde{M}$ which are linearly independent modulo M is also linearly independent in the ordinary sense.

The *dimension of* $\tilde{M}$ *modulo* M we define as the maximal number m of vectors in M which are linearly independent modulo M. We write

$$\dim \tilde{M} = m \qquad (\mathrm{mod}\ M).$$

If there are arbitrary many vectors in $\tilde{M}$ which are linearly independent vectors modulo M, then we write

$$\dim \tilde{M} = \infty \qquad (\mathrm{mod}\ M).$$

It is evident that the dimension of $\tilde{M}$ modulo M does not exceed the ordinary dimension of $\tilde{M}$.

We express the relation $f \in M$ also by

$$f \equiv 0 \qquad (\mathrm{mod}\ M)\ ;$$

and then the equation

$$f \equiv g \qquad (\mathrm{mod}\ M),$$

means that $f - g \in M$.

The dimension of $\tilde{M}$ modulo M defined above is the ordinary dimension of the quotient manifold $\tilde{M}/M$.

THEOREM: *Let A be an arbitrary symmetric operator with domain* D_A *and let* $\mathfrak{N}_{\bar{z}}$ *and* $\mathfrak{N}_z$ $(\Im z > 0)$ *be any pair of its deficiency subspaces. Then the domain* D_{A^*} *of the operator* A^* *has the following representation as the direct sum of three linear manifolds:*

$$D_{A^*} = D_A \oplus \mathfrak{N}_{\bar{z}} \oplus \mathfrak{N}_z.$$

Proof: Since the manifold D_{A^*} is linear and contains D_A, $\mathfrak{N}_{\bar{z}}$, and $\mathfrak{N}_z$,

$$D_{A^*} \supset D_A \oplus \mathfrak{N}_{\bar{z}} \oplus \mathfrak{N}_z.$$

We show that, conversely, each $f \in D_{A^*}$ can be represented in the form

$$f = f_0 + g_z + g_{\bar z}, \tag{1}$$

where[2] $f_0 \in D_A$, $g_z \in \mathfrak{N}_{\bar z}$ and $g_{\bar z} \in \mathfrak{N}_z$. Note that (1) yields

$$A^*f = Af_0 + zg_z + \bar z g_{\bar z}. \tag{1'}$$

Let $f \in D_{A^*}$. We decompose $A^*f - zf$ into its components in the orthogonal subspaces $\mathfrak{M}_z$ and $\mathfrak{N}_z$:

$$A^*f - zf = (Af_0 - zf_0) + (\bar z - z)\, g_{\bar z}.$$

But $A^*g_{\bar z} = \bar z g_{\bar z}$; therefore

$$A^*(f - f_0 - g_{\bar z}) = z(f - f_0 - g_{\bar z}).$$

We conclude that

$$f - f_0 - g_{\bar z} \in \mathfrak{N}_{\bar z},$$

i.e., that

$$f - f_0 - g_{\bar z} = g_z,$$

or

$$f = f_0 + g_{\bar z} + g_z.$$

To complete the proof of the theorem it remains to show that the representation (1) of each element $f \in D_{A^*}$ is unique. Assuming the contrary, let us suppose that

$$f_0 + g_z + g_{\bar z} = 0. \tag{2}$$

Applying the operator A^* to both sides of this equation, we get

$$Af_0 + zg_z + \bar z g_{\bar z} = 0. \tag{2'}$$

Multiply (2) by z and subtract from (2′) to obtain

$$Af_0 - zf_0 + (\bar z - z)\, g_{\bar z} = 0.$$

It follows from the orthogonality of the summands that $(\bar z - z)\, g_{\bar z} = 0$. In the same way we get $(\bar z - z)\, g_z = 0$. Hence

$$f_0 = g_{\bar z} = g_z = 0,$$

and the theorem is proved.

From formulas (1) and (1′) it follows that, for each $f \in D_{A^*}$,

$$\begin{aligned}(A^*f, f) &= (Af_0 + zg_z + \bar z g_{\bar z}, f_0 + g_z + g_{\bar z}) = \\ &= (Af_0, f_0) + [(Af_0, g_z) + z(g_z, f_0)] + [(Af_0, g_{\bar z}) + \bar z(g_{\bar z}, f_0)] + \\ &\qquad + [z(g_z, g_{\bar z}) + \bar z(g_{\bar z}, g_z)] + [z(g_z, g_z) + \bar z(g_{\bar z}, g_{\bar z})] = \\ &= (Af_0, f_0) + [\bar z(f_0, g_z) + z(g_z, f_0)] + [z(f_0, g_{\bar z}) + \bar z(g_{\bar z}, f_0)] + \\ &\qquad + [\bar z(g_{\bar z}, g_z) + z(g_z, g_{\bar z})] + [z\,\|g_z\|^2 + \bar z\,\|g_{\bar z}\|^2].\end{aligned}$$

---

[2] We recall (cf. item 2°, Section 78) that $\mathfrak{N}_z$ is the eigenmanifold of the operator A^* associated with the eigenvalue z. Therefore, elements from $\mathfrak{N}_{\bar z}$ we denote by g_z.

Since the first four summands are real,

$$(3) \qquad \Im\,(A^*f, f) = \Im z\,(\,\|g_z\|^2 - \|g_{\bar z}\|^2).$$

In correspondence with formula (3), the domain D_{A^*} consists of three (nonlinear) manifolds: Γ^+ (the set of all elements f for which $\Im\,(A^*f, f) > 0$), Γ^- (the set of elements f for which $\Im\,(A^*f, f) < 0$) and Γ^0 (the set of elements f, for which (A^*f, f) is real). The element

$$f = f_0 + g_z + g_{\bar z}$$

belongs to Γ^+, Γ^-, or Γ^0 according as

$$\|g_z\| > \|g_{\bar z}\|, \quad \|g_z\| < \|g_{\bar z}\|, \quad \text{or} \quad \|g_z\| = \|g_{\bar z}\|.$$

We derive next a representation similar to (1) for the domain $D_{\tilde A}$ of an arbitrary symmetric extension $\tilde A$ of the operator A.

In order to indicate the dependence on z of the subspaces F and G, which were introduced in Section 79, we write F_z and G_z. Thus $F_z \subset \mathfrak{N}_{\bar z}$ and $G_z \subset \mathfrak{N}_z$.

From the considerations of Section 79 it follows that

$$\begin{aligned} D_{\tilde A} &= (\tilde V - E)\,D_V = (\tilde V - E)(D_V \oplus F_z) = \\ &= (V - E)\,D_V \oplus (V' - E)\,F_z = D_A \oplus (V' - E)\,F_z \end{aligned}$$

and

$$D_{\tilde A} = D_A \oplus (V' + E)\,F_z,$$

if V' is replaced by V in the previous equation.

From $A^* \supset A$ it follows that if

$$(4) \qquad f = f_0 + g_z + V'g_z \quad (g_z \in F_z),$$

then

$$(4') \qquad \tilde A f = Af_0 + zg_z + \bar z V' g_z.$$

Formulas (1) and (4) will be called respectively the first and second Neumann formulas.

Let the deficiency indices of the operator A and its symmetric extension $\tilde A$ be (m, n) and $(m - p, n - p)$ where $m, n < \infty$. Then the second Neumann formula yields

$$\dim\ D_{\tilde A} = p \quad (\mathrm{mod}\ D_A).$$

We now illustrate the above theory on the differential operator P which was introduced in Section 49. The equation

$$P^*g = zg$$

has the form

$$\frac{dg}{dt} + izg = 0.$$

Its formal solution is the function

$$g(t) = Ce^{-izt}. \tag{5}$$

In the case with $-\infty < t < \infty$, this function belongs to L^2 only for $C = 0$. Hence, the deficiency indices of the differential operator P on the entire real axis equal (0, 0). In the case of the semi-axis ($0 \leqq t < \infty$) the deficiency indices are (0, 1), since the function (5) belongs to $L^2(0, \infty)$ for $\Im z < 0$ and does not belong to $L^2(0, \infty)$ for $\Im z > 0$. Finally in the case of the interval ($0 \leqq t \leqq 2\pi$) the deficiency indices are (1, 1), since the function (5) belongs to $L^2(0, 2\pi)$ for any z.

If we let in formula (4)

$$z = i, \quad g_z = e^t, \quad V'g_z = \theta g_{\bar{z}},$$

where $g_{\bar{z}} = e^{2\pi - t}$ and $\theta \, (|\theta| = 1)$ is a fixed constant and let $\arg \theta$ vary in the interval $[0, 2\pi]$, then we obtain all self-adjoint extensions $\widetilde{P}$ of the operator P (for the interval $[0, 2\pi]$) in the form

$$\varphi(t) = \varphi_0(t) + \alpha(e^t + \theta e^{2\pi - t}),$$
$$\widetilde{P}\varphi(t) = i\varphi_0'(t) + \alpha i(e^t - \theta e^{2\pi - t}).$$

Here $\varphi_0(0) = \varphi_0(2\pi) = 0$ and α is an arbitrary constant.

It is easy to verify that this result coincides with the result of Section 49; moreover the relation between the parameter here and the parameter there is given by

$$\theta = \frac{\beta - e^{2\pi}}{1 - e^{2\pi}\beta}.$$

81. Simple Symmetric Operators

A symmetric (respectively, isometric) operator A is *simple* if there does not exist a subspace invariant under A such that the restriction of A to this subspace is self-adjoint (respectively, unitary).

A symmetric operator A is simple if and only if its Cayley transform V is simple. This fact is a consequence of the following proposition.

THEOREM 1: *A subspace* G *reduces a symmetric operator A if and only if* G *reduces its Cayley transform V.*

Proof: Choose $h \in \mathrm{D}_A$ arbitrarily. Then

$$h = (V - E)f, \quad Ah = (\bar{z}V - zE)f \quad (f \in \mathrm{D}_V).$$

Assume that the subspace G reduces V and let P denote the operator of projection on G. Then

$$Pf \in \mathrm{D}_V, \quad VPf = PVf.$$

Therefore,

$$Ph = (V - E)\,Pf \in \mathrm{D}_A$$

and

$$APh = A(V - E)Pf = (\bar{z}V - zE)Pf = P(\bar{z}V - zE)f = PAh.$$

Thus, one of the assertions of the theorem is proved.

We now prove the other assertion. Choose $f \in \mathrm{D}_V$ arbitrarily. Then

$$f = (A - \bar{z}E)h, \quad Vf = (A - zE)h \quad (h \in \mathrm{D}_A).$$

Assume that the subspace G reduces the operator A. Then

$$Pf = (A - \bar{z}E)\,Ph \in \mathrm{D}_V.$$

It remains to verify that

$$VPf = PVf.$$

Since

$$VPf = V(A - \bar{z}E)Ph = (A - zE)Ph = P(A - zE)h = PVf,$$

the theorem is proved.

If an isometric operator V is not simple, so that it has unitary restrictions, then there exists a *maximal unitary restriction* of V (which is an extension of every unitary restriction of V). Indeed, the maximal unitary restriction of the operator V is the restriction of V to the closed linear envelope G_0 of all the subspaces G invariant under V such that the restriction of V to G is unitary.

Analogously, there exists a *maximal self-adjoint restriction* of a nonsimple symmetric operator.

LEMMA: *Let V be an isometric operator with equal deficiency indices, U_0 its maximal unitary restriction and U any unitary extension of the operator V. Then D_{U_0} is the orthogonal complement of the linear envelope* L *of the subspaces* $U^k(\mathrm{H} \ominus \mathrm{D}_V)$ $(\pm k = 0, 1, 2, \ldots)$.

Proof: The subspace $\mathrm{G} = \mathrm{H} \ominus \mathrm{L}$ is invariant under both U and U^{-1}. Therefore, by Theorem 3 of Section 42, the subspace G reduces the operator U. Let U' be the restriction of U to G. Since G is orthogonal to L, it is also orthogonal to $\mathrm{H} \ominus \mathrm{D}_V$. Hence,

$$\mathrm{G} \subset \mathrm{D}_V.$$

By similar reasoning,

$$\mathrm{G} \subset \Delta_V.$$

Hence, G reduces V and U' is a unitary restriction of V to G. Therefore,

$$\mathrm{G} \subset \mathrm{D}_{U_0}. \tag{1}$$

But, on the other hand,

$$D_{U_0} \subset D_V.$$

This implies that D_{U_0} is orthogonal to the subspace $H \ominus D_V$. Since

$$U^k D_{U_0} = U_0^k D_{U_0} = D_{U_0},$$

$U^k D_{U_0}$ is orthogonal to $H \ominus D_V$ for $k = 0, \pm 1, \pm 2, \ldots$. Hence, D_{U_0} is orthogonal to L, so that

$$D_{U_0} \subset G. \tag{2}$$

A comparison of (1) and (2) completes the proof of the lemma.

The following result is an immediate consequence of the lemma.

COROLLARY: *In a nonseparable space there exists no simple symmetric operator with equal and finite deficiency numbers.*

A further consequence of the lemma is the following.

THEOREM 2: *An isometric operator V with equal deficiency numbers is simple if and only if for each unitary extension U of V the linear envelope of the subspaces $U^k(H \ominus D_V)$ ($\pm k = 0, 1, 2, \ldots$) is dense in H.*[3]

From this theorem and Theorem 1 of Section 69 it follows that every unitary (self-adjoint) extension of a simple isometric (symmetric) operator with deficiency indices (1, 1) has a simple spectrum.

Analogously, one proves the following general result.

THEOREM 3: *The multiplicity of the spectrum of a unitary (self-adjoint) extension of a simple isometric (symmetric) operator with finite deficiency indices (m, m) does not exceed m.*

82. The Structure of Maximal Operators

Let the space H be separable and let $\{e_k\}_1^\infty$ be any orthonormal basis for H. We define linear operators V_+ and V_- by the formulas

$$V_+ e_k = e_{k+1} \qquad (k = 1, 2, 3, \ldots),$$
$$V_- e_k = e_{k-1} \qquad (k = 2, 3, 4, \ldots).$$

It is evident that V_+ and V_- are isometric operators with deficiency indices (0, 1) and (1, 0) respectively.

Since we wish to use the Cayley transforms A_+ and A_- of the operators V_+ and V_-, we must show that the manifolds $\Delta_{V+}(1)$ and $\Delta_{V-}(1)$ are dense in H. We present the proof only for $\Delta_{V+}(1)$. Letting

$$f = e_k + \frac{r-1}{r} e_{k+1} + \ldots + \frac{1}{r} e_{k+r-1} \quad (k = 1, 2, 3, \ldots),$$

[3] Translator's Note: In the German edition, there is the following additional statement: Such an operator V is simple if there exists a unitary operator U with the given properties.

we find that

$$(V_+ - E)f = \left(e_{k+1} + \frac{r-1}{r} e_{k+2} + \ldots + \frac{1}{r} e_{k+r}\right)$$

$$- \left(e_k + \frac{r-1}{r} e_{k+1} + \ldots + \frac{1}{r} e_{k+r-1}\right) =$$

$$= \frac{1}{r}(e_{k+1} + e_{k+2} + \ldots + e_{k+r}) - e_k.$$

But

$$\left\| \frac{1}{r}(e_{k+1} + e_{k+2} + \ldots + e_{k+r}) \right\|^2 = \frac{1}{r};$$

whence we conclude that the basis vectors $e_k (k = 1, 2, 3, \ldots)$ are limits of vectors of $\Delta_{V+}(1)$. Therefore, the manifold $\Delta_{V+}(1)$ is dense in H.

The symmetric operators A_+ and A_- are maximal with deficiency indices (0, 1) and (1, 0) respectively.

THEOREM 1: *The operators A_+ and A_- are irreducible.*

For the proof it is sufficient to show (cf. Section 81) that the operators V_+ and V_- are irreducible. We proceed to prove that V_+ is irreducible. Assume that the subspace F (and, hence, its orthogonal complement G) reduces V_+ and let $V_+\mathrm{F} = \mathrm{F}_+$ and $V_+\mathrm{G} = \mathrm{G}_+$ so that

$$\mathrm{F}_+ \subset \mathrm{F}, \quad \mathrm{G}_+ \subset \mathrm{G}.$$

In the last two relations, simultaneous equality is not possible, for $\mathrm{F}_+ = \mathrm{F}$ and $\mathrm{G}_+ = \mathrm{G}$ would imply that $V_+\mathrm{H} = \mathrm{H}$. Let, for example, $\mathrm{F} \ominus \mathrm{F}_+ \neq 0$ and

$$f \in \mathrm{F} \ominus \mathrm{F}_+ \qquad (f \neq 0).$$

Since $f \in \mathrm{F}$, f is orthogonal to G. Hence,

$$f \perp \mathrm{F}_+ \oplus \mathrm{G}_+. \tag{1}$$

From (1) it follows that $f \perp e_k$ $(k = 2, 3, \ldots)$. Therefore, $f = \alpha e_1 (\alpha \neq 0)$, whence $e_1 \in \mathrm{F} \ominus \mathrm{F}_+$ and $e_1 \in \mathrm{F}$.

Since F reduces V_+, F contains e_k $(k = 2, 3, \ldots)$ as well as e_1. So $\mathrm{F} = \mathrm{H}$ and the assertion concerning A_+ is proved. The proof for A_- is quite similar.

Thus, the operators A_+ and A_- are irreducible and, hence, are simple symmetric operators. The importance of these operators is indicated by the following theorem.

THEOREM 2: (J. von Neumann) *If a simple symmetric operator A in the space* H *has deficiency indices* (0, 1) (*respectively*, (1, 0)), *then the space*

H *is separable and the operator A is isomorphic to the operator A_+ (respectively, A_-).*

Proof: Let us assume for definiteness that the deficiency indices of the operator A are (0, 1). Let V be the Cayley transform of the operator A and let $e_1 \perp \Delta_V(\|e_1\|=1)$. We form the closed linear envelope M of the vectors $V^k e_1 = e_{k+1}$ $(k = 0, 1, 2, \ldots)$; M is a subspace with the orthonormal basis $\{e_k\}_1^\infty$.

Since M is invariant with respect to V and V^{-1}, M reduces V and, hence, H $\ominus$ M also reduces V. The restriction of V to M is evidently isomorphic to V_+ and, hence, has deficiency indices (0, 1). The restriction of V to H $\ominus$ M must be unitary since otherwise at least one of the deficiency indices of V would exceed the corresponding deficiency index of the operator V_+ and, therefore, of the operator A.

On the other hand, a simple unitary operator V has no proper unitary restriction. Thus, H $\ominus$ M $= 0$ and the operator V is isomorphic to V_+. Hence, its Cayley transform A is isomorphic to A_+.

With the aid of the operator A_+ (respectively, A_-) it is possible to construct a maximal operator with the deficiency indices $(0, \mathfrak{n})$, (respectively, $(\mathfrak{m}, 0)$). For this purpose, it is sufficient to form the direct sum of separable Hilbert spaces H_α, where α runs through a set of cardinality $\mathfrak{n}$ (respectively, $\mathfrak{m}$), and in each of these to realize the operator A_+ (respectively, A_-).

As J. von Neumann first showed, in this way it is possible to get any simple maximal operator, i.e., the following theorem holds.

THEOREM 3: *A simple symmetric operator A with deficiency indices $(0, \mathfrak{n})$ (or $(\mathfrak{m}, 0)$) can be decomposed into a direct sum of operators A_+ (respectively, A_-):*

$$A = \sum_{\alpha \in \mathbf{M}} \oplus A_+^\alpha \quad ,$$

where cardinality (M) $= \mathfrak{N}$; *or, correspondingly*

$$A = \sum_{\alpha \in \mathbf{M}} \oplus A_-^\alpha \quad ,$$

where cardinality (M) $= \mathfrak{m}$.

Proof: Let V be the Cayley transform of the operator A, $D_V = H$ and $H \ominus D_V = M_1$ (dim $M_1 = \mathfrak{n}$). We let

$$V^k M_1 = M_{k+1} \qquad (k = 0, 1, 2, \ldots)$$

and denote by M the closure of the linear envelope of the spaces $M_k (k = 1, 2, 3, \ldots)$.

In the same way as in Theorem 2, we get

$$M_i \perp M_k \qquad (i \neq k;\ i, k = 1, 2, 3, \ldots),$$

$$M = H.$$

Let $e_k^{(\alpha)}$ $(k = 1, 2, 3, \ldots)$ be a complete orthonormal system in M_k and let $\mathrm{H}^{(\alpha)}$ be the closure of the linear envelope of the vectors $e_1^{(\alpha)}, e_2^{(\alpha)}, e_3^{(\alpha)}, \ldots$. The spaces $\mathrm{H}^{(\alpha)}$ and $\mathrm{H}^{(\alpha')}$ are orthogonal for $\alpha \neq \alpha'$ and the direct sum of all these subspaces is H.

It is evident that each of the subspaces $\mathrm{H}^{(\alpha)}$ reduces V and that the restriction of V to $\mathrm{H}^{(\alpha)}$ is isomorphic to V_+. The proof of the theorem is completed by means of the Cayley transform.

Theorem 3 provides the answer to the question about the structure of simple maximal operators.

The operators A_+ and A_- (respectively, V_+ and V_-) will be called *elementary maximal operators.*

With the aid of these elementary maximal operators it is possible to construct an operator with given deficiency indices $(\mathfrak{m}, \mathfrak{n})$. To accomplish this, one need only construct, by the described above method, operators with deficiency indices $(\mathfrak{m}, 0)$ and $(0, \mathfrak{n})$ and then form their direct sum. However, an arbitrary simple operator with deficiency indices $(\mathfrak{m}, \mathfrak{n})$ in general can not be constructed from elementary maximal operators.

As an illustration of the proposition of the present section we consider again the differential operator P on the positive real semi-axis. This operator has deficiency indices (0, 1). In order to study the structure of the operator P we pass to its Cayley transform

$$V = (P - iE)(P + iE)^{-1}$$

and form the powers $V^k g$, where $g = e^{-t}$ is a deficiency element of P for the point $-i$. Note that e^{-t} is the Tchebysheff-Laguerre function of order zero with (cf. Section 11) the argument doubled:

$$e^{-t} = \psi_0(2t).$$

We shall show that

$$V\psi_k(2t) = \psi_{k+1}(2t) \qquad (k = 0, 1, 2, \ldots).$$

We represent $\psi_k(2t)$ in the form

$$\psi_k(2t) = if'(t) + if(t) \equiv (P + iE)f,$$

where $f(0) = 0$. It follows that

$$f(t) = -ie^{-t} \int_0^t \psi_k(2s)\, e^s\, ds,$$

and, therefore,

$$V\psi_k(2t) = (P - iE)f = -2e^{-t} \int_0^t \psi_k(2s)\, e^s\, ds + \psi_k(2t).$$

Thus, we must establish the identity

$$\psi_{k+1}(2t) = \psi_k(2t) - 2e^{-t}\int_0^t \psi_k(2s)\, e^s\, ds,$$

and this presents no difficulty. Since the Tchebysheff-Laguerre functions form a complete orthonormal system in $L^2(0, \infty)$, the operator V is isomorphic to V_+ and hence P is isomorphic to A_+.

Thus, Theorem 2 means that every simple operator with deficiency indices (0, 1) or (1, 0) is isomorphic to the differential operator P on the semi-axis $(0, \infty)$ or, correspondingly, $(-\infty, 0)$.

The operator P of differentiation on the axis $(-\infty, \infty)$ is evidently isomorphic to some self-adjoint extension of the operator $A_+ \oplus A_-$. Thus, (cf. Section 77) we can prove by different means the unitary equivalence of the differential operator P and the operator of multiplication by the independent variable.

83. Spectra of Self-Adjoint Extensions of Symmetric Operators

According to the general definition given in Section 43, the spectrum of a symmetric, but not self-adjoint, operator A contains the complement of the set of all points of regular type of A. However, the spectrum is not exhausted by this complement. For example, at least one of the open half-planes $\Im z > 0$ and $\Im z < 0$ belongs to the spectrum. Nevertheless, this complement forms a particularly important part of the spectrum.

We shall call the complement of the set of all points of regular type of a symmetric operator the *spectral kernel* of this operator. In order to classify the points in the spectral kernel of a symmetric operator A, we first denote by A_λ the restriction of the operator A to the subspace $\mathrm{H} \ominus \mathrm{G}_\lambda$, where G_λ is the eigenmanifold of A corresponding to λ if λ is an eigenvalue of A and G_λ is the null manifold otherwise.

In the spectral kernel we single out a particular subset. This is the set of all eigenvalues of the operator (it is empty if the operator is simple). We call this set the *discrete part* of the spectral kernel.

Turning now to the characterization of the remaining part of the spectral kernel of the operator A, we remark that for each λ the operator $A_\lambda - \lambda E$ has an inverse. The set of all λ for which the operator $(A_\lambda - \lambda E)^{-1}$ is unbounded obviously belongs to the spectral kernel. We call this set the continuous part of the spectral kernel. Thus, every point of the spectral kernel belongs either to the discrete part or to the continuous part or to both simultaneously.

For a self-adjoint operator the concepts of regular point and point of regular type coincide. Therefore, the spectral kernel of a self-adjoint operator coincides with the spectrum of this operator. Hence, the spectral kernel of a self-adjoint operator cannot be the empty set. For an arbitrary symmetric operator, such an assertion would be incorrect.

If $\tilde{A}$ is a symmetric (in particular, maximal or self-adjoint) extension of the operator A, then the spectral kernel of $\tilde{A}$ contains the spectral kernel of A. Moreover, each of the parts (discrete and continuous) of the spectral kernel of $\tilde{A}$ contains the corresponding part of the spectral kernel of A.

We now consider a special case, for which the continuous part of the spectral kernel of the operator A is invariant for symmetric extensions of this operator. This happens if the deficiency numbers of the operator A are finite. In this case, by the second Neumann formula (cf. Section 80), the manifold $(\tilde{A}_\lambda - \lambda E)\mathrm{D}_{\tilde{A}}$, where $\tilde{A}$ is any symmetric extension of the operator A, contains the manifold $(A_\lambda - \lambda E)\,\mathrm{D}_A$. However, the difference in dimensions is finite (i.e., the dimension of the first manifold modulo the second is finite). Therefore, the operator $(\tilde{A}_\lambda - \lambda E)^{-1}$ is bounded along with the operator $(A_\lambda - \lambda E)^{-1}$.

From these remarks follow the simple theorem.

THEOREM 1: *All self-adjoint extensions of an operator with equal and finite deficiency indices have the same continuous spectrum.*

Concerning the set of eigenvalues in the spectral kernel we have

THEOREM 2: *For an arbitrary extension of an operator with finite deficiency indices* (m, n) *to a self-adjoint operator, the multiplicity of an arbitrary eigenvalue is increased not more than* m *(in particular, the multiplicity of a new eigenvalue is not larger than* m*).*

Proof: Let $\tilde{A}$ be a self-adjoint extension of the operator A and let λ be an eigenvalue of multiplicity p of A. We assume that the multiplicity of λ as eigenvalue of the operator $\tilde{A}$ is equal to $p + q$ with $q > m$, contrary to the assertion of the theorem. We select a linear independent system of solutions $f_1, f_2, \ldots, f_p, f_{p+1}, \ldots, f_{p+q}$ of the equation $\tilde{A}f - \lambda f = 0$, so that $f_k \in \mathrm{D}_A$ for $k \leqq p$. Since the dimension of $\mathrm{D}_{\tilde{A}}$ modulo D_A is equal to m, there exists constants α_k such that

$$\alpha_1 f_{p+1} + \alpha_2 f_{p+2} + \ldots + \alpha_q f_{p+q} \in \mathrm{D}_A.$$

This equation implies that the multiplicity of λ as eigenvalue of the operator A is larger than p. Since this is a contradiction, the theorem is proved.

The following theorem is in some sense a converse of Theorem 2.

THEOREM 3: *If* λ *is a real point of regular type of a symmetric opera-*

tor A with finite deficiency indices (m, m), then there exists a self-adjoint extension $\tilde{A}$ of A for which the number λ is an eigenvalue of multiplicity m.

Proof: Let $\mathfrak{N}_\lambda$ denote the linear manifold of all solutions of the equations

$$A^*g - \lambda g = 0.$$

By the theorem on the invariance of the deficiency numbers in the field of regularity (cf. Section 78), the dimension of $\mathfrak{N}_\lambda$ is m.

The domain D_A of the operator A and the linear manifold $\mathfrak{N}_\lambda$ are linearly independent since, otherwise, the number λ would be an eigenvalue of A. Let

$$\mathrm{D} = \mathrm{D}_A \oplus \mathfrak{N}_\lambda. \tag{1}$$

Let $\tilde{A}$ be the operator which coincides with A^* on $\mathrm{D} = \mathrm{D}_{\tilde{A}}$. Then λ is an eigenvalue of A of multiplicity m.

We show now that the operator $\tilde{A}$ is self-adjoint. For this it is sufficient to prove that $\tilde{A}$ is symmetric, since it follows from (1) that

$$\dim \mathrm{D}_{\tilde{A}} = m \qquad (\mathrm{mod}\ \mathrm{D}_A).$$

If f and g are arbitrary elements of $\mathrm{D}_{\tilde{A}}$ and

$$f = f_1 + f_2 \qquad (f_1 \in \mathrm{D}_A,\ f_2 \in \mathfrak{N}_\lambda),$$
$$g = g_1 + g_2 \qquad (g_1 \in \mathrm{D}_A,\ g_2 \in \mathfrak{N}_\lambda),$$

then

$$(\tilde{A}f, g) = (Af_1, g_1) + (A^*f_2, g_1) + (Af_1, g_2) + (A^*f_2, g_2) =$$
$$= (Af_1, g_1) + \lambda(f_2, g_1) + \lambda(f_1, g_2) + \lambda(f_2, g_2),$$

and

$$(f, \tilde{A}g) = (f_1, Ag_1) + (f_1, A^*g_2) + (f_2, Ag_1) + (f_2, A^*g_2) =$$
$$= (f_1, Ag_1) + \lambda(f_1, g_2) + \lambda(f_2, g_1) + \lambda(f_2, g_2).$$

It follows that $\tilde{A}$ is symmetric.

In conclusion we give another theorem on the number of solutions of the equation

$$A^*g - \lambda g = 0$$

for real λ.

THEOREM 4: *If A is a symmetric operator with finite deficiency indices (m, m) and λ is a real number not belonging to the point spectrum of the operator A, then the number $m(\lambda)$ of the solutions of the equations*

$$A^*g - \lambda g = 0 \tag{2}$$

does not exceed the deficiency number m.

Proof: We define the domain $\mathrm{D}_{\tilde{A}}$ by formula (1), where $\mathfrak{N}_\lambda$ is the linear manifold of the solutions of equation (2). Then $\tilde{A} \subset A^*$. From the proof of the preceding theorem it follows that $\tilde{A}$ is a symmetric extension of A, so that

$$m(\lambda) = \dim \mathrm{D}_{\tilde{A}} \leqq m \qquad (\mathrm{mod}\ \mathrm{D}_A).$$

We shall refer to this last theorem in Appendix II.

84. The Formula of Krein for the Resolvent of the Self-Adjoint Extensions of a Symmetric Operator

In this section we consider symmetric operators with equal and finite deficiency indices.

Let A_1 and A_2 be two self-adjoint extensions of an operator A with finite deficiency indices (m, m):

$$A_1 \supset A, \qquad A_2 \supset A.$$

It is natural to call an operator C which satisfies the conditions

$$(1) \qquad A_1 \supset C, \qquad A_2 \supset C$$

a common part of the operators A_1 and A_2. It is evident that there exists an operator C which satisfies conditions (1) and which is an extension of every common part of A_1 and A_2. We call such an operator the *maximal common part* of A_1 and A_2. The maximal common part either is an extension of A or coincides with A and in the latter case we call the extensions A_1 and A_2 *relatively prime*. Two extensions A_1 and A_2 are relatively prime if and only if the conditions

$$(2) \qquad h \in \mathrm{D}_{A_1}, \quad h \in \mathrm{D}_{A_2}$$

imply that $h \in \mathrm{D}_A$.

If the maximal number of vectors which are linearly independent modulo D_A and which satisfy conditions (2) is equal to $p (0 \leqq p \leqq m)$, then the maximal common part A_0 of the operators A_1 and A_2 has deficiency indices $(m - p, m - p)$. In this case the operators A_1 and A_2 can be considered as relatively prime self-adjoint extensions of A_0.

The problem of the present section is the derivation of a formula which relates the resolvents of two self-adjoint extensions of an operator A. Let $\mathring{B}$ be a fixed self-adjoint extension, B an arbitrary self-adjoint extension, and let $\mathring{R}_z$ and R_z be their resolvents. Further, let λ be any common regular point of the operators $\mathring{B}$ and B (in particular, λ can be an arbitrary nonreal number).

In order not to exclude the case when $\mathring{B}$ and B are not relatively prime extensions of A, we shall consider them as relatively prime extensions of their maximal common part A_0 which has deficiency indices (r, r) where $0 < r \leqq m$.

We now let $\mathfrak{M}_\lambda = \Delta_{A_0}(\lambda)$ and $\mathfrak{N}_\lambda = \mathrm{H} \ominus \mathfrak{M}_\lambda$. Then the difference of the resolvents satisfies the formula

$$(3) \qquad (\mathring{R}_\lambda - R_\lambda) f \begin{cases} = 0 & \text{for } f \in \mathfrak{M}_\lambda, \\ \in \mathfrak{N}_{\bar{\lambda}} & \text{for } f \in \mathfrak{N}_\lambda. \end{cases}$$

This follows from the fact that, for each $h \in \mathfrak{M}_{\bar{\lambda}}$,

$$(\{\mathring{R}_\lambda - R_\lambda\} f, h) = (f, \{\mathring{R}_\lambda - R_\lambda\}^* h) = (f, \{\mathring{R}_{\bar{\lambda}} - R_{\bar{\lambda}}\} h) = (f, 0) = 0.$$

Now we choose any r linearly independent vectors $g_1(\bar{\lambda}), g_2(\bar{\lambda}), \ldots, g_r(\bar{\lambda})$ from $\mathfrak{N}_\lambda$ and r linearly independent vectors $g_1(\lambda), g_2(\lambda), \ldots, g_r(\lambda)$ from $\mathfrak{N}_{\bar{\lambda}}$. It follows from (3) that, for each $f \in \mathrm{H}$,

$$(4) \qquad (\mathring{R}_\lambda - R_\lambda) f = \sum_{k=1}^{r} c_k g_k(\lambda).$$

By (4), the constants c_k are linear functionals of f. Hence, there exist vectors $h_k(\lambda)$ such that

$$c_k = (f, h_k(\lambda)) \qquad (k = 1, 2, \ldots, r).$$

Since the vectors $g_1(\lambda), g_2(\lambda), \ldots, g_r(\lambda)$ are linearly independent, it follows from (3) and (4) that, for each $f \perp \mathfrak{N}_\lambda$,

$$(f, h_k(\lambda)) = 0 \qquad (k = 1, 2, \ldots, r).$$

Therefore,

$$h_k(\lambda) \in \mathfrak{N}_\lambda \qquad (k = 1, 2, \ldots, r),$$

so that each $h_k(\lambda)$ has a representation of the form

$$(5) \qquad h_k(\lambda) = \sum_{i=1}^{r} \overline{p_{ik}(\lambda)} g_i(\bar{\lambda}) \qquad (k = 1, 2, \ldots, r),$$

and (4) can be expressed by

$$(6) \qquad (\mathring{R}_\lambda - R_\lambda) f = \sum_{i,k=1}^{r} p_{ik}(\lambda)\,(f, g_i(\bar{\lambda}))\, g_k(\lambda).$$

We remark that the matrix function $(p_{ik}(\lambda)) = \mathfrak{P}(\lambda)$, which is defined on the set of all common regular points of the operators $\mathring{B}$ and B, is nonsingular. Indeed, if the determinant of $(p_{ik}(\lambda_0))$ were zero then, by (5), the vectors $h_k(\lambda_0)$ $(k = 1, 2, \ldots, r)$ would be linearly dependent, and this would imply the existence of a vector $h \neq 0$ such that

$$h \perp h_k(\lambda_0), \quad h \in \mathfrak{N}_{\lambda_0} \quad (k = 1, 2, \ldots, r).$$

Then it would follow from (4) that

$$(\mathring{R}_{\lambda_0} - R_{\lambda_0}) h = 0.$$

This would contradict the fact that $\mathring{B}$ and B are relatively prime extensions of the operator A_0.

In (6) we omit the element f and consider the expressions $(\cdot, g_i(\bar{\lambda}))g_k(\lambda)$ $(i, k = 1, 2, \ldots, r)$ as operators in order to obtain, for each common regular point λ of the operators $\mathring{B}$ and B, the formula

$$R_\lambda = \mathring{R}_\lambda - \sum_{i,k=1}^{r} p_{ik}(\lambda)\,(\cdot, g_i(\bar{\lambda}))\, g_k(\lambda). \tag{7}$$

Until now the choice of the vector functions $g_k(\lambda)$ and $g_i(\bar{\lambda})$ $(i, k = 1, 2, \ldots, r)$ have been arbitrary. At the same time the left member and, hence, also the right member of formula (6) is a regular analytic vector function of λ. We show next that $g_k(\lambda)$ $(k = 1, 2, \ldots, r)$ can be defined as a regular analytic function of λ and then we obtain a formula for the matrix function $\mathfrak{P}(\lambda)$ which corresponds to this choice.

With this purpose, we choose an arbitrary fixed value λ_0 and introduce the operator

$$U_{\lambda\lambda_0} = \frac{\mathring{B} - \lambda_0 E}{\mathring{B} - \lambda E} = E + (\lambda - \lambda_0)\,\mathring{R}_\lambda$$

with domain

$$(\mathring{B} - \lambda E)\, \mathrm{D}_{\mathring{B}} = \mathrm{H}$$

and range

$$(\mathring{B} - \lambda_0 E)\, \mathrm{D}_{\mathring{B}} = \mathrm{H}.$$

The operator $U_{\lambda\lambda_0}$ is defined by the formulas

$$\left.\begin{aligned} (\mathring{B} - \lambda E) f &= h, \\ (\mathring{B} - \lambda_0 E) f &= U_{\lambda\lambda_0} h, \end{aligned}\right\} \quad (f \in \mathrm{D}_{\mathring{B}}),$$

from which it follows that $U_{\lambda\lambda_0}$ determines a one-to-one correspondence of H to H.

In the special case with $\lambda = \bar{\lambda}_0$, the operator $U_{\lambda\lambda_0}$ becomes the Cayley transform of $\mathring{B}$ and it maps the deficiency subspace $\mathfrak{N}_{\bar{\lambda}_0}$ of A_0 into its deficiency subspace $\mathfrak{N}_{\lambda_0}$. We show that, in general,

$$U_{\lambda\lambda_0}\, \mathfrak{N}_{\bar{\lambda}_0} = \mathfrak{N}_{\bar{\lambda}}.$$

Let $g_1(\lambda_0), g_2(\lambda_0), \ldots, g_r(\lambda_0)$ be an arbitrary basis for $\mathfrak{N}_{\bar{\lambda}_0}$ (the vectors $g_k(\lambda_0)$ are not necessarily orthogonal or normalized). Now,

$$U_{\lambda\lambda_0} g_k(\lambda_0) \in \mathfrak{N}_{\bar{\lambda}}$$

since

$$\begin{aligned} A_0^* U_{\lambda\lambda_0} g_k(\lambda_0) &= A_0^* \{E + (\lambda - \lambda_0) \mathring{R}_\lambda\} g_k(\lambda_0) = \\ &= \lambda_0 g_k(\lambda_0) + (\lambda - \lambda_0) \mathring{B} \mathring{R}_\lambda g_k(\lambda_0) = \\ &= \lambda_0 g_k(\lambda_0) + (\lambda - \lambda_0)(E + \lambda \mathring{R}_\lambda) g_k(\lambda_0) = \\ &= \lambda \{E + (\lambda - \lambda_0) \mathring{R}_\lambda\} g_k(\lambda_0) \\ &= \lambda U_{\lambda\lambda_0} g_k(\lambda_0). \end{aligned}$$

Since the operator $U_{\lambda\lambda_0}$ is one-to-one, the vectors $U_{\lambda\lambda_0}\, g_k(\lambda_0)$ form a basis for $\mathfrak{N}_{\bar{\lambda}}$. Hence, we may define the vectors $g_k(\lambda)$ at each regular point λ of the operator $\mathring{B}$ by the formulas

$$g_k(\lambda) = U_{\lambda\lambda_0} g_k(\lambda_0) = g_k(\lambda_0) + (\lambda - \lambda_0) \mathring{R}_\lambda g_k(\lambda_0) \quad (k = 1, 2, \ldots, r).$$

With the aid of the functional equation of the resolvent, it is easy to verify that

$$g_k(\mu) = U_{\mu\lambda} g_k(\lambda) = g_k(\lambda) + (\mu - \lambda) \mathring{R}_\mu g_k(\lambda) \tag{8}$$

for any two regular points λ and μ of the operator $\mathring{B}$.

The value of the matrix function $\mathfrak{P}(\lambda)$ for any common regular point λ of $\mathring{B}$ and B is determined by its value $\mathfrak{P}(\lambda_0)$. In order to find the corresponding formula we make use of the functional equation of the resolvent:

$$R_\lambda = R_{\lambda_0} + (\lambda - \lambda_0) R_\lambda R_{\lambda_0}. \tag{9}$$

On the other hand, by (7)

$$\begin{aligned} R_\lambda &= \mathring{R}_\lambda - \sum_{i,k=1}^{r} p_{ik}(\lambda) (\cdot\,, g_i(\bar{\lambda}))\, g_k(\lambda), \\ R_{\lambda_0} &= \mathring{R}_{\lambda_0} - \sum_{i,k=1}^{r} p_{ik}(\lambda_0) (\cdot\,, g_i(\bar{\lambda}_0))\, g_k(\lambda_0). \end{aligned} \tag{10}$$

Substitute (10) into (9) and use the functional equation of the resolvent $\mathring{R}_\lambda$ to get

$$\begin{aligned} -\sum_{i,k=1}^{r} p_{ik}(\lambda) (\cdot\,, g_i(\bar{\lambda}))\, g_k(\lambda) &= -\sum_{i,k=1}^{r} p_{ik}(\lambda_0) (\cdot\,, g_i(\bar{\lambda}_0))\, g_k(\lambda_0) \\ &- (\lambda - \lambda_0) \sum_{i,k=1}^{r} p_{ik}(\lambda_0) (\cdot\,, g_i(\bar{\lambda}_0))\, \mathring{R}_\lambda g_k(\lambda_0) \\ &- (\lambda - \lambda_0) \sum_{i,k=1}^{r} p_{ik}(\lambda) (\mathring{R}_{\lambda_0} \cdot\,, g_i(\bar{\lambda}))\, g_k(\lambda) \\ &+ (\lambda - \lambda_0) \sum_{i,k,j,s=1}^{r} p_{ik}(\lambda) (g_s(\lambda_0), g_i(\bar{\lambda}))\, p_{js}(\lambda_0) (\cdot\,, g_j(\bar{\lambda}_0))\, g_k(\lambda). \end{aligned} \tag{11}$$

By means of (8), we express the sum of the second and third summands on the right in the form

$$\sum_{i,k=1}^{r} p_{ik}(\lambda_0)(\cdot\,, g_i(\bar{\lambda}_0))\{g_k(\lambda_0) - g_k(\lambda)\} + \sum_{i,k=1}^{r} p_{ik}(\lambda)(\cdot\,, g_i(\bar{\lambda}_0) - g_i(\bar{\lambda}))\, g_k(\lambda).$$

With this substitution, (11) yields

$$-\sum_{i,k=1}^{r} p_{ik}(\lambda_0)(\cdot\,, g_i(\bar{\lambda}_0))\, g_k(\lambda) + \sum_{i,k=1}^{r} p_{ik}(\lambda)(\cdot\,, g_i(\bar{\lambda}_0))\, g_k(\lambda) +$$

$$+ (\lambda - \lambda_0) \sum_{i,k,j,s=1}^{r} p_{ik}(\lambda)(g_s(\lambda_0), g_i(\bar{\lambda}))\, p_{js}(\lambda_0)(\cdot\,, g_j(\bar{\lambda}_0))\, g_k(\lambda) = 0.$$

Since the vectors $g_k(\lambda)$ are linear independent,

$$-\sum_{i=1}^{r} p_{ik}(\lambda_0)(\cdot\,, g_i(\bar{\lambda}_0)) + \sum_{i=1}^{r} p_{ik}(\lambda)(\cdot\,, g_i(\bar{\lambda}_0)) +$$

$$+ (\lambda - \lambda_0) \sum_{i,j,s=1}^{r} p_{ik}(\lambda)(g_s(\lambda_0), g_i(\bar{\lambda}))\, p_{js}(\lambda_0)(\cdot\,, g_j(\lambda_0)) = 0,$$

and, further, since the vectors $g_i(\bar{\lambda}_0)$ are linearly independent,

$$-p_{ik}(\lambda_0) + p_{ik}(\lambda) + (\lambda - \lambda_0) \sum_{j,s=1}^{r} p_{is}(\lambda_0)(g_s(\lambda_0), g_j(\bar{\lambda}))\, p_{jk}(\lambda) = 0$$

or, in matrix form,

$$\mathfrak{P}(\lambda) - \mathfrak{P}(\lambda_0) + (\lambda - \lambda_0)\, \mathfrak{P}(\lambda_0)\big((g_s(\lambda_0), g_j(\bar{\lambda}))\big)\, \mathfrak{P}(\lambda) = 0.$$

Multiply this equation on the right by $\mathfrak{P}^{-1}(\lambda)$ and on the left by $\mathfrak{P}^{-1}(\lambda_0)$ to obtain finally the relations

$$\mathfrak{Q}(\lambda) = \mathfrak{Q}(\lambda_0) + (\lambda - \lambda_0)\big((g_s(\lambda_0), g_j(\bar{\lambda}))\big)_{s,j=1}^{r}, \tag{12}$$

where $\mathfrak{Q}(\lambda) = \mathfrak{P}^{-1}(\lambda)$. It is not difficult to verify by means of (12) that, for any two common regular points λ and μ of the operators $\mathring{B}$ and B,

$$\mathfrak{Q}(\lambda) = \mathfrak{Q}(\mu) + (\lambda - \mu)\big((g_s(\mu), g_j(\bar{\lambda}))\big)_{s,j=1}^{r}.$$

85. Semi-Bounded Operators

A symmetric operator A is *bounded below* if

$$(Af, f) \geqq m\, \|f\|^2 \qquad (m > -\infty) \tag{1}$$

for each $f \in D_A$, and *bounded above* if

$$(Af, f) \leqq M\, \|f\|^2 \qquad (M < \infty) \tag{2}$$

for each $f \in D_A$. We shall assume that the numbers m and M in the inequalities (1) and (2) cannot be improved. Thus,

$$m = \inf_{f \in D_A} \frac{(Af, f)}{(f, f)}, \qquad M = \sup_{f \in D_A} \frac{(Af, f)}{(f, f)},$$

and we call m the *lower bound* and M the *upper bound* of A.

A special case of an operator semi-bounded from below is the positive operator introduced in Section 41. The study of semi-bounded operators can be reduced to the study of positive operators. Indeed, the operator A has the lower bound $m > -\infty$ if and only if the operator $A - mE$ is positive.

If A is a positive operator, then the negative semi-axis belongs to its field of regularity, since the inequality

$$(Af, f) \geqq 0$$

implies, for negative λ,

$$\|(A - \lambda E)f\|^2 = \|Af\|^2 - 2\lambda(Af, f) + \lambda^2\|f\|^2 \geqq \lambda^2\|f\|^2$$

and, hence,

$$\|(A - \lambda E)f\| \geqq |\lambda| \cdot \|f\|.$$

Therefore, by proposition 1° of Section 78, a semi-bounded operator has equal deficiency numbers and has self-adjoint extensions. If the negative semi-axis belongs to the field of regularity of a symmetric operator A then it is not necessarily true that A is positive,[4] unless A is self-adjoint. In the latter case, for each $f \in D_A$,

$$(Af, f) = \int_{-\infty}^{\infty} t\, d(E_t f, f) = \int_{0}^{\infty} t\, d(E_t f, f) \geqq 0.$$

The square of any self-adjoint operator is a positive operator. Conversely, any positive self-adjoint operator A can be represented in the form of the square of some self-adjoint operator B. Indeed, if

$$A = \int_0^\infty t\, dE_t,$$

then it is possible, for example, to let

$$B = \int_0^\infty \sqrt{t}\, dE_t.$$

[4] For example, for the differential operator on a finite interval, the field of regularity is the entire plane, but this operator is not semi-bounded.

If a semi-bounded operator has finite deficiency indices, then all of its self-adjoint extensions are also semi-bounded. Moreover, the following theorem holds.

THEOREM 1: *If A is a positive operator with the finite deficiency indices (m, m), then each of its self-adjoint extensions has only a finite number of negative eigenvalues, and the sum of the multiplicities of these eigenvalues does not exceed m.*

PROOF: Let $\tilde{A}$ be some self-adjoint extension of the operator A and

$$\tilde{A} = \int_{-\infty}^{\infty} t\, dE_t.$$

Further, let Δ_ε denote the interval $(-\infty, -\varepsilon)$. To establish the theorem it is sufficient to prove that, for each $\varepsilon > 0$, the dimension of the subspace $E(\Delta_\varepsilon)\mathrm{H}$ does not exceed the deficiency number m.

We assume that for some $\varepsilon > 0$

$$\dim E(\Delta_\varepsilon)\,\mathrm{H} > m. \tag{3}$$

Since

$$\dim \mathrm{D}_{\tilde{A}} = m \pmod{\mathrm{D}_A},$$

there exists a vector f_0 which is both in D_A and in the subspace $E(\Delta_\varepsilon)\,\mathrm{H}$. Then

$$(Af_0, f_0) = (\tilde{A}f_0, f_0) = \int_{-\infty}^{-\varepsilon} t\, d(E_t f_0, f_0) < 0.$$

Since this contradicts the fact that A is positive, the theorem is proved.

We present another theorem about self-adjoint extensions of semi-bounded operators with arbitrary deficiency indices.

THEOREM 2: *A semi-bounded operator A with lower bound m has a self-adjoint extension $\tilde{A}$ with lower bound not smaller than an arbitrarily pre-assigned number $m' < m$.*

Proof: First, we prove the theorem for $m = 1$ and $m' = 0$. In this case,

$$(f, f) \leqq (Af, f) \leqq \|Af\| \cdot \|f\|,$$

so that

$$\|Af\| \geqq \|f\|$$

and

$$\|Af - Ag\| \geqq \|f - g\|.$$

It follows that A^{-1} exists as a bounded operator defined on the subspace Δ_A. Repeating the argument of item 2° of Section 78, it is easy to see that

$$\mathfrak{N}_0 = \mathrm{H} \ominus \Delta_A$$

is the eigenmanifold of the operator A^* which belongs to the eigenvalue $\lambda = 0$. The subspace $\mathfrak{N}_0$ and the manifold D_A are linearly independent since if an element g belongs to both $\mathfrak{N}_0$ and D_A, then $Ag = 0$, which implies that $g = 0$.

We define an extension $\tilde{A}$ of the operator A on the domain

$$\mathrm{D}_{\tilde{A}} = \mathrm{D}_A \oplus \mathfrak{N}_0$$

by the formula

$$\tilde{A}h = Af$$

with

$$h = f + g, \quad f \in \mathrm{D}_A, \quad g \in \mathfrak{N}_0.$$

It is evident that $\tilde{A}$ is a symmetric extension of A, since if

$$h_1, h_2 \in \mathrm{D}_{\tilde{A}} \ (h_i = f_i + g_i,\ f_i \in \mathrm{D}_A,\ g_i \in \mathfrak{N}_0,\ i = 1, 2),$$

then

$$(\tilde{A}h_1, h_2) = (Af_1, f_2 + g_2) = (Af_1, f_2) =$$
$$= (f_1, Af_2) = (f_1 + g_1, Af_2) = (h_1, \tilde{A}h_2).$$

Furthermore, it is evident that the subspace $\mathfrak{N}_0$ reduces $\tilde{A}$ since $\mathfrak{N}_0 \subset \mathrm{D}_{\tilde{A}}$ and $\tilde{A}h = 0$ for $h \in \mathfrak{N}_0$. Both $\mathfrak{N}_0$ and its orthogonal complement Δ_A are reduced by the operator $\tilde{A}$. Let $\tilde{A}'$ and $\tilde{A}''$ be the restrictions of $\tilde{A}$ to Δ_A and $\mathfrak{N}_0$.

The range of the operator $\tilde{A}'$ is the subspace Δ_A and, hence, (cf. Section 41), $\tilde{A}'$ is self-adjoint. Since $\tilde{A}'' = 0$, the operator

$$\tilde{A} = \tilde{A}' \oplus \tilde{A}''$$

is also self-adjoint.

Furthermore, for

$$h = f + g \qquad (f \in \mathrm{D}_A, g \in \mathfrak{N}_0),$$

we have

$$(\tilde{A}h, h) = (Af, f + g) = (Af, f) \geqq \|f\|^2 \geqq 0,$$

so that the lower bound of the operator $\tilde{A}$ is not negative. Thus, the theorem is proved for the case with $m = 1$ and $m' = 0$. It is easy to see that the general case is reduced to the case considered by the linear transformation

$$A_1 = \frac{1}{m - m'} A - \frac{m'}{m - m'} E.$$

The theorem proved above was established by J. von Neumann. In this connection J. von Neumann made the conjecture that the theorem is true with $m' = m$. Later this conjecture was proved by Stone and Friedrichs, but these authors left open the question of the uniqueness of the extension with preservation of the lower bound.

A complete theory of the self-adjoint extensions of semi-bounded operators with preservation of the lower bound and also the application of this theory to differential equations is due to Krein.[5] A presentation of these results exceeds the scope of this book.

Concluding the present section, we prove a proposition which is a generalization of the theorem of Section 33. There we introduced the notion of inequality between projection operators. Now we introduce the concept of inequality between bounded self-adjoint operators. We write

$$B \geqq A$$

if $B - A$ is a positive operator.

LEMMA: *A monotone increasing sequence of positive self-adjoint operators*

$$0 < A_1 < A_2 < A_3 < \dots$$

with bounded norms

$$\| A_n \| \leqq N \qquad (n = 1, 2, 3, \dots)$$

converges strongly to some limit operator.

Proof: First we remark that if B is a positive self-adjoint operator with norm less than or equal to one, then $B^2 \leqq B$, i.e.,

$$(B^2 f, f) \leqq (Bf, f).$$

This fact follows immediately from the spectral representation of the operator B, but it can be established also without the application of the spectral representation in the following way. Since $\| B \| \leqq 1$,

$$(Bf, f) \leqq (f, f),$$

and, hence,

$$E - B \geqq 0.$$

[5] cf. M. G. Krein [3], [4], Vol. I.

Also

$$(E-B)\,B\,(E-B) \geqq 0, \quad B\,(E-B)\,B \geqq 0,$$

since

$$(\{E-B\}\,B\,\{E-B\}f,f) = (B\,\{E-B\}f,\,\{E-B\}f) \geqq 0,$$
$$(B\,\{E-B\}\,Bf,f) = (\{E-B\}\,Bf,\,Bf) \geqq 0.$$

But, on the other hand,

$$(E-B)\,B\,(E-B) + B\,(E-B)\,B = B - B^2.$$

Hence, we conclude that

$$B - B^2 \geqq 0.$$

Turning now to the proof of the lemma we let $N = 1$ without loss of generality. Since the numerical sequence $\{(A_n f, f)\}_1^\infty$, where f is a fixed element of H, is monotone and bounded, it converges to a finite limit. Therefore, for sufficiently large p_ϵ and for $n > m > p_\epsilon$

$$(\{A_n - A_m\}\,f, f) \leqq \epsilon.$$

On the other hand the identity

$$(\{A_n - A_m\}f,f) + (A_m f,f) = (A_n f,f)$$

and the positivity of the summands yield

$$\| A_n - A_m \| \leqq 1.$$

By the remark which was made in the beginning of the proof, we obtain

$$\|(A_n - A_m)f\|^2 = (\{A_n - A_m\}f,\,\{A_n - A_m\}f) =$$
$$= (\{A_n - A_m\}^2 f,f) \leqq (\{A_n - A_m\}f,f) \leqq \epsilon,$$

and the lemma is proved.

86. Some Remarks about the General Theory of Extensions

The presentation in the present chapter of the theory of extensions is based essentially on the assumption that the domain of the operator to be extended is dense in H. This assumption was included in Section 41 for the definition of the concept of a symmetric operator.

Some questions reduce, however, to the consideration of operators with domain not dense in H which satisfy the conditions

$$(Af, g) = (f, Ag) \qquad (f, g \in \mathrm{D}_A).$$

Such operators are also called symmetric. The theory of extensions of von Neumann does not apply to them. In particular, this theory leaves open even the question of the existence of self-adjoint extensions for bounded operators which are symmetric in the extended sense. This last question was resolved in the affirmative by Krein[6], who established the existence of self-adjoint extensions of a bounded operator with preservation of the norm, described these extensions and also gave a criterion for their uniqueness.

Recently Krasnoselski[7] considered the question of symmetric extensions of symmetric operators with nondense domain in complete generality. He showed that every closed symmetric operator has maximal symmetric extensions. If the deficiency indices of such an operator are equal (and only in that case) then it has self-adjoint extensions. The self-adjoint extensions are characterized by formulas similar to formulas (4) and (4') of Section 80.

In conclusion it is necessary to remark that for the construction of the theory of extensions in such generality one must assume that the operator has a closure. For operators with nondense domains, the existence of closure is not a consequence of symmetry.

The extension of a symmetric operator with a nondense domain for the case when the closure of the domain contains its range was studied first in a remarkable work of M. A. Naimark in connection with the construction of the theory of generalized spectral functions. The presentation of the theory of M. A. Naimark forms the main part of Appendix I.

---

[6] cf. M. G. Krein [3], [4], Vol. I.
[7] cf. M. A. Krasnoselski [2], Vol. I.

Appendix I

GENERALIZED EXTENSIONS AND GENERALIZED SPECTRAL FUNCTIONS OF SYMMETRIC OPERATORS

1. Generalized Resolution of the Identity. Naimark's Theorem[1]

A generalized resolution of the identity is defined as a one parameter family of operators F_t which satisfy the following conditions:

(A) Each difference $F_{t_2} - F_{t_1}$, where $t_2 > t_1$, is a bounded positive operator,

(B) $F_{t-0} = F_t$,

(C) $F_{-\infty} = 0$, $F_{\infty} = E$.

In contrast with the definition of the ordinary resolution of the identity (cf. Section 61) it is not assumed here that the operators F_t are projections. Furthermore, we do not assume here the earlier orthogonality condition,

$$F_u F_v = F_s \qquad (s = \min \{u, v\}), \tag{1}$$

since (1) and (A) would imply that F_t is a projection operator.

In view of the lemma of Section 85 on monotone sequences of operators, any family of operators F_t which satisfy condition (A) can be modified or "normalized" so that condition (B) is also satisfied.

The ordinary resolution of the identity is a special case of the generalized one. Sometimes the latter is called simply the resolution of the identity and the ordinary one is called the orthogonal resolution of the identity.

In terms of F_t we define a positive additive operator function

$$F(\Delta) = F_\Delta = F_{t_2} - F_{t_1}$$

of an interval $\Delta = (t_1, t_2)$, where $t_1 < t_2$.

A simple example of a generalized resolution of the identity is given by

$$F_t = \mu_1 E_t^{(1)} + \mu_2 E_t^{(2)},$$

[1] Cf. M. A. Naimark [1], [4], Vol. I.

where $E_t^{(1)}$ and $E_t^{(2)}$ are arbitrary orthogonal resolutions of the identity and μ_1 and μ_2 are any positive numbers, such that $\mu_1 + \mu_2 = 1$.

A more instructive example is given by the operator function F_t which is obtained by the following construction. Let E_t be an orthogonal resolution of the identity of the space H, let G be a subspace of H, and let P be the operator of projection of H onto G. We define

$$F_t = PE_t,$$

and consider F_t as restricted to G. It is easy to see that F_t satisfies conditions (A), (B), and (C) and, hence, is a resolution of the identity (not necessarily orthogonal) of the space G.

Naimark proved that every generalized resolution of the identity of the space H can be obtained by the method described above if one imbeds H in a particular space H^+. For the proof of the theorem of M. A. Naimark we need a certain important method for the construction of Hilbert spaces. We first present this method. For this purpose we introduce the concept of a positive-definite function. A scalar function $\Phi(\mathfrak{f}, \mathfrak{g})$ which is defined for each pair $\mathfrak{f}$, $\mathfrak{g}$ of elements of some set $\mathfrak{R}$ is called positive-definite if

$$\Phi(\mathfrak{f}, \mathfrak{g}) = \overline{\Phi(\mathfrak{g}, \mathfrak{f})}$$

for all $\mathfrak{f}$, $\mathfrak{g}$ in $\mathfrak{R}$ and if for arbitrary $\mathfrak{f}_1, \mathfrak{f}_2, \ldots, \mathfrak{f}_n$ $(n < \infty)$ in $\mathfrak{R}$ the quadratic form

$$\sum_{i,k=1}^{n} \Phi(\mathfrak{f}_i, \mathfrak{f}_k)\, \xi_i \bar{\xi}_k$$

is non-negative.

An example of a positive-definite function of a vector pair in Hilbert space is the scalar product, since

$$(f, g) = \overline{(g, f)},$$

and

$$\sum_{i,k=1}^{n} (f_i, f_k)\, \xi_i \bar{\xi}_k = \left\| \sum_{i=1}^{n} \xi_i f_i \right\|^2 \geqq 0.$$

On the other hand, if a positive-definite function of a pair of elements of an arbitrary set $\mathfrak{R}$ (no algebraic operations are defined in $\mathfrak{R}$) is given, then this set can be made into a Hilbert space. To be precise, $\mathfrak{R}$ can be imbedded in a Hilbert space H^+ such that the scalar product of $\mathfrak{f}$ and $\mathfrak{g}$ in $\mathfrak{R}$ is defined by

$$(\mathfrak{f}, \mathfrak{g}) = \Phi(\mathfrak{f}, \mathfrak{g}).$$

To construct H^+ we complete $\mathfrak{R}$ to a linear space $\hat{\mathfrak{R}}$, in which we introduce formally the finite sums

$$\hat{\mathfrak{f}} = \sum_{i=1}^{n} \xi_i \mathfrak{f}_i$$

for arbitrary $\mathfrak{f}_i \in \mathfrak{R}$ and arbitrary numbers $\xi_i (i = 1, 2, \ldots, n)$. For elements

$$\hat{\mathfrak{f}} = \sum_{i=1}^{n} \xi_i f_i \quad , \quad \hat{\mathfrak{g}} = \sum_{k=1}^{n} \eta_k \mathfrak{g}_k$$

in $\mathfrak{R}$ we define a scalar product by the formula

$$(\hat{\mathfrak{f}}, \hat{\mathfrak{g}}) = \sum_{i,k=1}^{n} \Phi(\mathfrak{f}_i, \mathfrak{g}_k)\, \xi_i \bar{\eta}_k.$$

The scalar product thus defined has all the properties listed in Section 2, Chapter 1, with the exception of perhaps one; namely, it is possible that

$$(2) \qquad (\hat{\mathfrak{f}}, \hat{\mathfrak{f}}) = 0 \qquad \text{for } \hat{\mathfrak{f}} \neq 0.$$

If we retrace the proof of the Cauchy-Bunyakovski inequality in Section 2, we see that this proof is valid for the scalar product defined above (strictly speaking, we should call it a *quasiscalar product*). Thus,

$$(3) \qquad |(\hat{\mathfrak{f}}, \hat{\mathfrak{g}})|^2 \leqq (\hat{\mathfrak{f}}, \hat{\mathfrak{f}})(\hat{\mathfrak{g}}, \hat{\mathfrak{g}}).$$

From this inequality it follows that the set $\hat{\mathfrak{N}}$ of all elements $\hat{\mathfrak{f}}$ for which $(\hat{\mathfrak{f}}, \hat{\mathfrak{f}}) = 0$ is a linear manifold.

In order to remove the defect indicated by (2), we introduce the quotient space

$$\mathrm{R}^+ = \hat{\mathfrak{R}}/\hat{\mathfrak{N}}.$$

Elements of the space R^+ are the sets in $\hat{\mathfrak{F}} \subset \hat{\mathfrak{R}}$ such that if $\hat{\mathfrak{f}}_0 \in \hat{\mathfrak{F}}$, then $\hat{\mathfrak{F}}$ consists of all the elements

$$\hat{\mathfrak{f}} = \hat{\mathfrak{f}}_0 + \hat{\mathfrak{g}},$$

where $\hat{\mathfrak{g}}$ is an arbitrary element of $\hat{\mathfrak{N}}$. Multiplication of an element $\hat{\mathfrak{F}}$ of R^+ by a scalar λ is defined such that the product $\lambda\hat{\mathfrak{F}}$ consists of all the elements $\lambda\hat{\mathfrak{f}}$ where $\hat{\mathfrak{f}} \in \hat{\mathfrak{F}}$. The sum in R^+ is defined analogously.

The scalar product of two elements $\hat{\mathfrak{F}}$ and $\hat{\mathfrak{G}}$ in R^+ is defined by the equation

$$(4) \qquad (\hat{\mathfrak{F}}, \hat{\mathfrak{G}}) = (\hat{\mathfrak{f}}, \hat{\mathfrak{g}}) \qquad (\hat{\mathfrak{f}} \in \hat{\mathfrak{F}},\ \hat{\mathfrak{g}} \in \hat{\mathfrak{G}}).$$

Using (3), it is easy to conclude that the scalar product $(\hat{\mathfrak{F}}, \hat{\mathfrak{G}})$ does not depend on the choice of the representatives $\hat{\mathfrak{f}}$ and $\hat{\mathfrak{g}}$ and, hence, that definition (4) is unambiguous. The scalar product $(\hat{\mathfrak{F}}, \hat{\mathfrak{G}})$ now has without exception all the properties listed in Section 2. The completion (cf. Section 3) of the inner product space R^+ is a Hilbert space which we denote by H^+. This space H^+ is the one sought.

Now we come to the theorem of Naimark.

THEOREM: *Let F_t be an arbitrary resolution of the identity for the space* H. *Then there exists a Hilbert space* H^+ *which contains* H *as a subspace and there exists an orthogonal resolution of the identity E_t^+ for the space* H^+ *such that*

$$F_t f = P^+ E_t^+ f$$

for each $f \in$ H where P^+ is the operator of projection on H.

Proof: Consider the set $\mathfrak{R}$ of all pairs $\mathfrak{p}$ of the form

$$\mathfrak{p} = \{\Delta, f\},$$

where Δ is an arbitrary real interval and f is an arbitrary vector of H. On $\mathfrak{R}$ we define a function $\Phi(\mathfrak{p}_1, \mathfrak{p}_2)$ such that if $\mathfrak{p}_1 = \{\Delta_1, f_1\}$ and $\mathfrak{p}_2 = \{\Delta_2, f_2\}$, then

$$\Phi(\mathfrak{p}_1, \mathfrak{p}_2) = (F_{\Delta_1 \cdot \Delta_2} f_1, f_2).$$

We show that the function $\Phi(\mathfrak{p}_1, \mathfrak{p}_2)$ is positive-definite. Indeed,

$$\Phi(\mathfrak{p}_1, \mathfrak{p}_2) = (F_{\Delta_1 \cdot \Delta_2} f_1, f_2) = (f_1, F_{\Delta_1 \cdot \Delta_2} f_2) = \overline{(F_{\Delta_1 \cdot \Delta_2} f_2, f_1)} = \overline{\Phi(\mathfrak{p}_2, \mathfrak{p}_1)}$$

and, on the other hand,

$$(5) \qquad \sum_{i,k=1}^{n} \Phi(\mathfrak{p}_i, \mathfrak{p}_k)\, \xi_i \bar{\xi}_k = \sum_{i,k=1}^{n} (F_{\Delta_i \cdot \Delta_k} f_i, f_k)\, \xi_i \bar{\xi}_k.$$

If the intervals Δ_i $(i = 1, 2, \ldots, n)$ are pairwise disjoint, then

$$(6) \qquad \sum_{i,k=1}^{n} (F_{\Delta_i \cdot \Delta_k} f_i, f_k)\, \xi_i \bar{\xi}_k = \sum_{i=1}^{n} (F_{\Delta_i} f_i, f_i)\, |\xi_i|^2 \geqq 0.$$

If the intervals $\Delta_i (i = 1, 2, \ldots, n)$ are pairwise disjoint and the intervals Δ_1 and Δ_2 coincide, then the sums in the right member of (5) fall into two parts. One part, with indices from 3 to n, is of the form (6), and the other part, with indices 1 and 2, satisfies

$$\sum_{i,k=1}^{2} (F_{\Delta_i \cdot \Delta_k} f_i, f_k)\, \xi_i \bar{\xi}_k = \sum_{i,k=1}^{2} (F_{\Delta_1} f_i, f_k)\, \xi_i \bar{\xi}_k = (F_{\Delta_i} \sum_{i=1}^{2} \xi_i f_i, \sum_{k=1}^{2} \xi_k f_k) \geqq 0.$$

The case with arbitrary intervals $\Delta_i (i = 1, 2, \ldots, n)$ can be reduced, with the aid of additional partitions, to the cases already considered. Hence, if $\Delta_1 \cap \Delta_2 = 0$, then

$$(F_{(\Delta_1+\Delta_2) \cdot \Delta_3} f, g) = (F_{\Delta_1 \cdot \Delta_3 + \Delta_2 \cdot \Delta_3} f, g) = (F_{\Delta_1 \cdot \Delta_3} f, g) + (F_{\Delta_2 \cdot \Delta_3} f, g).$$

Thus, $\Phi(\mathfrak{p}_1, \mathfrak{p}_2)$ is a positive-definite function on $\mathfrak{R}$.

Using the method described earlier we imbed $\mathfrak{R}$ in a Hilbert space H^+.

Not desiring to introduce new notations for those elements $\mathfrak{P}$ of the space H^+ which are subsets of $\mathfrak{R}$ by the construction described earlier, we agree on the following: if an element $\mathfrak{p}$ of $\mathfrak{R}$ belongs to $\mathfrak{P}$ then we write $\mathfrak{p}$ instead of $\mathfrak{P}$.

We indicate the scalar product in the space H^+ by the symbol $_+$, and have

$$(\mathfrak{p}_1, \mathfrak{p}_2)_+ = \Phi(\mathfrak{p}_1, \mathfrak{p}_2).$$

We now consider elements of H^+ of the form $\{I, f\}$, $I = [-\infty, \infty]$. By means of the equation

$$(\{I, f\}, \{I, g\})_+ = (F_I f, g) = (f, g),$$

we can identify the pair $\{I, f\}$ with the element f from H. The element $\sum_{k=1}^{n} \xi_k \{I, f_k\}$ of the space H^+ is identified with the element $\sum_{k=1}^{n} \xi_k f_k$ of the space H. Thus, H can be considered as a subspace of the space H^+.

We now solve the following problem: find the projection of the element $\{\Delta, f\}$ of the space H^+ on the subspace H. We denote the projection to be found by $\{I, g\}$. For each h of H,

$$(\{\Delta, f\} - \{I, g\}, \{I, h\})_+ = 0,$$

or

$$(\{\Delta, f\}, \{I, h\})_+ - (\{I, g\}, \{I, h\})_+ = (F_\Delta f, h) - (g, h) = (F_\Delta f - g, h) = 0,$$

so that

$$g = F_\Delta f,$$

i.e.

$$P^+\{\Delta, f\} = \{I, F_\Delta f\}. \tag{7}$$

The theorem will be proved if it is established that the operator function E_Δ^+, which is defined by

$$E_\Delta^+\{\Delta', f\} = \{\Delta \cap \Delta', f\} \tag{8}$$

for each element of the form $\{\Delta', f\} \in \mathrm{H}^+$ is an orthogonal resolution of the identity for the space H^+, since then (7) can be expressed in the form

$$P^+ E_\Delta^+ f = P^+ E_\Delta^+ \{I, f\} = P^+\{\Delta \cap I, f\} = P^+\{\Delta, f\} = \{I, F_\Delta f\} = F_\Delta f$$

for each $f \in \mathrm{H}$.

It is evident that E_Δ^+ is an additive operator function of an interval. Furthermore, the two equations

$$(E_\Delta^+)^2\{\Delta', f\} = E_\Delta^+\{\Delta \cap \Delta', f\} = \{\Delta \cap \Delta \cap \Delta', f\} = E_\Delta^+\{\Delta', f\},$$

and

$$(E_\Delta^+\{\Delta',f\}, \{\Delta'', g\})_+ = (\{\Delta \cap \Delta',f\}, \{\Delta'',g\})_+ =$$
$$= (F_{\Delta\cdot\Delta'\cdot\Delta''} f, g) = (F_{\Delta'\cdot\Delta\cdot\Delta''} f, g) = (\{\Delta',f\}, E_\Delta^+ \{\Delta'',g\})_+,$$

imply that E_Δ^+ is a projection operator. Finally, it is evident that $E_I^+\{\Delta',f\} = \{\Delta',f\}$.

Since the family of all elements of the form $\{\Delta',f\}$ is dense in H^+, the extension to H^+ by continuity of the operator E_Δ^+ defined by formula (8) is an orthogonal resolution of the identity for the space H^+. The theorem is proved.

2. Self-Adjoint Extensions to Larger Spaces and Spectral Functions of Symmetric Operators[2]

In the present section we apply the concept of the orthogonal sum of the spaces $H_1 \oplus H_2$. A special case ($H_1 \oplus H_1$) of this concept occurred in the definition of the graph of an operator (cf. Section 46). The orthogonal sum

$$H^+ = H_1 \oplus H_2. \tag{1}$$

of two arbitrary Hilbert spaces H_1 and H_2 is the set of all pairs $f^+ = \{f_1, f_2\}$ with $f_1 \in H_1$ and $f_2 \in H_2$. The algebraic operations and the metric are defined analogously to those introduced in Section 46.

It is evident that H^+ is a Hilbert space. If H_1 and H_2 are identified with the sets of all pairs $\{f_1, 0\}$ and $\{0, f_2\}$ with $f_1 \in H_1$ and $f_2 \in H_2$, then H_1 and H_2 can be regarded as mutually orthogonal subspaces of H^+. After the construction of H^+, formula (1) can be considered as a decomposition of H^+ into an orthogonal sum.

A generalization of the concept of a symmetric extension is basic for our further considerations. Let A be a symmetric operator which is defined in a space H and let H^+ be a Hilbert space which contains H. In supplement to the definition in Section 41, we shall call each symmetric (in particular, self-adjoint) operator B^+ which is defined in H^+ and is an extension of the operator A, a symmetric (in particular, self-adjoint) extension of the operator A.

The concept of an isometric (in particular, unitary) extension of an isometric operator is generalized analogously. It is evident that the Cayley transform of a symmetric extension B^+ of an operator A is an isometric

[2] Cf. M. A. Naimark [2], [1], Vol. I. Before Naimark constructed the general theory which is presented in this and the following section, separate results were obtained by different methods by A. I. Plesner [4], Vol. I, and N. I. Akhiezer [4], Vol. I.

(respectively, unitary) extension of the Cayley transform of the operator A. Conversely, the Cayley transform of an isometric (unitary) extension of an isometric operator is a symmetric (respectively, self-adjoint) extension of the Cayley transform of this operator.

For $\mathrm{H}^+ = \mathrm{H}$ we obtain the ordinary symmetric and isometric extensions which have been considered heretofore.

Let B^+ be an arbitrary symmetric extension of the operator A. Then the following relation holds:

$$\mathrm{D}_A \subset \mathrm{D}_{B+} \cap \mathrm{H} \subset \mathrm{D}_{B+}.$$

It is convenient to classify the symmetric extensions B^+ of an operator A by the following scheme.

Extension of type I: $\mathrm{D}_A \neq \mathrm{D}_{B+} \cap \mathrm{H} = \mathrm{D}_{B+}$; this extension coincides with the ordinary one.

Extension of type II: $\mathrm{D}_A = \mathrm{D}_{B+} \cap \mathrm{H} \neq \mathrm{D}_{B+}$.

Extension of type III. $\mathrm{D}_A \neq \mathrm{D}_{B+} \cap \mathrm{H} \neq \mathrm{D}_{B+}$.

Thus, the ordinary symmetric extensions (i.e., those which do not extend beyond the space) are extensions of the first type, and symmetric extensions which do go beyond the space are of types II and III. It is evident that a maximal operator has only symmetric extensions of type II.

If a symmetric extension B^+ of an operator A is reduced by a subspace $\mathrm{G}^+ \subset \mathrm{H}^+ \ominus \mathrm{H}$, then we shall always exclude G^+ from H^+ (i.e., we replace the space H^+ by the space $\mathrm{H}^+ \ominus \mathrm{G}^+$ and the operator B^+ by its restriction to $\mathrm{H}^+ \ominus \mathrm{G}^+$). Under this condition, a self-adjoint operator admits no symmetric extensions.

We now turn to the fundamental theorem of the present section.

THEOREM 1: *Every symmetric operator A defined in a Hilbert space with arbitrary deficiency indices* $(\mathfrak{m}, \mathfrak{n})$ *can be extended to a self-adjoint operator B^+ which is defined in a space* $\mathrm{H}^+ \supset \mathrm{H}$.

Proof: In some space H' of sufficiently large dimension we construct a symmetric operator A' with deficiency indices $(\mathfrak{n}, \mathfrak{m})$. (One can, for example, choose H' isomorphic to H and A' isomorphic to $(-1)\,A$). We now construct the space

$$\mathrm{H}^+ = \mathrm{H} \oplus \mathrm{H}',$$

and we introduce in this space the symmetric operator

$$A^+ = A \oplus A'.$$

It is evident that A^+ is a symmetric extension of A of type II. For the proof of the theorem it is sufficient to establish that the operator A^+ can be extended to a self-adjoint operator. Hence, we must verify that A^+ has equal deficiency indices. For $\Im z \neq 0$,

$$\mathfrak{M}_z^+ = (A^+ - zE^+)\,\mathrm{D}_{A+} = (A^+ - zE^+)(\mathrm{D}_A \oplus \mathrm{D}_{A'}) = \\ = (A - zE)\,\mathrm{D}_A \oplus (A' - zE')\,\mathrm{D}_{A'} = \mathfrak{M}_z \oplus \mathfrak{M}_z',$$

or, passing to the orthogonal complement of the corresponding manifolds $\mathfrak{M}_z^+$, $\mathfrak{M}_z$ and $\mathfrak{M}_z'$ in the spaces H^+, H and H',

$$\mathfrak{N}_z^+ = \mathfrak{N}_z \oplus \mathfrak{N}_z'.$$

It follows that the deficiency indices of the operator A^+ are $(\mathfrak{m} + \mathfrak{n}, \mathfrak{m} + \mathfrak{n})$. Thus, the theorem is proved.

If we choose the space H' and the operator A' in a different manner, then we obtain different self-adjoint extensions of A.

The operator B^+ constructed in the proof of Theorem 1 is, generally speaking, an extension of type III of the initial symmetric operator A. However, it is always possible to require that B^+ be an extension of type II of the operator A. For this, it is sufficient to construct a unitary operator U^+ such that

$$U^+\mathfrak{N}_z = \mathfrak{N}_{\bar{z}}',$$

and to define B^+ as the Cayley transform of U^+.

By the theorem just proved, we show now that an arbitrary symmetric operator has an integral representation similar to the one for self-adjoint operators which was derived in Chapter VI. Thus, let A be a symmetric operator in H. We extend[3] it to a self-adjoint operator B^+ and we pass from the space H to a larger space H^+. Let $E^+(\Delta)$ be the spectral function of the operator B^+ and P^+ the operator of projection of H^+ on H. Finally, let

$$F(\Delta) = P^+E^+(\Delta).$$

For arbitrary elements $f \in \mathrm{D}_{B+}$ and $g \in \mathrm{H}^+$,

$$(B^+f, g) = \int_{-\infty}^{\infty} t\,d\,(E_t^+ f, g),$$

$$\| B^+f \|^2 = \int_{-\infty}^{\infty} t^2 d\,(E^+f, f).$$

In particular, if $f \in \mathrm{D}_A$ and $g \in \mathrm{H}$, then these formulas can be written in

[3]We do not assume here that the extension is defined by the special procedure used in the proof of Theorem 1.

the form

$$(2) \qquad (Af, g) = \int_{-\infty}^{\infty} t\, d(F_t f, g),$$

$$(3) \qquad \| Af \|^2 = \int_{-\infty}^{\infty} t^2 d(F_t f, f).$$

Thus, we obtain an integral representation of an arbitrary symmetric operator which is similar to the integral representation of a self-adjoint operator.

In view of similarity of the formulas (2) and (3) to the representation obtained in Section 66 for a self-adjoint operator, we now give the following definition which generalizes the concept of the spectral function.

DEFINITION: *If A is a symmetric operator and F_t is a resolution of the identity such that formulas* (2) *and* (3) *hold for arbitrary $f \in D_A$ and $g \in H$, then F_t is a spectral function of the operator A.*

Before we compare formulas (2) and (3) with the integral representation of a self-adjoint operator obtained earlier, we show that the method used in the derivation of these formulas is general. Namely, the following theorem holds.

THEOREM 2: *Every spectral function of a symmetric operator A which is defined in H has the form*

$$F_t = P^+ E_t^+,$$

where E_t^+ is the spectral function of some self-adjoint extension B^+ of the operator A, obtained with the aid of an extension of H to $H^+ \supset H$, and P^+ is the projection operator of H^+ on H.

Proof: By means of the theorem of Naimark we construct the space H^+ and, in it, an orthogonal resolution of the identity E_t^+ such that

$$(4) \qquad F_t = P^+ E_t^+.$$

We show that the operator B^+, defined by

$$\int_{-\infty}^{\infty} td\, E_t^+ f$$

for all $f \in H^+$ which satisfy the inequality

$$\int_{-\infty}^{\infty} t^2 d(E_t^+ f, f) < \infty,$$

is a self-adjoint extension of the operator A. From Theorem 1 of Section 66 it follows that the operator B^+ is self-adjoint.

If $f \in \mathrm{D}_A$, then $f \in \mathrm{D}_{B+}$, since

$$\int_{-\infty}^{\infty} t^2 d(E_t^+ f, f) = \int_{-\infty}^{\infty} t^2 d(F_t f, f) = \| Af \|^2 < \infty.$$

Furthermore, for $f \in \mathrm{D}_A$ and $g \in \mathrm{H}$, it follows from the equation

$$(Af, g) = \int_{-\infty}^{\infty} td(F_t f, g) = \int_{-\infty}^{\infty} t\, d(E_t^+ f, g) = (B^+ f, g)$$

that

$$Af = P^+ B^+ f \qquad (f \in \mathrm{D}_A). \tag{5}$$

But, for $f \in \mathrm{D}_A$,

$$\| Af \|^2 = \int_{-\infty}^{\infty} t^2 d(F_t f, f) = \int_{-\infty}^{\infty} t^2 d(E_t^+ f, f) = \| B^+ f \|^2. \tag{6}$$

From (5) and (6) it follows that

$$Af = B^+ f \qquad (f \in \mathrm{D}_A).$$

It can be shown that the operator B^+ is reduced by no subspace of $\mathrm{H}^+ \ominus \mathrm{H}$. Indeed, if a subspace G of $\mathrm{H}^+ \ominus \mathrm{H}$ reduces the operator B^+ then it also reduces the resolution of the identity E_t^+ and the exclusion of G from H^+ will reduce to the exclusion of the restriction of the operator E_t^+ to G. This does not affect formula (4). Thus, the theorem is proved.

If a spectral function F_t of a symmetric operator is represented in the form (4), where E_t^+ is the spectral function of a self-adjoint extension B^+ of the operator A, then the spectral function F_t is *generated by* the self-adjoint extension B^+. Thus, every self-adjoint extension of the operator A generates some spectral function of this operator, and conversely, every spectral function of the operator A is generated by one of its self-adjoint extensions.

We present now several facts which will clarify the interrelation between the new and old definitions of a spectral function and which indicate to what extent there is an analogy between the representation (2), (3) and the spectral representation of a self-adjoint operator.

1° The spectral function of a self-adjoint operator defined in Chapter VI is its unique spectral function in the sense of the definition of the present section. A self-adjoint operator does not have other spectral functions since, by our theorem, every such spectral function must be generated by a self-adjoint extension and a self-adjoint operator does not have such extensions.

2° From formula (3) it follows that the integral

$$\int_{-\infty}^{\infty} t^2 d(F_t f, f) \tag{7}$$

converges for each $f \in \mathrm{D}_A$. The converse statement, generally speaking, is not correct. It is possible to assert only that if the integral (7) converges for some $f \in \mathrm{H}$ then

$$\int_{-\infty}^{\infty} t^2 d(E_t^+ f, f) < \infty. \tag{8}$$

Thus the vector f belongs to the manifold $\mathrm{D}_{B+} \cap \mathrm{H}$. This coincides with D_A if and only if B^+ is an extension of type II of the operator A.

Thus, only those spectral functions of the operator A which are generated by self-adjoint extensions of type II of A characterize the domain D_A as the set of vectors for which the integral (7) converges. In particular, the domains of maximal operators are characterized by the spectral functions and by inequality (8), since such operators admit only extensions of type II.

In the general case the spectral function F_t of the operator A, generated by a self-adjoint extension B^+, characterizes the domain of the operator $C = P^+B^+P^+$. This is a symmetric extension of type I of A with a maximal domain D_C which satisfies the condition

$$\mathrm{D}_A \subset \mathrm{D}_C \subset \mathrm{D}_{B+}. \tag{9}$$

With this point of view it would be natural to consider the resolution of the identity F_t not as a spectral function of A but of C and, in correspondence with this, to include in the definition the requirement that the resolution of the identity F_t, apart from the representations (2), (3), define still the domain D_A by the inequality (8). However, it seems impractical to include in the definition an additional requirement concerning the characterization of the domain D_A. Thus, the definition admits that a single resolution of the identity may be the spectral function of two different operators (for example, the operators A and $C = P^+B^+P^+$). In particular, the spectral function of a self-adjoint extension of type I of a given operator A with equal deficiency numbers is the spectral function of the same operator A and of each operator C such that $C \subset B$.

3° From the preceding it follows that every symmetric operator corresponds to some set of spectral functions and that one and the same spectral function corresponds to some set of symmetric operators.

In connection with this, there arises the question: is every resolution of the identity the spectral function of a symmetric operator? In the case of the orthogonal resolution of the identity a positive answer to this question was given by Theorem 1, Section 66.

If, however, F_t is a nonorthogonal resolution of the identity then it is impossible to assert that the set of vectors for which the integral on the right side of equation (3) exists will be dense in the space H. In fact, as Naimark observed, there exist resolutions of the identity such that the integral on the right side of equation (3) does not converge for any vector $f \neq 0$. As such an example,[4] we consider the operator function defined on the space H by the equations

$$F_t = \begin{cases} 0 & t \leqq 0 \\ e^{-\frac{1}{t}}E & t > 0. \end{cases}$$

It is evident, the operator function F_t satisfies conditions A, B, C in section 1° and, hence, is a resolution of the identity. However this resolution of the identity is not a spectral function of any symmetric operator, since the integral

$$\int_{-\infty}^{\infty} t^2 d(F_t f, f) = (f, f) \int_0^{\infty} t^2 de^{-\frac{1}{t}} = (f, f) \int_0^{\infty} e^{-\frac{1}{t}} dt$$

diverges for each vector $f \neq 0$.

4° In Chapter VI it was shown that for a self-adjoint operator the representation (2) holds in the "strong" sense. In the case of the nonorthogonal resolution of the identity the transition from the "weak" representation (2) to the "strong" representation

$$Af = \int_{-\infty}^{\infty} t dF_t f \tag{10}$$

is impossible. Therefore, we regard equation (10) only as a symbolic entry of the formulas (2) and (3). The concept of the integral in the strong sense in the case of nonorthogonal resolutions of the identity is not introduced.

5° From the property of the orthogonality of the spectral function E_t of a self-adjoint operator, it follows that for each finite real interval Δ, the vector $E(\Delta)f$ with $f \in$ H belongs to D_A and

$$(AE(\Delta)f, g) = \int_{\Delta} t\, d(E_t f, g)$$

---

[4]For this example we thank M. A. Krasnoselski.

for each $g \in H$. This important fact which we applied repeatedly in Chapter VI does not hold for the integral representation of symmetric operator considered here (even if we restricted ourselves to spectral functions generated by extensions of type II).

However the following proposition is correct: if A is a symmetric operator, if F_t is its spectral function, if Δ is a finite real interval, and if g is an arbitrary vector in H, then for each $h \in H$,

$$F(\Delta)\, g \in D_A{}^*$$

and

$$(A^*F(\Delta)\, g, h) = \int_\Delta t\, d\,(F_t g, h).$$

Indeed, if B^+ is the extension which generates F_t and if P^+ is the operator of projection of H^+ on H, then for each $f \in D_A$

$$(11) \quad (Af, F(\Delta) g) = (B^+ f, P^+ E^+(\Delta) g) = (B^+ f, E^+(\Delta) g) = \\ = \int_\Delta t\, d(E_t^+ f, g) = \int_\Delta t\, d(E_t^+ f, P^+ g) = \int_\Delta t\, d(F_t f, g).$$

But the integral

$$\int_\Delta t\, d\,(F_t h, g)$$

is a bilinear functional of h and g on H and, therefore, has a representation (h, Cg) where C is a bounded operator. It follows that

$$(Af, F(\Delta)\, g) = (f, g^*),$$

and, hence,

$$F(\Delta)\, g \in D_{A^*}.$$

Therefore, (11) yields

$$(f, A^*F(\Delta)\, g) = \int_\Delta t\, d\,(F_t f, g).$$

Since D_A is dense in H, this equation holds for an arbitrary vector $h \in H$.

3. Spectral Functions of Symmetric Operators and Generalized Resolvents

We now consider an operator function $\mathbf{R}_z$, which is related to the spectral function of a symmetric operator in the same way that the resolvent of a self-adjoint operator is related to its spectral function.

DEFINITION: *Let A be a symmetric (but not self-adjoint) operator. Let B^+ be a self-adjoint extension of A such that B^+ is defined in a space* $\mathrm{H}^+ \supset \mathrm{H}$. *Let R_z^+ be the resolvent of B^+. Finally, let P^+ be the operator of projection of* H^+ *on* H. *The operator* $\mathbf{R}_z$ *defined on* H *by*

$$\mathbf{R}_z = P^+ R_z^+$$

for each nonreal z, is called a generalized resolvent of the operator A.

If the operator B^+ and $\mathbf{R}_z$ are related as indicated in the definition, then we say that the generalized resolvent $\mathbf{R}_z$ is generated by the self-adjoint extension B^+.

The generalized resolvents of an operator A (with equal deficiency numbers) generated by its self-adjoint extensions of type I are called orthogonal; these generalized resolvents are also the ordinary resolvents of the self-adjoint extensions which generate them.

The following theorem establishes an integral representation for a generalized resolvent, analogous to the one obtained in Chapter VI for an ordinary resolvent of a self-adjoint operator.

THEOREM 1: *In order that the operator function* $\mathbf{R}_z(\Im z \neq 0)$ *be a generalized resolvent of the symmetric operator A, it is necessary and sufficient that it have the representation*

$$(1) \qquad (\mathbf{R}_z f, g) = \int_{-\infty}^{\infty} \frac{d(F_t f, g)}{t - z} \qquad (f, g \in \mathrm{H}),$$

where F_t is some spectral function of the operator A.

Proof: For the necessity of the representation (1) we may take the spectral function F_t corresponding to that self-adjoint extension B^+ which generates the generalized resolvent $\mathbf{R}_z$. For the sufficiency we may take the generalized resolvent $\mathbf{R}_z$ generated by that self-adjoint extension B^+ corresponding to the spectral function F_t.

If the generalized resolvent $\mathbf{R}_z$ and the spectral function F_t are related by formula (1), then it is said that they correspond to each other. The well known inversion formula between the set of the spectral functions of a given symmetric operator and the set of its generalized resolvents determines a one-to-one correspondence.

We remark that for every vector of the subspace

$$\mathfrak{M}_z = \Delta_A(z) \qquad (\Im z \neq 0),$$

the values of all generalized resolvents of the operator A coincide. Indeed, if

$$(2) \qquad (A - zE)f = g,$$

then

(3) $\quad \mathbf{R}_z g = \mathbf{R}_z(A - zE)f = P^+ R_z^+ (B^+ - zE^+)f = P^+ f = f.$

If the generalized resolvent $\mathbf{R}_z$ is generated by an extension B^+ with the resolvent R_z^+, then for $f, g \in \mathrm{H}$

$$(\mathbf{R}_z f, g) = (P^+ R_z^+ f, g) = (R_z^+ f, g) = (f, R_{\bar{z}}^+ g) = (f, P^+ R_{\bar{z}}^+ g) = (f, \mathbf{R}_{\bar{z}} g),$$

and this yields

(4) $$\mathbf{R}_{\bar{z}} = \mathbf{R}_z^*.$$

If A is a maximal symmetric operator and

$$\mathfrak{M}_z = \mathrm{H}$$

for $\Im z > 0$, then its generalized resolvents coincide for $\Im z > 0$, since by (2) and (3),

$$\mathbf{R}_z g = (A - zE)^{-1} g$$

for $\Im z > 0$ and $g \in \mathrm{H}$. According to (4), the coincidence of the operators $\mathbf{R}_z$ for $\Im z > 0$ implies that they coincide for $\Im z < 0$. Thus, a maximal operator has a unique generalized resolvent and hence possesses a unique spectral function.

On the other hand, a generalized resolvent of a non-maximal symmetric operator A is not uniquely defined. In order to ascertain this, it is sufficient to take two different maximal symmetric extensions C' and C'' of A of type I and to extend them to self-adjoint operators B'^+ and B''^+. It is evident that the generalized resolvents $\mathbf{R}_z'$ and $\mathbf{R}_z''$ of the operator A generated by the extensions B'^+ and B''^+ do not coincide.

Since a non-maximal symmetric operator has different generalized resolvents, its spectral function is not defined uniquely.

In view of an earlier fact (Section 88, 1°) about spectral functions of self-adjoint operator, the following proposition is valid.

THEOREM 2: *A symmetric operator has a unique spectral function if and only if it is maximal. This unique spectral function is orthogonal if and only if the operator is self-adjoint.*

It is easy to see that the set of all spectral functions (respectively, generalized resolvents) of a symmetric operator A is convex. This means that if F_t' and F_t'' ($\mathbf{R}_z'$ and $\mathbf{R}_z''$) are two spectral functions (respectively, generalized resolvents) of the operator A, then for $\alpha' + \alpha'' = 1$, $\alpha' > 0$, $\alpha'' > 0$ the operator function $\alpha' F_t' + \alpha'' F_t''$ (respectively, $\alpha' \mathbf{R}_z' + \alpha'' \mathbf{R}_z''$) is also a spectral function (respectively, generalized resolvent) of A.

In connection with this we indicate one method of construction of generalized resolvents (spectral functions) which are generated by self-adjoint extensions of type II by means of operations inside the space H. Let A be a symmetric operator defined in H, and let A' and A'' be any two of its maximal extensions of type I. For definiteness we assume that

$$\Delta_{A'}(z) = \Delta_{A''}(z) = \mathrm{H}$$

for $\Im z > 0$.

We define the operators $\mathbf{R}'_z$ and $\mathbf{R}''_z$ by the equations

$$R'_z = \begin{cases} (A' - zE)^{-1} & (\Im z > 0), \\ (\mathbf{R}'_{\bar{z}})^* & (\Im z < 0), \end{cases}$$

$$R''_z = \begin{cases} (A'' - zE)^{-1} & (\Im z > 0), \\ (\mathbf{R}''_{\bar{z}})^* & (\Im z < 0). \end{cases}$$

The operators $\mathbf{R}'_z$ and $\mathbf{R}''_z$ are the generalized resolvents of the maximal symmetric operators A' and A'' respectively. By means of $\mathbf{R}'_z$ and $\mathbf{R}''_z$ we determine the spectral functions F'_t and F''_t of A' and A'', respectively, and form the operator function

$$F_t = \alpha' F'_t + \alpha'' F''_t \qquad (\alpha' + \alpha'' = 1, \alpha' > 0, \alpha'' > 0).$$

Since the set of the spectral functions is convex, F_t is a spectral function of A.

We show now how to choose the maximal extensions A' and A'' in order that the spectral function F_t is generated by an extension of type II or, equivalently, in order that the spectral function F_t and the inequality

$$\int_{-\infty}^{\infty} t^2 d(F_t f, f) < \infty \tag{5}$$

determine the domain D_A. It is easy to see that a vector $f \in \mathrm{H}$ satisfies the condition (5) if and only if

$$f \in \mathrm{D}_{A'} \cap \mathrm{D}_{A''}.$$

It remains to prove that the maximal extensions A' and A'' of the operator A can be chosen such that the intersection of the domains $\mathrm{D}_{A'}$ and $\mathrm{D}_{A''}$ is the domain D_A. By the second Neumann formula (Section 80)

$$\mathrm{D}_{A'} = \mathrm{D}_A + \Gamma',$$
$$\mathrm{D}_{A''} = \mathrm{D}_A + \Gamma'',$$

where Γ' is the set of all vectors of the form

$$g + U'g \qquad (g \in \mathfrak{N}_{\bar{z}}),$$

and U' is an isometric operator which maps $\mathfrak{N}_{\bar{z}}$ into $U'\mathfrak{N}_{\bar{z}} \subset \mathfrak{N}_z (\Im z > 0$, if it is assumed for definiteness that $\mathfrak{m} \leqq \mathfrak{n})$; Γ'' is defined analogously.

Note that $D_{A'} \cap D_{A''} \supset D_A$. Let us assume that D_A is a proper subset of $D_{A'} \cap D_{A''}$ and choose a vector $h \notin D_A$ which belongs to both $D_{A'}$ and $D_{A''}$. Then

$$h = f' + g' + U'g',$$
$$h = f'' + g'' + U''g'',$$

where $f', f'' \in D_A$ and $g', g'' \in \mathfrak{N}_{\bar{z}}$. These representations of h yield

$$(f'' - f') + (g'' - g') + (U''g'' - U'g') = 0.$$

The summands in the left member of the equation belong to the manifolds D_A, $\mathfrak{N}_{\bar{z}}$ and $\mathfrak{N}_z$ respectively. Since these manifolds are linearly independent,

$$f' = f'', \quad g' = g'' = g, \qquad (6)$$
$$U'g = U''g.$$

From the last equation it follows that if

$$D_{A'} \cap D_{A''} = D_A,$$

then the isometric operators U' and U'' must be defined such that equation (6) holds for no vector of $\mathfrak{N}_{\bar{z}}$ which is different from zero. This condition can always be achieved, for example, by choosing U' arbitrarily and letting $U'' = -U'$.

Thus, the constructed operator functions

$$F_t = \alpha' F_t' + \alpha'' F_t''$$

and

$$\mathbf{R}_z = \alpha' \mathbf{R}_z' + \alpha'' \mathbf{R}_z''$$

for arbitrary $\alpha', \alpha''(\alpha' + \alpha'' = 1,\ \alpha' > 0, \alpha'' > 0)$ are spectral functions and generalized resolvents, respectively, of the operator A, which are generated in the sense of Section 2 by a self-adjoint extension B^+ of type II of A.

However, it is not true that such a method yields all spectral functions of a symmetric operator.[5]

From the given method of construction of spectral functions it follows, among other things, that Theorem 2 remains correct even if the spectral functions are required to be generated by self-adjoint extensions of type II of the operator A. In order to prove this it is necessary to show that a non-maximal symmetric operator A has different spectral functions which are generated by self-adjoint extensions of type II of A. But this is a consequence of the fact that the generalized resolvents

$$\alpha \mathbf{R}_z' + \beta \mathbf{R}_z''$$

[5] In this connection, see M. A. Naimark [5], Vol. I, and I. M. Glazman [2], Vol. I.

and

$$\beta\mathbf{R}'_z + \alpha\mathbf{R}''_z$$

for the indicated selections of the maximal extensions A' and A'' coincide only for vectors in $\Delta_A(z)$ and, therefore, generate different spectral functions.

In conclusion, in this and the following section, we illustrate facts and methods with a differential operator. Let P_0 be the operator of differentiation which acts in $L^2(0, \infty)$. The deficiency indices of the operator P_0 are (0, 1).

To obtain a generalized self-adjoint extension of the operator P_0 by the method of Theorem 1, we introduce the operator P' of differentiation in $L^2(-\infty, 0)$ which is defined by the formula

$$P'_0 = i\frac{d}{dt}$$

for each absolutely continuous function $\varphi(t)$ such that $\varphi(t)$ and its derivative $\varphi'(t)$ belong to $L^2(-\infty, 0)$ and such that $\varphi(t)$ satisfies the boundary condition $\varphi(0) = 0$. It is evident that the deficiency indices of the operator P'_0 are (1, 0).

We form the orthogonal sums

$$L^2(-\infty, \infty) = L^2(-\infty, 0) \oplus L^2(0, \infty),$$

$$P_0^+ = P'_0 \oplus P_0.$$

Obviously, the operator P_0^+ is defined by the formula

$$P_0^+\varphi = i\frac{d}{dt}\varphi(t)$$

for each function $\varphi(t)$ which is absolutely continuous in the intervals $(-\infty, 0)$ and $(0, \infty)$ such that $\varphi(t)$ and $\varphi'(t)$ belong to $L^2(-\infty, \infty)$ and such that $\varphi(t)$ satisfies the boundary condition $\varphi(0) = 0$.

It is easy to see that one obtains the domain $D_{(P_0^+)^*}$ of the adjoint operator $(P_0^+)^*$ if one omits the boundary condition $\varphi(0) = 0$. Therefore, each of the equations

$$(P_0^+)^*g + ig = 0,$$

$$(P_0^+)^*g - ig = 0$$

has a unique solution. These solutions are defined, respectively, by the formulas

$$g_1(t) = \begin{cases} 0 & (t<0) \\ e^{-t} & (t \geqq 0) \end{cases} \qquad g_2(t) = \begin{cases} e^t & (t<0) \\ 0 & (t \geqq 0). \end{cases}$$

Thus, the deficiency indices of the operator P_0^+ are (1, 1).

The operator P of differentiation on the entire axis $(-\infty, \infty)$ considered in Section 49 is evidently a self-adjoint extension of the operator P_0^+. Thus, the operator of differentiation on the entire axis is a generalized self-adjoint extension of the operator of differentiation on the semi-axis.

4. The Formula of Krein for Generalized Resolvents

In connection with the results of the preceding section, there is the problem of describing the set of all spectral functions of a given symmetric operator. Since the set of all spectral functions and the set of all resolvents of a symmetric operator have a one-to-one correspondence determined by the formula

$$(\mathbf{R}_\lambda f, g) = \int_{-\infty}^{\infty} \frac{d(F_t f, g)}{t-\lambda},$$

this problem is equivalent to the problem of describing all resolvents. The latter problem was solved by Krein[6] for the case of equal and finite deficiency numbers. We present here the result of Krein for the case of deficiency indices (1, 1).

Let A be a symmetric operator with deficiency indices (1, 1), let $\mathring{A}$ be a fixed self-adjoint extension of type I of A, let $\mathring{R}_\lambda$ be the resolvent of the operator $\mathring{A}$, and finally let $\mathbf{R}_\lambda$ be an arbitrary generalized resolvent of the operator A such that

$$\mathbf{R}_\lambda = P^+ R_\lambda^+.$$

Here R_λ^+ is the orthogonal resolvent of some self-adjoint extension A^+ of the operator A which is defined in a space $\mathrm{H}^+ \supset \mathrm{H}$ and P^+ is the operator of projection of H^+ on H.

We let as usual

$$\mathfrak{M}_\lambda = \Delta_A(\lambda) \qquad (\Im\lambda \neq 0),$$
$$\mathfrak{N}_\lambda \neq \mathrm{H} \ominus \mathfrak{M}_\lambda.$$

In the case considered the dimension of the subspace $\mathfrak{N}_\lambda$ is one.

For the difference of the resolvents we have, as in Section 84,

$$(1) \qquad \begin{cases} (\mathring{R}_\lambda - \mathbf{R}_\lambda) f = 0 & \text{for} \quad f \in \mathfrak{M}_\lambda, \\ (\mathring{R}_\lambda - \mathbf{R}_\lambda) f \in \mathfrak{N}_{\bar{\lambda}} & \text{for} \quad f \in \mathfrak{N}_\lambda. \end{cases}$$

[6]Cf. M. G. Krein [1], Vol. I. A formula for generalized resolvents of an operator with deficiency indices (1, 1) was obtained by Naimark [1], Vol. I. A further generalization of the formulas of Naimark and Krein was given by Strauss [1], Vol. I.

The last relation follows from the fact that for $f \in \mathfrak{N}_\lambda$ and $h \in \mathfrak{M}_{\bar\lambda}$

$$((\mathring{R}_\lambda - \mathbf{R}_\lambda)f, h) = (\mathring{R}_\lambda f, h) - (R^+_\lambda f, h) =$$
$$= (f, \mathring{R}_{\bar\lambda} h) - (f, R^+_{\bar\lambda} h) = (f, (\mathring{R}_{\bar\lambda} - \mathbf{R}_{\bar\lambda}) h) = (f, 0) = 0.$$

From (1), repeating the argument of Section 84, we get

(2) $$\mathbf{R}_\lambda f = \mathring{R}_\lambda f - \frac{(f, g(\bar\lambda))\, g(\lambda)}{Q(\lambda)},$$

where

(2′) $$g(\lambda) = g(\lambda_0) + (\lambda - \lambda_0)\, \mathring{R}_\lambda g(\lambda_0).$$

Here $g(\lambda_0)$ is a vector in $\mathfrak{N}_{\bar\lambda_0}$ with norm one. We assume for definiteness that the fixed point λ_0 lies in the upper half-plane (it is possible to choose $\lambda_0 = i$). However, generally speaking, the function $Q(\lambda)$ does not satisfy the relation

(3) $$Q(\lambda) = Q(\lambda_0) + (\lambda - \lambda_0)(g(\lambda_0), g(\bar\lambda)),$$

which we obtained in Section 84 for the case of orthogonal resolvents.

Let us turn to the clarification of the nature of the function $Q(\lambda)$. First, we find $Q(\lambda_0)$, for which we let $\lambda = \lambda_0$ in (2) and pass from the resolvent to the Cayley transform

$$\mathring{U}_{\lambda_0} = \frac{\mathring{A} - \bar\lambda_0 E}{\mathring{A} - \lambda_0 E}, \qquad U^+_{\bar\lambda_0} = \frac{A^+ - \bar\lambda_0 E^+}{A^+ - \lambda_0 E^+}.$$

After an elementary calculation, we get

$$P^+ U^+_{\lambda_0} f = \mathring{U}_{\lambda_0} f - \frac{\lambda_0 - \bar\lambda_0}{Q(\lambda_0)} (f, g(\bar\lambda_0))\, g(\lambda_0).$$

In this formula, we let $f = g(\bar\lambda_0)$. Then we have

(4) $$P^+ U^+_{\lambda_0} g(\bar\lambda_0) = g(\lambda_0) - \frac{\lambda_0 - \bar\lambda_0}{Q(\lambda_0)} g(\lambda_0).$$

Now, since the element $\psi = U^+_{\lambda_0} \varphi$ belongs to $\mathfrak{M}_{\bar\lambda_0}$ if and only if φ belongs to $\mathfrak{M}_{\lambda_0}$,

$$0 = (g(\bar\lambda_0), \varphi) = (U^+_{\lambda_0} g(\bar\lambda_0), U^+_{\lambda_0} \varphi) = (U^+_{\lambda_0} g(\bar\lambda_0), \psi) = (P^+ U^+_{\lambda_0} g(\bar\lambda_0), \psi);$$

which implies that

(5) $$P^+ U^+_{\lambda_0} g(\bar\lambda_0) = \theta g(\lambda_0).$$

In (5), θ is a constant which does not exceed one in absolute value, and which is determined by the extension A^+; and θ equals one in absolute value if and only if A^+ is an extension of type I.

Our considerations yield an important consequence: if

$$\mathbf{R}_\lambda = \mathring{R}_\lambda$$

for any one nonreal value $\lambda = \lambda_1$, *then the equation holds for all nonreal* λ. Indeed, replacing λ_0 in formulas (2) and (2′) by λ_1 ($\Im\lambda_1 > 0$) we find from formula (4) that $Q(\lambda_1) = \infty$, which implies that the parameter θ in formula (5) equals one. Therefore, $\mathbf{R}_\lambda$ is an orthogonal resolvent; and two orthogonal resolvents of one and the same operator which coincide at one point are obviously identical.

Comparing (4) and (5), we find that

$$Q(\lambda_0) = \frac{\lambda_0 - \bar{\lambda}_0}{1-\theta},$$

or

$$Q(\lambda_0) = i\Im\lambda_0 + \tau. \tag{6}$$

Here τ is a new parameter which is related to θ by the formula

$$\tau = i(\Im\lambda_0)\frac{1+\theta}{1-\theta},$$

which maps the unit circle of the θ-plane into the upper half τ-plane.

If $\mathbf{R}_\lambda$ runs through the set of orthogonal resolvents, then, since (3) is valid in this case, formula (2) takes the form

$$\mathbf{R}_\lambda f = \mathring{R}_\lambda f - \frac{(f, g(\bar{\lambda}))\, g(\lambda)}{\tau + Q_1(\lambda)}, \tag{7}$$

where

$$Q_1(\lambda) = i\Im\lambda_0 + (\lambda - \lambda_0)(g(\lambda_0), g(\bar{\lambda})). \tag{7'}$$

The parameter τ is real ($-\infty < \tau \leqq \infty$). Formula (7) defines a one-to-one correspondence between the set of all orthogonal resolvents and the values of the parameter τ.

In the general case where $\mathbf{R}_\lambda$ is an arbitrary resolvent, we let in formula (2)

$$Q(\lambda) = \tau(\lambda) + Q_1(\lambda).$$

Then

$$\mathbf{R}_\lambda f = \mathring{R}_\lambda f - \frac{(f, g(\bar{\lambda}))\, g(\lambda)}{\tau(\lambda) + Q_1(\lambda)}. \tag{8}$$

We show that the function $\tau(\lambda)$ is holomorphic and has a non-negative imaginary part in the upper half-plane. With this aim, we remark first that the equation

$$\mathbf{R}_\lambda g(\bar{\lambda}) = \mathring{R}_\lambda g(\bar{\lambda}),$$

and also the equation

$$(\mathbf{R}_\lambda g(\bar{\lambda}), g(\lambda)) = (\mathring{\mathbf{R}}_\lambda g(\bar{\lambda}), g(\lambda))$$

hold at no nonreal point. Therefore, we let $f = g(\bar{\lambda})$ in (8) and form the inner product of both sides with $g(\lambda)$. We obtain

$$\frac{1}{\tau(\lambda) + Q_1(\lambda)} = ((\mathring{R}_\lambda - \mathbf{R}_\lambda) g(\bar{\lambda}), g(\lambda)).$$

But, since $Q_1(\lambda)$ and the right member are holomorphic and the latter is different from zero, the function $\tau(\lambda)$ is holomorphic for all nonreal values λ. We prove next that if $\Im\, \tau(\lambda') = 0$ at any one point of the upper half-plane then $\tau(\lambda) =$ constant. It follows, in particular, that the inequality $\Im\, \tau(\lambda) < 0$ is impossible at any point of the upper half-plane. Thus, let $\Im \tau(\lambda') = 0$ $(\Im \lambda' > 0)$. We choose $\lambda = \lambda'$ in both formulas (7) and (8) and take in formula (7) the number $\tau(\lambda')$ as the constant τ. Then the right members of these formulas will be identical for each f. Hence, at the point $\lambda = \lambda'$ the generalized resolvent $\mathbf{R}_\lambda$ coincides with an orthogonal resolvent R_λ and as we showed above, these resolvents must coincide.

We form the inner product of both sides of equation (8) with f and obtain the formula

$$(\mathbf{R}_\lambda f, f) = \frac{p_0(\lambda) + p_1(\lambda)\,\tau(\lambda)}{q_0(\lambda) + q_1(\lambda)\,\tau(\lambda)}.$$

For fixed f and λ the functions $p_0(\lambda)$, $p_1(\lambda)$, $q_0(\lambda)$, $q_1(\lambda)$ will have completely determined values, not depending on the choice of the resolvent $\mathbf{R}_\lambda$, which are determined by the function $\tau(\lambda)$. Since orthogonal resolvents are obtained when $\tau(\lambda)$ is a real constant and also when $\tau(\lambda)$ runs through a particular set of holomorphic functions with non-negative imaginary part, the point $w = (\mathbf{R}_\lambda f, f)$ belongs to some circular region. The boundary of this region, the circle $C(f; \lambda)$ is generated by the point w, when $\mathbf{R}_\lambda$ is an orthogonal resolvent. In other words, the circle $C(f; \lambda)$ is described by the point

$$w = \frac{p_0(\lambda) + p_1(\lambda)\,\tau}{q_0(\lambda) + q_1(\lambda)\,\tau},$$

when τ runs through the real axis.

Since the set of all resolvents of a given operator is convex, the point $w = (\mathbf{R}_\lambda f, f)$ runs through the region $K(f; \lambda)$ with boundary $C(f; \lambda)$, when $\mathbf{R}_\lambda$ runs through the set of all resolvents of the operator A.

By our considerations we have now established that: *every resolvent*

$\mathbf{R}_\lambda$ *of a symmetric operator* A *with the deficiency indices* (1, 1) *is representable in the form*

$$\mathbf{R}_\lambda = \mathring{R}_\lambda - \frac{(\cdot, g(\bar{\lambda}))\, g(\lambda)}{\tau(\lambda) + Q_1(\lambda)}, \tag{9}$$

where $\tau(\lambda)$ *belongs to the class* N *of all functions which are defined in the half-plane* $\Im \lambda > 0$ *and are holomorphic and in this half-plane have a non-negative imaginary part.*

We now prove the converse proposition: *each function* $\tau(\lambda)$ *of the class* N *generates by means of formula* (9) *a resolvent of the operator* A. Let $\tau(\lambda) \in N$ and let S_λ be an operator defined by the equation

$$S_\lambda = \mathring{R}_\lambda - \frac{(\cdot, g(\bar{\lambda}))\, g(\lambda)}{\tau(\lambda) + Q_1(\lambda)}. \tag{9'}$$

Since the circle $C(f; \lambda)$ lies in the upper half-plane, but the point $(S_\lambda f, f)$ lies in the circular region $K(f; \lambda)$, the scalar product $(S_\lambda f, f)$ is a holomorphic function in the half-plane $\Im \lambda > 0$ which evidently belongs to the class N.

Furthermore, since, for orthogonal resolvents R_λ, the scalar product $(R_\lambda f, f)$ satisfies the inequality

$$|(R_\lambda f, f)| \leqq \frac{(f, f)}{\Im \lambda} \qquad (\Im \lambda > 0),$$

the scalar product $(S_\lambda f, f)$, which lies in the circular region $K(f; \lambda)$, satisfies the inequality

$$|(S_\lambda f, f)| \leqq \frac{(f, f)}{\Im \lambda}.$$

Using this inequality and repeating the argument of Section 65, we obtain a representation of the scalar product $(S_\lambda f, f)$ in the form

$$(S_\lambda f, f) = \int_{-\infty}^{\infty} \frac{d\omega(t; f)}{t - \lambda}. \tag{10}$$

Here $\omega(t; \lambda) = \omega(t)$ is a nondecreasing left-continuous function which tends to zero as $t \to -\infty$ and satisfies the condition

$$\omega(t; f) \leqq (f, f) \qquad (-\infty < t \leqq \infty).$$

From formula (10), repeating the calculation of Section 65, we obtain a representation of the operator S_λ in the form

$$(S_\lambda f, g) = \int_{-\infty}^{\infty} \frac{d(F_t f, g)}{t - \lambda}.$$

Here F_t is a nondecreasing left-continuous operator function which tends to zero as $t \to -\infty$ and satisfies the condition

$$(F_t f, f) \leqq (f, f) \qquad (-\infty < t \leqq \infty). \tag{11}$$

In contrast with Section 65, the operator function S_λ does not satisfy the Hilbert functional equation. Hence, we cannot derive the orthogonality relation

$$F_u F_v = F_s \qquad (s = \min\{u, v\}). \tag{12}$$

Therefore, we fail to have the relation used in the end of Section 65 in the proof of the property

$$\lim_{t \to \infty} F_t f = f \qquad (f \in \mathrm{H}). \tag{13}$$

In order to prove that the operator function F_t is a generalized resolution of the identity, we must prove property (13), without the aid of (12).

According to the lemma on monotone increasing sequences of operators (Section 85), the limit on the left side of equation (13) exists. Hence, it is sufficient to show that as $t \to \infty$ the operator F_t tends weakly to E from the left, i.e., that

$$\lim_{t \to \infty} (F_t f, g) = (f, g)$$

for all $f, g \in \mathrm{H}$. Evidently this relation holds if and only if

$$\lim_{t \to \infty} (F_t f, f) = (f, f) \tag{14}$$

for all $f \in \mathrm{H}$. Since, by (11), the norm of the operator function F_t is bounded ($\| F_t \| \leqq 1$), it is sufficient to verify equation (14) for some dense set of vectors in H. As such a set we take the domain D_A of the operator A.

By (10), equation (14) is equivalent to the equation

$$\lim_{\eta \to \infty} i\eta (S_{i\eta} f, f) = -(f, f) \tag{15}$$

for all $f \in \mathrm{H}$. Since the point $(S_{i\eta} f, f)$ lies inside the circle $C(f; \lambda)$ it is evident that

$$| i\eta (S_{i\eta} f, f) + (f, f) | \leqq \max_{-\infty < \tau \leqq \infty} | i\eta (R^\tau_{i\eta} f, f) + (f, f) |.$$

Here R^τ_λ denotes the resolvent of type I which corresponds to the parameter τ. It remains to prove that

$$| i\eta (R^\tau_{i\eta} f, f) + (f, f) |$$

tends to zero uniformly with respect to τ $(-\infty < \tau \leqq \infty)$. This is true since, for $f \in \mathrm{D}_A$ and $\eta > 0$,

$$|i\eta(R^\tau_{i\eta}f,f) + (f,f)| = \left| \int_{-\infty}^{\infty} \frac{i\eta}{t - i\eta} d(E^\tau_t f,f) + \int_{-\infty}^{\infty} d(E^\tau_t f,f) \right| =$$

$$= \left| \int_{-\infty}^{\infty} \frac{t}{t-i\eta} d(E^\tau_t f,f) \right| \leqq \sqrt{\int_{-\infty}^{\infty} t^2 d(E^\tau_t f,f)} \sqrt{\int_{-\infty}^{\infty} \frac{1}{|t-i\eta|^2} d(E^\tau_t f,f)} \leqq$$

$$\leqq \frac{1}{\eta} \| Af \| \cdot \|f\|.$$

Thus, relation (13) is proved so that the operator function F_t is indeed a generalized resolution of the identity and according to the theorem of Naimark it has the representation

$$F_t = P^+ E_t^+.$$

Let us introduce the self-adjoint operator

$$A^+ f = \int_{-\infty}^{\infty} t \, d\, E_t^+ f.$$

Evidently

$$S_\lambda = P^+ R_\lambda^+,$$

where R_λ^+ is the resolvent of the operator A^+. To complete the proof of the theorem it remains to show that A^+ is an extension of A:

$$A^+ \supset A,$$

or that

$$R_\lambda^+ f = R_\lambda f$$

for $f \in \mathfrak{M}_\lambda$.

From formula (9′) it follows that

$$P^+ R_\lambda^+ f = S_\lambda f = R_\lambda f$$

for $f \in \mathfrak{M}_\lambda$. Therefore,

$$R_\lambda^+ f = R_\lambda f + h,$$

where $h \perp \mathrm{H}$. We show that $h = 0$. For $f \in \mathfrak{M}_\lambda$,

$$A R_\lambda f = f + \lambda R_\lambda f,$$

$$A^+(R_\lambda f + h) = f + \lambda R_\lambda f + \lambda h.$$

Letting $g = R_\lambda f$ we obtain

$$(A^+(g + h), g + h) = (Ag + \lambda h, g + h) = (Ag, g) + \lambda(h, h),$$

which is possible only for $h = 0$, since $\Im\lambda \neq 0$.

Hence, the theorem is completely proved.

5. Quasi-Self-Adjoint Extensions and the Characteristic Function of a Symmetric Operator

In the present section we consider another class of extensions of symmetric operators with finite and equal deficiency numbers which was introduced by Lifschitz.[7]

A *quasi-self-adjoint extension* of a symmetric operator A with the deficiency indices (m, m) $(m < \infty)$ is an arbitrary linear operator B which satisfies the conditions

$$A \subset B \subset A^*, \tag{1}$$

$$\dim D_B = m \qquad (\text{mod } D_A), \tag{2}$$

but is not a self-adjoint extension of the operator A.

For simplicity we restrict ourselves to the case of operators with deficiency indices (1, 1). In this case condition (2) is a consequence of condition (1) and can be omitted. We shall assume that the operator A is simple (Section 81).

Next we present an example of a quasi-self-adjoint extension of a symmetric operator. Let P be the operator of differentiation in the space $L^2(0, a)$ with the boundary condition

$$\varphi(0) = \varphi(a) = 0.$$

In Chapter IV we proved that the domain of an arbitrary self-adjoint extension P_θ of the operator P is determined by the boundary condition

$$\varphi(a) = \theta\, \varphi(0) \qquad (|\,\theta\,| = 1), \tag{3}$$

and, conversely, each such condition with $|\,\theta\,| = 1$ determines the domain of a self-adjoint extension of the operator P.

If in equation (3) θ is replaced by an arbitrary complex number ρ such that $|\,\rho\,| \neq 1$, then the operator P_ρ $(P_\rho \subset P^*)$ which is defined on set of the functions $\varphi(t)$ which satisfy the condition

$$\varphi(a) = \rho\, \varphi(0) \qquad (|\,\rho\,| \neq 1) \tag{4}$$

is a quasi-self-adjoint extension of the operator P. It is evident that each quasi-self-adjoint extension of the operator P is given by condition (4) (one of them with $\rho = \infty$).

For quasi-self-adjoint extensions B of an operator A it is possible to introduce the Cayley transform S which is defined on the manifold

$$D_s = \Delta_B(\lambda)$$

[7]Cf. M. S. Lifschitz [2], [3], Vol. I.

by the formulas

$$\text{(5)} \qquad \varphi = (B - \lambda E) f$$

$$(f \in \mathrm{D}_B, \Im \lambda \neq 0)$$

$$\text{(6)} \qquad S \varphi = (B - \bar{\lambda} E) f.$$

This definition is ambiguous if the vector f, which is defined by the vector φ, is not unique, i.e., if λ is an eigenvalue of the operator B. But if λ is an eigenvalue of the operator B, then $\bar{\lambda}$ is not an eigenvalue, since otherwise

$$g_\lambda \in \mathrm{D}_B, \; g_{\bar{\lambda}} \in \mathrm{D}_B$$

and

$$B = A^*,$$

which would contradict condition (1).

Therefore, we can assume in formula (5) that λ is not an eigenvalue of the operator B. For definiteness we choose $\lambda = -i$ and then write formulas (5) and (6) in the form

$$\text{(5')} \qquad \varphi = (B + iE) f$$

$$(f \in \mathrm{D}_B)$$

$$\text{(6')} \qquad S\varphi = (B - iE) f.$$

It is evident that the operator S is defined in the whole space and is an extension of the Cayley transform V of the operator A. However, in contrast with the case of self-adjoint extensions, the operator S is not unitary.

We show that the orthogonal complement of the manifold

$$\mathrm{D}_V = \mathfrak{M}_{-i} = \Delta_A(-i)$$

is transformed by the operator S either into the orthogonal complement of the manifold

$$\Delta_V = \mathfrak{M}_i = \Delta_A(i),$$

or into the zero manifold. Indeed, if

$$g \perp \mathfrak{M}_{-i}$$

and

$$g = (B + iE)\, h \qquad (h \in \mathrm{D}_B),$$

then for each $f \in \mathrm{D}_A$

$$((B + iE)\, h, (A + iE) f) = 0.$$

On the other hand, using the inclusion $B \subset A^*$, we get for each $f \in D_A$

$$(Sg, (A - iE)f) = ((B - iE)h, (A - iE)f) =$$
$$= (Bh, Af) - i(h, Af) + i(Bh, f) + (h, f) =$$
$$= (Bh, Af) - i(Bh, f) + i(h, Af) + (h, f) =$$
$$= ((B + iE)h, (A + iE)f) = 0,$$

so that $Sg \perp \mathfrak{M}_i$ (in particular, it is possible that $Sg = 0$).

If g_1 denotes a vector orthogonal to the manifold $\mathfrak{M}_i$ such that $\| g_1 \| = \| g \|$, then, on the basis of the proof,

$$Sg = \kappa g_1, \tag{7}$$

where the number κ does not have absolute value one, since otherwise the operator B, which obviously is expressed in terms of S by means of the formulas

$$f = \tfrac{1}{2i}(E - S)\varphi, \tag{8}$$

$$(\varphi \in H)$$

$$Bf = \tfrac{1}{2}(E + S)\varphi \tag{9}$$

would be self-adjoint.

The operator $S \equiv S_\kappa$ is called a *quasi-unitary extension* of the isometric operator V. It is easy to verify that $S_\kappa^* = S_{\bar{\kappa}}$.

In general, we define a quasi-unitary extension of a given isometric operator V with deficiency indices (m, m) $(m < \infty)$ as an arbitrary linear but not unitary operator $S \supset V$ which is defined in the whole space and which transforms the orthogonal complement of the manifold D_V into a sub-space orthogonal to Δ_V (or into the null element).

It is evident that each quasi-unitary extension S of the operator V generates by formulas (8) and (9) a quasi-self-adjoint extension B of the operator A.

For the case of deficiency indices $(1, 1)$ there exists a one-to-one correspondence between the set of all self-adjoint extensions B of the operator A (respectively of quasi-unitary extensions S of the operator V) and the set of all complex numbers κ not equal to one in absolute value. In view of this relation, we denote the corresponding extension by B_κ (respectively, S_κ).

We turn now to the study of the spectrum of a quasi-self-adjoint extension B of the operator A with deficiency indices $(1, 1)$. We recall that by the general definition of Section 43, the number λ is called a regular point of the linear operator T if the operators $(T - \lambda E)^{-1}$ exists, is bounded, and is defined in the whole space. The spectrum of the operator T is defined as the complement of the set of its regular points.

THEOREM 1: *The spectrum of a quasi-self-adjoint extension B of a simple operator A with deficiency indices* (1, 1) *consists of the spectral kernel* (*cf.* Section 83) *of the operator A and the eigenvalues. The set of the eigenvalues lies in its entirety either in the upper or in the lower half-plane. We omit the special case*[8] *when the whole half-plane (upper or lower) consists of eigenvalues, and, as a consequence, the set of eigenvalues can have only real limit points in addition to the eigenvalues themselves.*

Proof: We denote the spectral kernel of the operator A by Λ. For each $\lambda \in \Lambda$, the operator $(A-\lambda E)^{-1}$ is unbounded and hence it is not possible for the operator $(B-\lambda E)^{-1}$ to be bounded. Hence Λ is contained in the spectrum of the operator B.

We assume now that $\lambda \notin \Lambda$. If the operator $(B - \lambda E)^{-1}$ does not exist, then λ is an eigenvalue of the operator B. If the operator $(B - \lambda E)^{-1}$ exists then λ is a regular point of the operator B. Indeed the operator $(B - \lambda E)^{-1}$ cannot be unbounded, since the operator $(A - \lambda E)^{-1}$ is bounded and $\Delta_B(\lambda)$ differs from $\Delta_A(\lambda)$ by not more than one dimension. It remains to show that $\Delta_B(\lambda) = \mathrm{H}$. Assuming the contrary, we find that $\Delta_B(\lambda) = \Delta_A(\lambda)$. Now if f is a vector in D_B which does not belong to D_A, then the vector

$$(B - \lambda E)f = f^*,$$

which belongs to $\Delta_B(\lambda) = \Delta_A(\lambda)$, can be represented in the form

$$f^* = (A - \lambda E)f' = (B - \lambda E)f' \qquad (f' \in \mathrm{D}_A).$$

Hence

$$(B - \lambda E)(f - f') = 0,$$

which contradicts the existence of the operator $(B - \lambda E)^{-1}$. Thus, it is proved that λ is a regular point, and the spectrum of the operator B consists of the spectral kernel Λ of the operator A and the eigenvalues.

We introduce the Cayley transform S_κ of the operator $B = B_\kappa$, and then we notice without difficulty that the point λ runs through the spectrum of the operator B as the point

$$\zeta = \frac{\lambda - i}{\lambda + i} \tag{10}$$

runs through the spectrum of the operator S_κ, and conversely. Moreover, formula (10) establishes a one-to-one correspondence between the eigenvalues of the operator B and of the operator S_κ.

On the basis of what has been said, it is sufficient to determine the

[8]This case will be treated later (cf. the corollary of Theorem 3).

eigenvalues of the operator S_κ. Since the absolute values of the eigenvalues of the operators S_κ are greater or less than one according as $|\kappa|>1$ or $|\kappa|<1$, all eigenvalues of the operator B lie in the upper half-plane (if $|\kappa|<1$) or in the lower half-plane (if $|\kappa|>1$).

We assume for definiteness that $|\kappa|<1$. In what follows g denotes a unit vector orthogonal to the manifold $\mathfrak{M}_i$, $\mathring{U}$ denotes a unitary extension of U and g^* is the vector defined by

$$g^* = \mathring{U}g.$$

We represent the eigenvector of the operator S_κ which corresponds to the number ζ, in the form $\varphi + \alpha g$ with $\varphi \in D_V$. Thus,

$$S_\kappa(\varphi + \alpha g) = \zeta(\varphi + \alpha g).$$

Then

$$S_\kappa\varphi + \alpha\kappa g^* = \zeta\varphi + \alpha\zeta g,$$

or

$$\mathring{U}\varphi - \zeta\varphi = \alpha(\zeta g - \kappa g^*),$$

and this yields

$$\frac{1}{\alpha}\varphi = \zeta(\mathring{U} - \zeta E)^{-1}g - \kappa(\mathring{U} - \zeta E)^{-1}g^*. \tag{11}$$

We form the scalar product of both members of (11) with g and obtain an equation satisfied by the eigenvalues of the operator S_κ:

$$\zeta((\mathring{U} - \zeta E)^{-1}g, g) - \kappa((\mathring{U} - \zeta E)^{-1}g^*, g) = 0. \tag{12}$$

Reversing the order of calculations we find that every root of equation (12) is an eigenvalue of the operator S_κ.

Since the left member of equation (12) is a regular function for $|\zeta| \neq 0$, the assertion of the theorem concerning the limit points of the discrete spectrum of the operator B is proved with the exception of the case when the left side of equation (12) becomes zero for $|\zeta|<1$.

In this last case each point ζ inside the unit circle is an eigenvalue of the operator S_κ.

We show now that the special case mentioned in the formulation of the theorem can occur. Let H be a separable Hilbert space with an orthonormal basis $\{e_k\}_{-\infty}^{\infty}$ and let $\tilde{U}$ be the unitary operator defined by the formulas

$$\tilde{U}e_k = e_{k-1}.$$

Further, let $\tilde{V} \subset \tilde{U}$ and $D_{\tilde{V}} \perp e_0$ and, hence, $\Delta_{\tilde{V}} \perp e_{-1}$. The operator $\tilde{V}$ is an isometric operator with deficiency indices (1, 1).

Now let S_0 be the quasi-unitary extension of the operator $\tilde{V}$ defined by the condition

$$S_0 e_0 = 0.$$

It is easy to verify that for each ζ inside the unit circle, the vector

$$\sum_{k=0}^{\infty} \zeta^k e_k$$

is an eigenvector of the operator S which corresponds to the eigenvalue ζ.

Later, we shall see that this special case is unique in the following sense. A simple symmetric operator with deficiency indices (1, 1) having a quasi-self-adjoint extension with a point spectrum which forms an entire half-plane is isomorphic to the Cayley transform of the operator $\tilde{V}$.

We turn now to the transformation of equation (12). If $\mathring{F}_t$ is the spectral function of the operator $\mathring{U}$, then

$$2\zeta((\mathring{U}-\zeta E)^{-1}g, g) = 2\zeta\int_0^{2\pi} \frac{1}{e^{is}-\zeta} d(\mathring{F}_s g, g) = \tag{13}$$

$$= \int_0^{2\pi} \frac{e^{is}+\zeta}{e^{is}-\zeta} d(\mathring{F}_s g, g) - 1,$$

and

$$2((\mathring{U}-\zeta E)^{-1}g^*, g) = 2((\mathring{U}-\zeta E)^{-1}\mathring{U}g, g) = \tag{14}$$

$$= 2\int_0^{2\pi} \frac{e^{is}}{e^{is}-\zeta} d(\mathring{F}_s g, g) = \int_0^{2\pi} \frac{e^{is}+\zeta}{e^{is}-\zeta} d(\mathring{F}_s g, g) + 1.$$

From the last formulas it follows in particular that the scalar product $((\mathring{U}-\zeta E)^{-1}g^*, g)$ does not vanish for $|\zeta| < 1$. Hence, equation (12) can be represented in the form

$$w(\zeta) - \kappa = 0,$$

where

$$w(\zeta) = \frac{\zeta((\mathring{U}-\zeta E)^{-1}g, g)}{((\mathring{U}-\zeta E)^{-1}g^*, g)} \qquad (g^* = \mathring{U}g). \tag{15}$$

On the basis of formulas (13) and (14) the function $w(\zeta)$ can be represented in the form

$$w(\zeta) = \frac{\Phi(\zeta)-1}{\Phi(\zeta)+1}, \tag{15a}$$

where

$$\Phi(\zeta) = \int_0^{2\pi} \frac{e^{is} + \zeta}{e^{is} - \zeta} d(\mathring{F}_s g, g). \tag{15b}$$

From these representations it follows (Section 59) that the function $w(\zeta)$ is regular in the unit circle, maps it into itself and satisfies the normalization condition $w(0) = 0$.

Following Lifschitz we call the function $w(\zeta)$ *the characteristic function of the isometric operator* V and the function

$$\omega(\lambda) = w\left(\frac{\lambda - i}{\lambda + i}\right) \tag{15c}$$

the characteristic function of the symmetric operator A.

The function $\omega(\lambda)$ is regular in the upper half-plane, maps it into a subset of the unit circle, and satisfies the normalization condition $\omega(i) = 0$.

In order to justify the above definitions, it is necessary to show that the function $w(\zeta)$ is defined in principle by the operator V although (cf. formula (15)) it depends formally on the choice of the unitary extension $\mathring{U}$. With this aim we express the operator $\mathring{U}$ in formula (15) in terms of its Cayley transform:

$$\mathring{U} = (\mathring{A} - iE)(\mathring{A} + iE)^{-1}.$$

After an elementary calculation we obtain for the characteristic function $\omega(\lambda)$ the formula

$$\omega(\lambda) = \frac{\lambda - i}{\lambda + i} \frac{(\{E + (\lambda - i)\mathring{R}_\lambda\} g, g^*)}{(\{E + (\lambda - i)\mathring{R}_\lambda\} g, g)}. \tag{16}$$

Here $\mathring{R}_\lambda$ is the resolvent of the operator $\mathring{A}$.

In Section 84 it was established that the vector

$$\{E + (\lambda - i)\mathring{R}_\lambda\} g$$

belongs to the deficiency subspace $\mathfrak{N}_{\bar{\lambda}}$, which is one-dimensional in the case under consideration. Hence, for another choice of the extension $\mathring{A}$ this vector is replaced by a scalar multiple of itself. And this does not alter $\omega(\lambda)$. As far as the vector g^* is concerned, it is multiplied by a number with absolute value one.

Thus, the characteristic function of a symmetric (isometric) operator is determined by this operator to within a multiplicative constant with absolute value one. We do not distinguish between two characteristic functions which differ from one another by such a factor.

Now formula (16) can be written in the form

$$\omega(\lambda) = \frac{\lambda - i}{\lambda + i} \frac{(g_\lambda, g^*)}{(g_\lambda, g)}, \tag{17}$$

where g_λ is an arbitrary solution of the equation

$$A^* h - \lambda h = 0.$$

As an example, we calculate the characteristic function of the operator P of differentiation on the interval $[0, a]$ with the boundary conditions

$$\varphi(0) = \varphi(a) = 0.$$

In this case

$$g = \frac{\sqrt{2}}{\sqrt{e^{2a} - 1}} e^t, \ g^* = \frac{\sqrt{2}}{\sqrt{1 - e^{2a}}} e^{-t}, \ g_\lambda = e^{-i\lambda t},$$

and, by formula (17),

$$\omega(\lambda) = \frac{e^a - e^{-ia\lambda}}{1 - e^a e^{-ia\lambda}}. \tag{18}$$

In addition to $w(\zeta)$ and $\omega(\lambda)$ we introduce the functions

$$w(\zeta; \kappa) = \frac{w(\zeta) - \kappa}{\bar{\kappa} w(\zeta) - 1},$$

$$\omega(\lambda; \kappa) = \frac{\omega(\lambda) - \kappa}{\bar{\kappa}\omega(\lambda) - 1},$$

and call them the *characteristic functions* of the quasi-unitary extension S of the operator V and, correspondingly, of the quasi-self-adjoint extension B_κ of the operator A. The characteristic functions $w(\zeta; \kappa)$ and $\omega(\lambda; \kappa)$ are normed by the conditions

$$w(0; \kappa) = \kappa, \qquad \omega(i; \kappa) = \kappa.$$

With the aid of the characteristic function $\omega(\lambda)$ the spectral kernel of the operator A is defined, as shown in the following proposition.

THEOREM 2: *In order that the real number λ_0 be a point of regular type of a simple symmetric operator A with deficiency indices* (1, 1) *it is necessary and sufficient that both of the following conditions be satisfied:*

1° *the function $\omega(\lambda)$ is regular in a neighborhood of λ_0;*

2° $|\omega(\lambda)| = 1$ in some interval $\lambda_0 - \epsilon < \lambda < \lambda_0 + \epsilon$.

Proof: Let λ_0 be a point of regular type of the operator A and V the Cayley transform of A. It is evident that there exists a self-adjoint extension $\mathring{A} \supset A$ for which the point λ_0 is regular, because if two self-adjoint extensions $\mathring{A}_1$ and $\mathring{A}_2$ have the same eigenvalue λ_0, then

$$\mathring{A}_1 f_1 = \lambda_0 f_1, \ \mathring{A}_2 f_2 = \lambda_0 f_2 \quad (f_1, f_2 \notin \mathrm{D}_A).$$

Then

$$A^* f_1 = \lambda_0 f_1, \ A^* f_2 = \lambda_0 f_2,$$

and hence

$$f_1 = \alpha f_2,$$

so that

$$\mathrm{D}_{\mathring{A}_1} = \mathrm{D}_{\mathring{A}_2}$$

and

$$\mathring{A}_1 = \mathring{A}_2.$$

In formula (15) let $\mathring{U}$ be the Cayley transform of $\mathring{A}$. Then from formulas (15a), (15b), (15c) it follows that the function $\omega(\lambda)$ has properties 1° and 2°.

Conversely, let $\omega(\lambda)$ have properties 1° and 2°. Then choosing the extension $\mathring{U}$ in formula (15) such that

$$w(\zeta_0) \neq 1, \qquad \left(\zeta_0 = \frac{\lambda_0 - i}{\lambda_0 + i}\right),$$

the function

$$\Phi(\zeta) = \frac{1 + w(\zeta)}{1 - w(\zeta)}$$

is regular in a neighborhood of the point $\zeta_0 = e^{is_0}$. Furthermore, on some arc of the unit circle containing the point ζ_0, $|\Phi(\zeta)| = 1$. Therefore, after the application of the inversion formula to the representation (15b) it follows that ζ_0 is a point of constancy of the function $(\mathring{F}_s g, g)$ and, since the vector g generates the operator $\mathring{U}$, ζ_0 is a regular point of this operator (cf. Section 69). Thus, λ_0 is a regular point of the operator $\mathring{A}$ and, hence, does not belong to the spectral kernel of A.

From the proof of Theorem 2 we obtain the following refinement of Theorem 1: *all finite limit points of the set of eigenvalues of any quasi-self-adjoint extension B_κ of the given operator A belongs* (*with the exception of the case mentioned in Theorem* 1) *to the spectral kernel of the operator A.*

If $\zeta_k (k = 1, 2, 3, \ldots)$ are the roots of the characteristic function $w(\zeta; \kappa)$ of a quasi-unitary extension S_κ of the operator V for $\kappa \neq 0$, then $w(\zeta; \kappa)$ can be represented in the form

$$w(\zeta; \kappa) = e^{-G(\zeta)} \prod_{k=1}^{\infty} \frac{\zeta_k - \zeta}{1 - \bar{\zeta}_k \zeta} \frac{|\zeta_k|}{\zeta},$$

where $G(\zeta)$ is a regular function with non-negative real part for $|\zeta| < 1$. If we represent the function $G(\zeta)$ in the form (Section 69)

$$G(\zeta) = \int_0^{2\pi} \frac{e^{is} + \zeta}{e^{is} - \zeta} d\rho(s) + i\beta,$$

where $\rho(s)$ is a nondecreasing function of bounded variation then we obtain

$$(19) \quad w(\zeta;\kappa) = Ce^{\int_0^{2\pi} \frac{\zeta + e^{is}}{\zeta - e^{is}} d\rho(s)} \prod_{k=1}^{\infty} \frac{1 - \frac{\zeta}{\zeta_k}}{1 - \bar{\zeta}_k \zeta} |\zeta_k| \quad (|C| = 1).$$

From Theorems 1 and 2, it follows that the point spectrum of a quasi-unitary extension S_κ of the operator V consists of the points ζ_k and the rest of the spectrum consists of points of increase of the function $\rho(s)$ and limit points of the set of roots ζ_k.[9]

If the spectral kernel of the operator is empty, then the eigenvalues of an arbitrary quasi-self-adjoint extension B_κ of the operator A have no finite limit points.

In particular, it can happen that the spectrum of some quasi-self-adjoint extension of the operator A is void. An example is the quasi-self-adjoint extension of the differential operator P with the boundary condition

$$\varphi(0) = 0.$$

Later (cf. Theorem 4) we see that this special case is unique if one does not consider isomorphic operators as distinct.

We prove now a general theorem which shows that the characteristic function determines an operator up to an isomorphism.

THEOREM 3: *In order that simple symmetric (isometric) operators with deficiency indices (1, 1) be unitary equivalent, it is necessary and sufficient that their characteristic functions coincide.*

Proof: We give the proof for isometric operators. Let the operators V and $\tilde{V}$ which are defined in the spaces H and $\tilde{\mathrm{H}}$ respectively be unitary equivalent, i.e., let

$$\tilde{V} = \mathfrak{U} V \mathfrak{U}^{-1},$$

where $\mathfrak{U}$ is an isometric operator which maps H on $\tilde{\mathrm{H}}$. Taking some unitary extension U of the operator V we construct a unitary extension $\tilde{U}$ of the

[9] Starting from the representation (19) for the characteristic function $w(\zeta; x)$ M. S. Lifschitz investigated the question of the invariant subspaces of the operators S_κ (cf. M. S. Lifschitz (4), Vol. I).

operator $\tilde{V}$ by the formula

$$\tilde{U} = \mathfrak{U}\, U\, \mathfrak{U}^{-1}.$$

Furthermore, we choose a unit vector $g \in \mathrm{H}$ orthogonal to the manifold D_V and we let

$$\tilde{g} = \mathfrak{U} g.$$

For these choices of the vector $\tilde{g}$ and of the operator $\tilde{U}$, the formulas

$$w(\zeta) = \frac{\zeta((U - \zeta E)^{-1} g, g)_H}{((U - \zeta E)^{-1} U g, g)_H}, \tag{20}$$

$$\tilde{w}(\zeta) = \frac{\zeta((\tilde{U} - \zeta \tilde{E})^{-1} \tilde{g}, \tilde{g})_{\tilde{H}}}{((\tilde{U} - \zeta \tilde{E})^{-1} \tilde{U} \tilde{g}, \tilde{g})_{\tilde{H}}} \tag{21}$$

imply that

$$\tilde{w}(\zeta) = w(\zeta).$$

Conversely, let the characteristic functions $\tilde{w}(\zeta)$ and $w(\zeta)$ of the operators $\tilde{V}$ and V, which are defined by formulas (20) and (21), coincide. Let

$$\tilde{\Phi}(\zeta) = \frac{1 + \tilde{w}(\zeta)}{1 - \tilde{w}(\zeta)}, \quad \Phi(\zeta) = \frac{1 + w(\zeta)}{1 - w(\zeta)},$$

and apply the inversion formula to the representations of the functions $\tilde{\Phi}(\zeta)$ and $\Phi(\zeta)$ to obtain

$$(\tilde{E}_t \tilde{g}, \tilde{g})_{\tilde{H}} = (E_t g, g)_H.$$

On the other hand, the simplicity of the operators $\tilde{V}$ and V implies the simplicity of the spectra of their unitary extensions $\tilde{U}$ and U; here the deficiency vectors $\tilde{g}$ and g are the generating elements for $\tilde{U}$ and U (Section 81).

Thus, the operators $\tilde{U}$ and U are reduced to the same canonical form—to the operator of multiplication by e^{it} in the space L^2_σ with the distribution function

$$\sigma(t) = (\tilde{E}_t \tilde{g}, \tilde{g})_{\tilde{H}} = (E_t g, g)_H.$$

and, hence, are isomorphic (Section 69). From the isomorphism of the operators $\tilde{U}$ and U it immediately follows that the operators $\tilde{V}$ and V are isomorphic. Thus, the theorem is proved.

REMARK: Each simple symmetric operator with deficiency indices (1, 1) which has a quasi-self-adjoint extension with point spectrum which covers an entire half-plane is isomorphic to the Cayley transform of the operator $\tilde{V}$ defined on page 151.

Indeed the characteristic function of such an operator must be identically zero, so that all such operators are isomorphic.

The following theorem gives us an interesting abstract characterization of the operator of differentiation on a finite interval.

THEOREM 4: *Each simple symmetric operator with deficiency indices* (1, 1) *which admits a quasi-self-adjoint extension without spectrum is isomorphic to the operator of differentiation on a finite interval.*

Proof: Let A be an operator with the property indicated in the formulation of the theorem, let B_κ be a self-adjoint extension without spectrum and let $\omega(\zeta; \kappa)$ be the characteristic function of the operator B_κ. Furthermore, let V be the Cayley transform of the operator A and let $w(\zeta; \kappa)$ be the characteristic function of the Cayley transform S_κ of the operator B_κ.

Since B_κ has no eigenvalues, the function $w(\zeta; \kappa)$, which maps the unit circle on a subset of itself, never vanishes. Therefore,

$$w(\zeta; \kappa) = e^{-G(\zeta)},$$

where the function $G(\zeta)$ is regular in the unit circle and has there a nonnegative real part. Replace ζ by

$$\frac{\lambda - i}{\lambda + i}$$

to obtain

$$\omega(\lambda; \kappa) = e^{iH(\lambda)}.$$

Here $H(\lambda)$ is a function which is regular in the upper half-plane and maps it into a subset of itself.

According to Section 59, the function $H(\lambda)$ can be represented in the form

$$H(\lambda) = \alpha + \mu\lambda + \int_{-\infty}^{\infty} \frac{1 + i\lambda}{t - \lambda}\, d\sigma(t),$$

where α is real, $\mu \geqq 0$ and $\sigma(t)$ is a nondecreasing function of bounded variation.

Since the spectral kernel of the operator A is empty, the function $H(\lambda)$ is also regular and real on the entire real axis. Therefore, by the Stieltjes inversion formula (Section 59)

$$\sigma(t) = \text{const.},$$

and

$$H(\lambda) = \alpha + \mu\lambda.$$

Thus, except possibly for a constant factor of absolute value one,

$$\omega(\lambda; \kappa) = e^{i\mu\lambda}.$$

This yields

$$\omega(\lambda) = \frac{\kappa - \omega(\lambda; \kappa)}{1 - \bar{\kappa}\,\omega(\lambda; \kappa)} = \frac{\kappa - e^{i\mu\lambda}}{1 - \bar{\kappa} e^{i\mu\lambda}}.$$

But since

$$\omega(i) = 0,$$

we have

$$\omega(\lambda) = \frac{e^{\mu} - e^{-i\mu\lambda}}{1 - e^{\mu} e^{-i\mu\lambda}}.$$

Comparing this formula with formula (18) we see that the characteristic function of the operator A coincides with the characteristic function of the operator of differentiation on the interval $[0, \mu]$. By Theorem 3 the operator A is unitarily equivalent to the operator of differentiation, which was to be proved.

In connection with Theorem 3, there arises the question of the existence of a symmetric (isometric) operator with given characteristic function. For isometric operators this question can be answered in the affirmative:

THEOREM 5: *Let $w(\zeta)$ be a function which is regular in the unit circle, maps it in a subset of itself and satisfies the normalization condition $w(0)=0$. Then there exists an isometric operator V for which $w(\zeta)$ is the characteristic function.*

Proof: We define a function $\Phi(\zeta)$ by means of the equation

$$w(\zeta) = \frac{\Phi(\zeta) - 1}{\Phi(\zeta) + 1}.$$

By Section 59, $\Phi(\zeta)$ can be represented in the form

$$\Phi(\zeta) = \int_0^{2\pi} \frac{e^{is} + \zeta}{e^{is} - \zeta}\, d\sigma(s),$$

where $\sigma(s)$ is a nondecreasing function with total variation one.

We introduce the space $L^2_\sigma(0, 2\pi)$ and define in it the unitary operator of multiplication by e^{it}:

$$\mathring{U} f(t) = e^{it} f(t).$$

Let V be the isometric operator which coincides with the operator $\mathring{U}$ on the hyperplane D_V orthogonal to the function $g(t) \equiv 1$. It is easy to verify that V satisfies the conditions of the theorem.

Indeed, since the spectral function $\mathring{E}_t$ of the operator $\mathring{U}$ is defined by the equation

$$\mathring{E}_t f(s) = \begin{cases} f(s) & (s \leqq t), \\ 0 & (s > t), \end{cases}$$

we have

$$((\mathring{U} - \zeta E)^{-1} g, g) = \int_0^{2\pi} \frac{d\sigma(s)}{e^{is} - \zeta},$$

$$((\mathring{U} - \zeta E)^{-1} \mathring{U} g, g) = \int_0^{2\pi} \frac{e^{is} d\sigma(s)}{e^{is} - \zeta},$$

and

$$\frac{\zeta((\mathring{U} - \zeta E)^{-1} g, g)}{((\mathring{U} - \zeta E)^{-1} \mathring{U} g, g)} = \frac{\Phi(\zeta) - 1}{\Phi(\zeta) + 1} = w(\zeta),$$

which proves the theorem (cf. formula (15)).

If, for an isometric operator V with a given characteristic function $w(\zeta)$, the manifold $\Delta_V(1)$ is not dense in H, then it is impossible to pass from V to its Cayley transform A. Therefore, in this case there exists no symmetric operator with characteristic function

$$\omega(\lambda) = w\left(\frac{\lambda - i}{\lambda + i}\right). \tag{22}$$

In order that the manifold $\Delta_V(1)$ be dense in H it is necessary to impose an additional restriction on the function (22). This matter is resolved in the following theorem.

THEOREM 6: *Let $\omega(\lambda)$ be any function which is regular in the upper half-plane, maps it into the interior of the unit circle and satisfies the normalization condition $\omega(i) = 0$. Then $\omega(\lambda)$ is the characteristic function of some simple symmetric operator if and only if*

$$\lim_{\lambda \to \infty} \lambda\{\omega(\lambda) - e^{i\alpha}\} = \infty \qquad (0 < \epsilon \leqq \arg \lambda \leqq \pi - \epsilon)$$

for each α such that $0 \leqq \alpha < 2\pi$.

Proof: We show that the manifold $\Delta_V(1)$ is not dense in H if and only if there exists a unitary extension of the operator V with an eigenvalue equal to one.

The sufficiency of the conditon is evident since U is a unitary extension of the operator V and $U\psi = \psi$ $(\psi \neq 0)$, so that for each $\varphi \in D_V$,

$$((V - E)\varphi, \psi) = ((U - E)\varphi, \psi) = (\varphi, (U^{-1} - E)\psi) = 0.$$

For the proof of the necessity of the condition we assume that

$$((V - E)\varphi, \psi) = 0 \qquad (\psi \neq 0)$$

for each $\varphi \in D_V$. We represent each vector $f \in H$ in the form $f = \varphi + \gamma g$, and obtain for an arbitrary quasi-unitary extension S_κ of the operator V the equation

$$(23) \quad ((S_\kappa - E)f, \psi) = (V\varphi - \varphi + \gamma\kappa g^* - \gamma g, \psi) = \gamma\{\kappa(g^*, \psi) - (g, \psi)\}.$$

Note that $(g^*, \psi) \neq 0$, since otherwise the vector ψ would have the representation $\psi = V\varphi_1 (\varphi_1 \in D_V, \varphi_1 \neq 0)$ so that

$$((V - E)\varphi, V\varphi_1) = 0 \qquad (\varphi \in D_V),$$

and, consequently,

$$(\varphi, (E - V)\varphi_1) = 0.$$

It follows that

$$V\varphi_1 - \varphi_1 = \gamma_1 g,$$

or

$$V\varphi_1 = \varphi_1 + \gamma_1 g.$$

But the orthogonality of the summands on the right side of the last equation and the condition $\| V\varphi_1 \| = \| \varphi_1 \|$ imply that $V\varphi_1 = \varphi_1 (\varphi_1 \neq 0)$, which contradicts the fact that the operator V is simple. Thus $(g^*, \psi) \neq 0$ and we can substitute

$$\kappa = \frac{(g, \psi)}{(g^*, \psi)}$$

in (23). It follows that, for each $f \in H$,

$$((S_\kappa - E)f, \psi) = 0,$$

or

$$(f, (S_\kappa^* - E)\psi) = 0.$$

Since $S_\kappa^* = S_{\bar{\kappa}}$, this equation implies that $S_{\bar{\kappa}}\psi = \psi\ (\psi \neq 0)$ which is possible only for $|\kappa| = 1$. Hence $S_{\bar{\kappa}}$ is the desired unitary extension of the operators V with an eigenvalue equal to one.

Now the proof of the theorem is reduced to the determination of necessary and sufficient conditions which the characteristic function of an isometric operator V must satisfy in order that none of its unitary extensions has an eigenvalue equal to one.

Let $\mathring{U}$ be the unitary extension of the operator V which appears in formula (15) and let $U^{(\alpha)}$ be an arbitrary unitary extension of the same operator such that $U^{(\alpha)}g = e^{i\alpha}g^*$. The deficiency element g is a generating element for each of the operators $U^{(\alpha)}$. Hence, the operator $U^{(\alpha)}\ (0 \leq \alpha < 2\pi)$

does not have an eigenvalue equal to one if and only if the distribution function $(F_t^{(\alpha)}g, g)$ has no jump at $t = 0$, where $F_t^{(\alpha)}$ is the spectral function of the operator $U^{(\alpha)}$ $(0 \leqq \alpha < 2\pi)$.

[10]On the other hand it follows from the formulas (15a), (15b), (15c) for the characteristic function $w(\zeta)$ that

$$\frac{e^{i\alpha}}{e^{i\alpha} - w(\zeta)} = \int_0^{2\pi} \frac{e^{it}}{e^{it} - \zeta}\, d(F_t^{(\alpha)}g, g).$$

This equation yields the value μ of the jump of the function $(F_t^{(\alpha)}g, g)$ at the point $t = 0$, namely

$$\mu = \lim_{\zeta \to 1} \frac{(1 - \zeta)\, e^{i\alpha}}{e^{i\alpha} - w(\zeta)}.$$

Using the indicated derivations, it is not difficult to complete the proof of the theorem. This task is left to the reader.

---

[10]Translator's Note: This sentence and the one following do not appear in the original (Russian) edition. They are taken from the authorized German edition.

Appendix II

DIFFERENTIAL OPERATORS

1. Self-Adjoint Differential Expressions

We begin with the presentation of some facts concerning real differential expressions (or, let us agree to call them *differential operations*) which are self-adjoint in the sense of Lagrange.

In analysis courses it is established that a self-adjoint differential operation of the second order

$$l_0D^2 + l_1D + l_2D^0 \left(l_k = l_k(t), D^k = \frac{d^k}{dt^k}\right)$$

under the assumption of the k-fold differentiability of the coefficients $l_k(t)$ can be represented in the form

$$- Dp_0D + D^0p_1 D^0,$$

where $p_0(t)$ is differentiable. However, it is possible to consider such operations without the assumption of the differentiability of the function $p_0(t)$. In this case the operator can be applied only to differentiable functions $\varphi (t)$ such that the product $p_0\varphi'$ is absolutely continuous.

We turn now to self-adjoint differential expressions of order $2n$. We assume as the canonical form of such an operation the expression

$$(1) \quad l = p_nD^0 - D\{p_{n-1}D - D[p_{n-2}D^2 - \ldots - D(p_1 D^{n-1} - Dp_0D^n)\ldots]\}.$$

If the coefficient $p_{n-k}(t)$ is k times ($k=0, 1, \ldots, n$) differentiable, then this is a self-adjoint ordinary differential operation, as may be verified immediately. Since we do not assume the differentiability of the coefficients, we call operation (1) a *quasi-differential operation.*

Let (a, b) be an open finite or infinite interval in which the differential operation (1) is considered. We assume that the coefficients $p_k(t)$ are measurable in this interval and satisfy the conditions

$$(2) \qquad \int_\alpha^\beta \frac{dt}{|p_0(t)|} < \infty, \int_\alpha^\beta |p_k(t)|\, dt < \infty, (k = 1, 2, \ldots, n)$$

in each closed subinterval $[\alpha, \beta]$ of (a, b).

If the interval (a, b) is finite and if condition (2) is satisfied for $\alpha=a$, $\beta=b$, then the operation (1) is called *regular*. If the interval (a, b) is infinite, or it is finite but conditions (2) are not satisfied for $\alpha=a$ or $\beta=b$, then the operation (1) is called *singular*. The left end point a is called *singular* if $a=-\infty$ or $a>-\infty$, but (2) fails with $\alpha=a$. Singularity of the right end point b is defined analogously. If an end point is not singular, then it is called *regular*. Without loss of generality, we consider singular differential operations with one singular end point on the semi-axis $(0, \infty)$ and with two singular end points on the entire axis $(-\infty, \infty)$.

For convenience we introduce, in place of the derivative $D^k\varphi = \varphi^{(k)}$, the so-called quasi-derivative $D^{[k]}\varphi = \varphi^{[k]}$ which is defined by the following formulas:

$$D^{[k]} = D^k \qquad (k = 0,1,2,\ldots, n-1),$$

$$D^{[n]} = p_0 D^n,$$

$$D^{[n+k]} = p_k D^{n-k} - D D^{[n+k-1]} \qquad (k = 1,2,\ldots,n).$$

Now operation (1) can be expressed in the form

$$l = D^{[2n]}.$$

Let D* denote the class of all functions $\varphi(t) \in L^2(a, b)$ for which each quasi-derivative $\varphi^{[k]}(t)$ $(k = 0, 1, \ldots, 2n-1)$ is absolutely continuous and the quasi-derivative $\varphi^{[2n]}(t)$ belongs to $L^2(a, b)$. It is evident that D* is the maximal linear manifold in $L^2(a, b)$, on which the operation l has a natural meaning and can be considered as an operator in $L^2(a, b)$. We denote this operator by L^*, so that $\mathrm{D}^* = \mathrm{D}_{L^*}$. We shall see later the purpose of this notation.

From (1) it is easy to obtain for each pair of functions $\varphi, \psi \in \mathrm{D}^*$ the so-called *Lagrange identity*

$$l[\varphi]\bar{\psi} - \varphi l[\bar{\psi}] = \frac{d}{dt}[\varphi, \psi]_t,$$

where $[\varphi, \psi]_t$ is the bilinear form

$$[\varphi, \psi]_t = \sum_{k=1}^{n} \{\varphi^{[k-1]}(t)\overline{\psi^{[2n-k]}(t)} - \varphi^{[2n-k]}(t)\overline{\psi^{[k-1]}(t)}\}.$$

We shall use the Lagrange identity in the form

$$\int_\alpha^\beta l[\varphi(t)]\overline{\psi(t)}\,dt - \int_\alpha^\beta \varphi(t)\, l[\overline{\psi(t)}]\,dt = [\varphi, \psi]_\alpha^\beta,$$

where $[\alpha, \beta]$ is an arbitrary closed subinterval of (a, b) and

$$[\varphi, \psi]_\alpha^\beta = [\varphi, \psi]_\beta - [\varphi, \psi]_\alpha.$$

Since each of the integrals

$$\int_\alpha^\beta l\,[\varphi(t)]\,\overline{\psi(t)}\,dt, \quad \int_\alpha^\beta \varphi(t)\,\overline{l[\psi(t)]}\,dt$$

exists for $\alpha = a$ and $\beta = b$, the bilinear form $[\varphi, \psi]_t$ has a finite value at the end points of the interval (a, b), whether these end points are regular or singular; for the value of the bilinear form at a singular point we take the limit $[\varphi, \psi]_t$ as t tends to this point.

Let l be a quasi-differential operation, regular or singular in the interval (a, b), and let $g(t)$ be a complex-valued measurable function defined in this interval. It is natural to define a *solution* of the quasi-differential equation

$$l\,[y] - \lambda y = g(t) \tag{3}$$

for a given value of the parameter λ as any function $\varphi(t)$ such that $\varphi(t)$ and its quasi-derivatives up to the $(2n-1)$th order are absolutely continuous and such that this equation is satisfied almost everywhere in (a, b).

The following existence theorem holds for quasi-differential equations and can be proved easily by using *Picard's method* of successive approximations.

THEOREM: *The quasi-differential equation* (3) *has a solution. Moreover, it has only one solution which satisfies the Cauchy conditions*

$$\varphi^{[k]}(t_0) = \alpha_k \qquad (k = 0, 1, 2, \ldots, 2n-1),$$

where t_0 is an arbitrary interior point of the interval (a, b) or a regular end point.

We note that if the operation l is regular in the interval (a, b), then the solution of equation (3) belongs to $L^2(a, b)$. If, in addition, $g(t) \in L^2(a, b)$ then, by equation (3) itself, this solution belongs to D^*.

From the existence theorem it follows, in particular, that the linear manifold of solutions of the homogeneous equation

$$l\,[u] - \lambda u = 0 \tag{4}$$

has dimension $2n$. Solutions $u_1, u_2, \ldots, u_{2n}$ of equation (4) are linearly independent if and only if the Wronskian determinant

$$w\,[u_1, u_2, \ldots, u_{2n}] = \operatorname{Det}\,(u_i^{[k-1]}(t))$$

does not vanish. An arbitrary system of $2n$ linearly independent solutions of equation (4) is called *fundamental*.

If the functions

$$u_1(t), u_2(t), \ldots, u_{2n}(t) \tag{5}$$

form a fundamental system of solutions of the equation (4), then any solution of the non-homogeneous equation (3) can be represented in the form

$$\varphi(t) = \sum_{k=1}^{2n} u_k(t) \int_{t_0}^{t} v_k(s) g(s)\, ds + \sum_{k=1}^{2n} c_k u_k(t).$$

This result can be easily verified if one modifies properly the classical method of variation of parameters. Here, as in the classical case, it is established that the functions

$$v_1(t), v_2(t), \ldots, v_{2n}(t) \tag{6}$$

form a fundamental system of solutions of equation (4). It is called the *adjoint* system of the system (5).

For what follows we shall need the following simple

LEMMA: *If the operation l is regular in the interval (a, b), then equation* (3) *under the boundary conditions*

$$y^{[k]}(a) = y^{[k]}(b) = 0 \quad (k = 0, 1, 2, \ldots, 2n-1)$$

has a solution if and only if the right member $g(t)$ is orthogonal to the $2n$-dimensional manifold of solutions of the homogeneous equation (4).

Proof: Let $\varphi(t)$ be the solution of (3) which satisfies the conditions

$$\varphi^{[k]}(b) = 0 \quad (k = 0, 1, 2, \ldots, 2n-1).$$

Applying the Lagrange identity to the function $\varphi(t)$ and some function $u_i(t)$ of a fundamental system for equation (4), we obtain

$$\sum_{k=1}^{n} \{u_i^{[k-1]}(a)\, \overline{\varphi^{[2n-k]}(a)} - u_i^{[2n-k]}(a)\, \overline{\varphi^{[k-1]}(a)}\} = \int_a^b u_i(s)\, \overline{g(s)}\, ds. \tag{7}$$

Imposing the fundamental system to the initial conditions

$$u_i^{[k-1]}(a) = \begin{cases} 0 & (i \neq k), \\ 1 & (i = k), \end{cases}$$

we obtain

$$
(8) \qquad \overline{\varphi^{[r]}(a)} = \begin{cases} -\int_a^b u_{2n-r}(s)\overline{g(s)}\,ds & (r = 0, 1, \ldots, n-1), \\ \int_a^b u_{2n-r}(s)\overline{g(s)}\,ds & (r = n, n+1, \ldots, 2n-1), \end{cases}
$$

which yields the assertion of the lemma.

2. Regular Differential Operators

Let l be a regular quasi-differential operation defined on the interval (a, b). If φ and ψ are arbitrary functions of D^*, then the difference

$$(L^*\varphi, \psi) - (\varphi, L^*\psi) = [\varphi, \psi]_a^b$$

in general is not equal to zero and hence L^* is not a symmetric operator. In order that the right member of the above relation vanish, it is necessary to impose an additional condition on φ and ψ and, moreover, to replace the domain D by a subset of D. In any case it is sufficient to require that each of the functions φ and ψ satisfy the relations

$$(1) \qquad y^{[k]}(a) = y^{[k]}(b) = 0 \quad (k = 0, 1, \ldots, 2n-1).$$

We denote by D the set of all functions in D* which satisfy the $4n$ conditions (1).

It is natural to expect that the operator L, with domain $D_L = D$, and which coincides with L^* in this domain, is a symmetric operator. Since

$$(L\varphi, \psi) = (\varphi, L\psi)$$

for all functions $\varphi, \psi \in D$, it is necessary only to establish that the manifold D is dense in $L^2(a, b)$. This fact is obvious if l is a differential operation; since in that case the manifold D contains, for example, all polynomials.

For the proof that D is dense in $L^2(a, b)$ in the case of a quasi-differential operation l, we note that, by the lemma of Section 1, one can decompose $L^2(a, b)$ into an orthogonal sum

$$(2) \qquad L^2(a, b) = \Delta_L \oplus \mathfrak{N}_0,$$

where $\mathfrak{N}_0$ is the $2n$-dimensional manifold of solutions of the homogeneous equation

$$l[u] = 0.$$

Let us assume that $(h, \varphi) = 0$ for all $\varphi \in \mathrm{D}$. We must show that $h = 0$. With this aim, we denote by ψ an arbitrary solution of the equation

$$l[\psi] = h.$$

By the Lagrange identity,

$$(\psi, l[\varphi]) = (l[\psi], \varphi) = (h, \varphi) = 0.$$

Since $l[\varphi] = L^*\varphi = L\varphi \in \Delta_L$, then by (2) $\varphi \in \mathfrak{N}_0$, i.e.,

$$l[\psi] = 0.$$

Hence, $h = 0$, which was to be proved.

We now prove that the operator L^* is adjoint to L, which justifies the relation used. For the moment we denote the operator adjoint to L by M. Let $\psi \in \mathrm{D}_M$ and let

$$M\psi = \chi.$$

Let ψ_0 denote any solution of the equation

$$l[y] = \chi.$$

By the Lagrange identity,

$$(\varphi, \chi) = (\varphi, l[\psi_0]) = (l[\varphi], \psi_0) = (L\varphi, \psi_0).$$

for each $\varphi \in \mathrm{D}$. On the other hand, by the definition of the adjoint operator,

$$(\varphi, \chi) = (\varphi, M\psi) = (L\varphi, \psi),$$

and, hence,

$$(L\varphi, \psi - \psi_0) = 0.$$

Now, since $\varphi \in \mathrm{D}_L$ is arbitrary, it follows from the decomposition (2) that

$$\psi - \psi_0 \in \mathfrak{N}_0.$$

This relation shows that $\psi \in \mathrm{D}^*$ and, therefore, that

$$M\psi = \chi = l[\psi_0] = l[\psi].$$

And so we have also shown that

$$M \subset L^*.$$

On the other hand, for arbitrary $\varphi \in \mathrm{D}$ and $\psi \in \mathrm{D}^*$,

$$(L\varphi, \psi) - (\varphi, L^*\psi) = [\varphi, \psi]_a^b = 0.$$

Hence,

$$L^* \subset M,$$

which implies that

$$M = L^*.$$

Hence, our assertion is proved.

The reader can easily verify that L^{**} coincides with L, whence it follows that the operator L is closed.

Since the equation

$$L^*\psi - \lambda\psi = 0$$

has (for each λ) $2n$ linearly independent solutions, the deficiency indices of the operator L are $(2n, 2n)$. Every symmetric operator generated by the operation l is an extension of L. Therefore, it is natural to call *L a regular quasi-differential operator of order $2n$ with minimal domain.*

3. Self-Adjoint Extensions[1] of a Regular Differential Operator

To describe all self-adjoint extensions of an operator L, it is sufficient to indicate the domains of all these extensions. We show that the domain of each self-adjoint extension of the operator L can be defined with the aid of certain boundary conditions, and we characterize all such conditions.

Let $\tilde{L}$ denote some self-adjoint extension of L. In order that a function φ in D_{L^*} also belong to $D_{\tilde{L}}$ it is necessary and sufficient that

$$(L^*\varphi, \psi) = (\varphi, \tilde{L}\psi),$$

or, by the Lagrange identity, that

$$[\varphi, \psi]_a^b = 0 \qquad (1)$$

for all $\psi \in D_{\tilde{L}}$. Since

$$\dim D_{\tilde{L}} = 2n \pmod{D_L},$$

there exists $2n$ functions $w_1, w_2, \ldots, w_{2n}$ in $D_{\tilde{L}}$ such that each function $\psi \in D_{\tilde{L}}$ can be represented in the form

$$\psi = \psi_0 + \sum_{k=1}^{2n} \alpha_k w_k \qquad (\psi_0 \in D_L).$$

By condition (1) of section (2),

$$[\varphi, \psi_0]_a^b = 0$$

for each function φ in D_{L^*}. Therefore, condition (1), which contains an "arbitrary" function ψ, is equivalent to the set of $2n$ conditions

$$[\varphi, w_k]_a^b = 0 \qquad (k = 1, 2, \ldots, 2n), \qquad (2)$$

[1]In the present and following sections, self-adjoint extensions are always of the first kind. For the general form of boundary conditions which characterize self-adjoint extensions of differential operators cf. M. G. Krein [4], Vol. I, and also A. A. Graff [1], Vol. I.

where the functions $w_1, w_2, \ldots, w_{2n}$ belong to $D_{\tilde{L}}$ and satisfy the conditions

$$[w_i, w_k]_a^b = 0 \quad (i, k = 1, 2, \ldots, 2n). \tag{3}$$

Equation (2) can be considered as a system of boundary conditions and equation (3) as a property of the coefficients of these conditions. The boundary conditions (2) can be written in the form of $2n$ linearly independent equations

$$\sum_{k=1}^{2n} \alpha_{ik}\varphi^{[k-1]}(a) + \sum_{k=1}^{2n} \beta_{ik}\varphi^{[k-1]}(b) = 0 \quad (i = 1, 2, \ldots, 2n), \tag{2'}$$

where

$$\begin{aligned} \alpha_{ik} &= \overline{w_i^{[2n-k]}(a)}, \ \alpha_{i,\,n+k} = -\overline{w_i^{[n-k]}(a)}, \\ \beta_{ik} &= -\overline{w_i^{[2n-k]}(b)}, \ \beta_{i,\,n+k} = \overline{w_i^{[n-k]}(b)}, \end{aligned} \quad (k = 1, 2, \ldots, n). \tag{4}$$

Then equation (3) takes the form

$$\begin{aligned} &\sum_{r=1}^{n} \alpha_{i,r}\,\bar{\alpha}_{k,\,2n-r+1} - \sum_{r=1}^{n} \alpha_{i,\,2n-r+1}\,\bar{\alpha}_{k,\,r} = \\ &= \sum_{r=1}^{n} \beta_{i,\,r}\bar{\beta}_{k,\,2n-r+1} - \sum_{r=1}^{n} \beta_{i,\,2n-r+1}\bar{\beta}_{k,\,r} \end{aligned} \quad (i, k = 1, 2, \ldots, 2n). \tag{3'}$$

Thus, the domain of each self-adjoint extension $\tilde{L}$ of L consists of all the functions $\varphi \in D_{L^*}$ which satisfy the $2n$ boundary conditions of the form (2′) with coefficients having the properties (3′).

We assume now a system of $2n$ boundary conditions of the form (2′) with coefficients which satisfy conditions (3′) and we prove that such a system determines in the sense mentioned above the domain of a self-adjoint extension of the operator L.

Let us assume for the moment that there exists $2n$ functions $w_1, w_2, \ldots, w_{2n}$ in D_{L^*} which satisfy conditions (4). Then condition (2′) and equation (3′) take the forms (2) and (3), respectively. We must show that condition (2), with the functions $w_k \in D_{L^*}$ which satisfy relations (3), determine a self-adjoint extension of L. With this aim we denote by $\tilde{D}$ the set of all functions $\varphi \in D_{L^*}$ which satisfy conditions (2) and by D' the set of all functions of the form

$$\psi = \psi_0 + \sum_{k=1}^{2n} \alpha_k w_k,$$

where $\psi_0 \in D_L$ and $\alpha_1, \alpha_2, \ldots, \alpha_{2n}$ are arbitrary constants. Condition (2) is equivalent to

$$[\varphi, \psi]_a^b = 0 \tag{5}$$

for every $\psi \in D'$. Since every function in D' satisfies this condition, relation

(3) holds. Hence, $D' \subset D$. But since both manifolds $\tilde{D}$ and D' are $2n$-dimensional modulo D_L, $\tilde{D} = D'$. Therefore, $\tilde{D}$ $(D_L \subset \tilde{D} \subset D_{L^*})$ is the set of all functions $\varphi \in D_{L^*}$ for which equation (5) is satisfied for all $\psi \in \tilde{D}$. By the lemma in Section 41, the operator $\tilde{L} \subset L^*$ with domain $\tilde{D}$ is a self-adjoint extension of L.

It remains to show that there exist functions in D_{L^*} which, together with their quasi-derivatives up to and including the $(2n-1)$th order, assume prescribed values at both end points of the interval $[a, b]$. Let $g_1(t)$ be any function which is orthogonal to the manifold of solutions of the homogeneous equation (5) of Section 1. Let $\omega_1(t)$ denote the solution of the equation

$$l[y] - \lambda y = g_1(t),$$

which satisfies the given conditions at the point a. By formula (8) of Section 1, $\omega^k(b) = 0$ $(k = 0, 1, 2, \ldots, 2n-1)$. We construct an analogous function $\omega_2(t)$ which satisfies the given conditions at the right end point. Then the sum $\omega_1(t) + \omega_2(t)$ is a function in D_{L^*} which satisfies all of the required conditions.

Thus, we have proved the following theorem.

THEOREM: *A system of linearly independent boundary conditions* (2′) *determines a self-adjoint extension of L if and only if there are* $2n$ boundary *conditions and the coefficients in* (2′) *have property* (3′).

By means of well-known calculations, it is easy to construct the Green's function $\tilde{G}(t, s)$ of an arbitrary self-adjoint extension $\tilde{L}$ of L. The function $\tilde{G}(t, s)$ is a Hilbert-Schmidt kernel and the integral operator $\tilde{K}$ determined by this kernel is related to the operator $\tilde{L}$ by the formula

$$\tilde{L} = \tilde{K}^{-1}.$$

It follows that the spectrum of the operator $\tilde{L}$ consists only of eigenvalues with the unique limit point infinity.

In conclusion, we note that if one omits equation (3′) then the boundary conditions (2′) determine a quasi-self-adjoint extension of the operator L in the sense of Section 5 of Appendix I.

4. Singular Differential Operators[2]

We begin with the case in which the interval is $(0, \infty)$, i.e., the right end point of the interval is singular. We assume the left end point to be regular.

[2]Sections 4, 5, and 6 of the present appendix are excerpts of the dissertation of I. M. Glazman [1], [3], Vol. I.

Let us define the operator L' by the equation

$$L'\varphi = l[\varphi]$$

on the manifold $D_{L'}$ of all functions in D^* which vanish outside a finite interval and which satisfy the conditions

$$\varphi(0) = \varphi^{[1]}(0) = \ldots = \varphi^{[2n-1]}(0) = 0. \tag{1}$$

Since the operator L' is obviously symmetric, it has a closure, which we denote by L and call the singular quasi-differential operator, with *minimal domain* generated by the operation l. The following theorem justifies this designation.

THEOREM 1: *The operator L^* is adjoint to the operator L.*

Proof: As in Section 2, we denote by M the operator adjoint to L. The relation

$$D_{L^*} \subset D_M$$

and the equation

$$M\psi = l[\psi]$$

for $\psi \in D_{L^*}$ are obvious. Hence, it remains only to prove that

$$D_M \subset D_{L^*}.$$

With this aim, as in Section 2, we choose a function $\psi \in D_M$ and let

$$M\psi = \kappa.$$

Furthermore, let ψ_0 denote some solution of the equation

$$l[y] = \chi.$$

Now it remains to prove that

$$\psi(t) = \psi_0(t) + u(t),$$

where $u(t)$ is some solution of the equation

$$l[y] = 0. \tag{2}$$

Now we choose a function $g(t) \in L^2(0, \infty)$ which vanishes for $t \geqq a$ and is orthogonal to the $2n$ dimensional manifold of solutions of equation (2) in the interval $0 \leqq t \leqq a$. Then, by Section 1, there exists a function $\varphi(t)$ which equals zero for $t \geqq a$ and satisfies the equation

$$l[\varphi] = g$$

and also condition (1). Obviously, $\varphi \in D_{L'} \subset D_L$ and, hence,

$$(g, \psi) = (L\varphi, \psi) = (\varphi, M\psi) = (\varphi, \chi).$$

But, on the other hand,

$$(g, \psi_0) = (l[\varphi], \psi_0) = (\varphi, l[\psi_0]) = (\varphi, \chi).$$

Therefore,

$$(g, \psi - \psi_0) = 0.$$

Since $g(t)$ is arbitrary,

$$\psi(t) - \psi_0(t) = u(t),$$

where $u(t)$ is a solution of equation (2).

Remark: Each function in D_L satisfies condition (1). Indeed, let $\psi(t)$ be a function in D_{L^*} which vanishes for $t \geqq a$. Then, for $\varphi \in D_L$, the equations

$$(L\varphi, \psi) = (\varphi, L^*\psi),$$
$$(l[\varphi], \psi) = [\varphi, \psi]_0^a + (\varphi, l[\psi])$$

imply that

$$[\varphi, \psi]_0^a = 0$$

and, hence,

$$[\varphi, \psi]_0 = 0.$$

Since the values $\psi^{k-1}(0)$ $(k = 1, 2, \ldots, 2n)$ are arbitrary, this means that the function $\varphi(t)$ satisfies condition (1).

We come now to the question about deficiency numbers of the operator L. They are identical since the coefficients of the operator l are real.

THEOREM 2: *The deficiency number m of a quasi-differential operator of order $2n$ on an interval with a singular end point satisfies the inequality*

(3) $$n \leqq m \leqq 2n.$$

Proof: The right inequality follows immediately from Theorem 1. For the proof of the left inequality we use the first Neumann formula (Section 80)

$$D_{L^*} = D_L \oplus \mathfrak{N}_\lambda \oplus \mathfrak{N}_{\bar{\lambda}}.$$

By the equality of the dimensions of the subspaces $\mathfrak{N}_\lambda$ and $\mathfrak{N}_{\bar{\lambda}}$ it is sufficient to prove that

$$\dim D_{L^*} \geqq 2n \quad (\text{mod } D_L).$$

With this aim, we prove the existence of $2n$ linearly independent functions which, together with their linear envelope, lie in D_{L^*} but (with the exception of the function zero) outside of D_L. It is not difficult to see that it is possible to take as such functions any linearly independent functions $\psi_1(t), \psi_2(t), \ldots, \psi_{2n}(t)$ from D_{L^*} which satisfy the condition

$$\text{Det}\,(\psi_i^{[k-1]}(0)) \neq 0.$$

Then from the assumption

$$\psi = \sum_{i=1}^{2n} \alpha_i \psi_i \in \mathrm{D}_L$$

follows (cf. the remark to Theorem 1) the equation

$$\psi(0) = \psi^{[1]}(0) = \ldots = \psi^{[2n-1]}(0) = 0,$$

and, hence

$$\psi(t) = 0.$$

From Theorems 1 and 2 it follows that the number of solutions of the quasi-differential equation

$$l[u] - \lambda u = 0 \qquad (\mathfrak{J}\lambda \neq 0),$$

which belong to $L^2(0, \infty)$ does not depend on λ and is not less than half of the order of the operation l.

The integer m in inequality (3) can be chosen arbitrarily between n and $2n$. It is interesting to note that for $l_1 = -i(1+t)\,\mathrm{D}\,(1+t)$ and $l_2 = \mathrm{D}^2 - 1$ the operation

$$l = l_1^{m-n} l_2^{2n-m} l_1^{m-n} \ (n \leqq m \leqq 2n)$$

generates an operator of order $2n$, on the interval $(0, \infty)$, with minimal domain and with deficiency indices (m, m). The verification of this fact is left to the reader.[3]

The investigations relating to the case of the interval $(-\infty, \infty)$ with two singular end points is similar to the one given above.

The operator L with minimal domain in this case is defined as the closure of the operator L', but it is necessary to omit condition (1). Theorem 1, which characterizes the operator L^*, is preserved. Theorem 2 is replaced by the following theorem.

THEOREM 3: *Let $L^{(-)}$ and $L^{(+)}$ be two singular quasi-differential operators with minimal domains, generated by the operator l on the intervals $(-\infty, 0)$ and $(0, \infty)$, respectively. Under these conditions the deficiency number m of the operator L is defined by the formula*

$$m = m^{(-)} + m^{(+)} - 2n,$$

where $m^{(-)}$ and $m^{(+)}$ are the deficiency numbers of the operators $L^{(-)}$ and $L^{(+)}$ respectively.

Proof: Let D_0 denote the linear manifold of functions $\varphi \in \mathrm{D}_{L'}$ such that

$$\varphi(0) = \varphi^{[1]}(0) = \ldots = \varphi^{[2n-1]}(0) = 0,$$

[3]In this connection cf. I. Glazman [1], Vol. I.

and let L_0' denote the restriction of the operator L' to D_0. Let L_0 be the closure of L_0'. Obviously

$$\dim D_L = 2n \quad (\text{mod } D_{L_0}).$$

But, by a remark in Section 80 (page 100), $\dim D_L$ (mod D_{L_0}) equals the difference of the deficiency numbers of L_0 and L. Hence, the deficiency number of the operator L_0 equals

$$m_0 = m + 2n.$$

On the other hand, the operator L_0 is reduced by the subspaces $L^2(-\infty, 0)$ and $L^2(0, \infty)$. Hence,

$$L_0 = L^{(-)} \oplus L^{(+)},$$

so that

$$m_0 = m^{(-)} + m^{(+)}$$

and

$$m = m^{(-)} + m^{(+)} - 2n,$$

which was to be proved.

5. Self-Adjoint Extensions of a Singular Differential Operator

Self-adjoint extensions of a singular operator, analogous to self-adjoint extensions of a regular operator, can be characterized by means of a system of boundary conditions that have a structure more complicated than the corresponding conditions of Section 3. Let us consider the case when only one end point of the interval is singular.

The general theory of extensions (cf. Section 80) yields

THEOREM 1: *Let the deficiency indices of the operator L be* (m, m) $(n \leqq m \leqq 2n)$ *and let the functions*

$$u_k(t; \lambda) \in L^2(0, \infty) \quad (k = 1, 2, \ldots, m)$$

form an orthonormal system of solutions of the equation

$$l[u] - \lambda u = 0$$

for some fixed non-real value λ. *Under these conditions there exists a one-to-one correspondence between the class of all self-adjoint extensions* L_θ *of the operator L and the class of all unitary matrices* $\theta = (\theta_{ik})$ *of order m. This correspondence is defined by the formula*

$$D_{L\theta} = D_L \oplus \Gamma_\theta, \tag{1}$$

where Γ_θ *is the linear envelope of the functions*

$$w_i^{(\theta)}(t; \lambda) = u_i(t; \lambda) + \sum_{k=1}^{m} \theta_{ik} \overline{u_k(t; \lambda)}.$$

Proof: It is sufficient to note that the functions $u_k(t; \lambda)$ form an orthonormal system of solutions of the equation

$$L^*g - \bar{\lambda}g = 0.$$

We now concern ourselves with the clarification of the boundary conditions which characterize the functions in D_{L_θ}.

THEOREM 2: *Let L_θ be a self-adjoint extension of the operator L with deficiency indices (m, m), defined in the sense of Theorem 1 by a unitary matrix θ. Furthermore, let the functions $w_i^{(\theta)}(t; \lambda)$ $(i = 1, 2, \ldots, m)$ have the same meaning as in Theorem 1. Then the domain D_{L_θ} of the operator L_θ consists of all the functions $\varphi(t)$ in D_{L^*} which satisfy the m conditions*

$$(2) \qquad [\varphi, w_i^{(\theta)}]_0^\infty = 0 \qquad (i = 1, 2, \ldots, m).$$

Proof: By the Lagrange identity, the functions $\varphi(t) \in D_{L^*}$ belong to D_{L_θ} if and only if the condition

$$(3) \qquad [\varphi, \psi]_0^\infty = 0$$

is satisfied for all $\psi(t)$ in D_{L_θ}. It remains to show that the system of conditions (3), which contain the arbitrary functions $\psi(t) \in D_{L^*}$ is reduced for $\varphi(t) \in D_{L^*}$ to a system of m conditions (2).

By formula (1), the function $\psi(t)$ has the representation

$$\psi(t) = \psi_0(t) + \sum_{i=1}^{m} c_i w_i^{(\theta)}(t; \lambda) \qquad (\varphi_0(t) \in D_L).$$

Therefore,

$$[\varphi, \psi]_0^\infty = [\varphi, \varphi_0]_0^\infty + \sum_{i=1}^{m} c_i [\varphi, w_i^{(\theta)}]_0^\infty.$$

Since $\varphi_0(t) \in D_L$,

$$[\varphi, \varphi_0]_0^\infty = 0$$

for $\varphi(t) \in D_{L^*}$. Since the constants $c_i (i = 1, 2, \ldots, m)$ are arbitrary, conditions (3) and (2) are equivalent.

We call equations (2), which involve passage to a limit, boundary conditions. The following theorem gives a case for which the boundary conditions involve no passage to a limit.

THEOREM 3: *If the deficiency indices of the operator L are (n, n), then the system of boundary conditions which define the self-adjoint extension L_θ of the operator L has the form*

$$(4) \qquad [\varphi, w_i^{(\theta)}]_0 = 0 \qquad (i = 1, 2, \ldots, n).$$

Proof: It suffices to show that, under the conditions of the theorem,

$$[\varphi, \psi]_\infty = 0$$

for all $\varphi(t), \psi(t) \in D_{L^*}$. With this aim, we choose in D_{L^*} the functions

$$\psi_i(t) \ (i = 1, 2, \ldots, 2n), \ \psi_i(t) = 0 \text{ for } t > a > 0,$$

so that

$$\text{Det}\,(\psi_i^{[k-1]}(0)) \neq 0.$$

The linear envelope of the functions $\psi_i(t)$ and D_L have only the zero elements in common. On the other hand, by the assumption on the deficiency indices of L,

$$\dim D_{L^*} = 2n \ (\text{mod } D_L).$$

Hence, for each function $\varphi(t) \in D_{L^*}$ there exists constants $\alpha_1, \alpha_2, \ldots, \alpha_{2n}$ such that

$$\varphi(t) - \sum_{i=1}^{2n} \alpha_i \psi_i(t) = \varphi_0(t) \in D_L.$$

Therefore, since the functions $\psi_i(t)$ $(i = 1, 2, \ldots, 2n)$ vanish for $t > a$ and $\varphi_0^{k-1}(0) = 0$ $(k = 1, 2, \ldots, 2n)$,

$$[\varphi, \psi]_\infty = [\varphi_0 + \sum_{i=1}^{2n} \alpha_i \psi_i, \psi]_\infty = [\varphi_0, \psi]_\infty = [\varphi_0, \psi]_0^\infty = 0.$$

We must note here that, in the general case, the boundary condition (2) depends essentially on the functions $w_i^{(\theta)}(i=1, 2, \ldots, m)$, so that it is not defined until the operator L is given. However, if the deficiency indices of the operator L are (n, n), then the boundary condition (4) can be freed from dependence on the functions $w_i^{(\theta)}(t; \lambda)$ by the method of Section 3 and then (4) does not depend on L.

In the case of the deficiency indices $(2n, 2n)$ the number of boundary conditions which determine a self-adjoint extension of a singular operator is equal to $2n$ as in the case of a regular operator. In view of this circumstance and some other considerations (cf. Theorem 2 of the following section) we call a singular operator with the deficiency indices $(2n, 2n)$ *quasi-regular*.

For the differential operator of the second order, the only possible cases are $m=2$ and $m=1$. This was first observed by Weyl who called $m=2$ the case of the *limit circle*, and $m=1$, the case of the *limit point*. We shall clarify the meaning of this terminology in Section 9.

Since the construction of the boundary conditions for the case of an interval with two singular end points can be reduced without difficulty to the method used in the present section, it is omitted.

In conclusion, we consider briefly the question of self-adjoint extensions L_0 which are real with respect to the operator of complex-conjugation in $L^2(0, \infty)$ (cf. Section 45). The operator L^* is real with respect to the operator of complex conjugation, since the coefficients which generate the quasi-differential operators are real. So L_θ is real if and only if the manifold D_{L_θ} contains the conjugate $\overline{\varphi(t)}$ of each function $\varphi(t)$ in D_{L_θ}. By means of this fact and formula (1), it is easy to establish that the operator L_θ is real if and only if the unitary matrix $\theta = (\theta_{ik})$ is symmetric,[4] i.e.,

$$\theta_{ik} = \theta_{ki} \quad (i, k = 1, 2, \ldots, m).$$

In the following section we shall need the existence of only one real extension. This follows from formula (1) if we choose for θ the unit matrix.

6. The Resolvents of Self-Adjoint Extensions

In the present section we show that the resolvents of self-adjoint extensions of a quasi-differential operator L are integral operators and we characterize the class of the kernels of these integral operators for distinct deficiency indices of the operator L. We begin with two lemmas.

LEMMA 1: *The resolvent of a self-adjoint operator which is defined in $L^2(0, \infty)$ and real with respect to the operation of complex conjugation on all functions $g(t) \in L^2(0, \infty)$ which vanish outside of a finite interval, has the form*

$$R_\lambda g = \int_0^\infty K(t, s; \lambda)\, g(s)\, ds.$$

The kernel $K(t, s; \lambda)$ is continuous in s and t for

$$t > 0, \; s > 0 \quad (t \neq s),$$

and

$$K(t, s; \lambda) = K(s, t; \lambda).$$

The proof follows immediately from the relation $R_\lambda^* = IR_\lambda I$ (cf. Section 45) and the fact that the function $g(t)$ is arbitrary.

LEMMA 2: *Let $\mathbf{L}$ be a homogeneous and additive (but not necessarily bounded) functional defined for all functions $g(t)$ of $L^2(0, \infty)$ which vanish outside of a finite interval. Suppose that $\mathbf{L}$ is bounded on $L^2(0, a)$ for each finite $a > 0$. Then $\mathbf{L}$ has the representation*

$$\mathbf{L}(g) = \int_0^\infty g(s)\, \overline{h(s)}\, ds. \tag{1}$$

Here $h(t)$ is a function determined by the functional $\mathbf{L}$ and belongs to $L^2(0, a)$ for each finite $a > 0$.

---

[4]In particular, it follows that in the case with the deficiency indices (1, 1) every self-adjoint extension is real.

The proof follows from the theorem of Riesz on the representation of a linear functional (cf. Section 16), according to which

$$\mathbf{L}(g) = \int_0^a g(s)\overline{h_a(s)}\,ds \quad (h_a(s) \in L^2(0, a)) \tag{2}$$

for each $g(t) \in L^2(0, a)$. The functions $h_a(t)$ which appears in this representation have the following property. If $a_1 < a_2$ then

$$h_{a_1}(t) = h_{a_2}(t)$$

for almost all $t < a_1$. This follows from the fact that for each function $g(t)$ in $L^2(0, a_1)$, there are the two representations

$$\mathbf{L}(g) = \int_0^{a_1} g(s)\overline{h_{a_1}(s)}\,ds,$$

$$\mathbf{L}(g) = \int_0^{a_2} g(s)\overline{h_{a_2}(s)}\,ds = \int_0^{a_1} g(s)\overline{h_{a_2}(s)}\,ds.$$

Hence,

$$\int_0^{a_1} [h_{a_2}(s) - h_{a_1}(s)]\overline{g(s)}\,ds = 0.$$

The fact just proved permits us to write formula (2) in the form (1).

THEOREM 1: *The resolvent $\tilde{R}_\lambda$ of each self-adjoint extension $\tilde{L}$ of a quasi-differential operator L with deficiency indices (m, m) at the points λ of regular type, is an integral operator.*

Proof: First we give the proof for an arbitrary real self-adjoint extension, after which we show that if the theorem is valid for one of the self-adjoint extensions, then it is valid for all self-adjoint extensions. Let $\tilde{L}_0$ be a real self-adjoint extension of L and let $\tilde{R}_\lambda^0$ be the corresponding resolvent (where λ is a fixed point of regular type of the operator L).

Note that the element $\tilde{R}_\lambda^0 g$ (for $g(t) \in L^2(0, \infty)$) must be in the manifold of solutions of the inhomogeneous quasi-differential equation

$$l[y] - \lambda y = g. \tag{3}$$

Consider the corresponding homogeneous equation,

$$l[u] - \lambda u = 0. \tag{4}$$

We choose a fundamental system of solutions $u_i(t; \lambda)$ $(i = 1, 2, \ldots, 2n)$ such that the first m of these functions belong to $L^2(0, \infty)$. (Then the solutions $u_{m+1}(t; \lambda), \ldots, u_{2n}(t; \lambda)$ together with their non-trivial linear combinations lie outside of $L^2(0, \infty)$). Here we do not assume as we did in the preceding section, that the solutions $u_i(t; \lambda)$ $(i=1, 2, \ldots, m)$ be orthonormal.

By Section 1 of Appendix II, the general solution of equation (3) has the form

$$\varphi(t) = \sum_{k=1}^{2n} u_k(t; \lambda) \int_0^t v_k(s; \lambda) g(s)\, ds + \sum_{k=1}^{2n} c_k u_k(t; \lambda). \tag{5}$$

Here the system $v_k(t; \lambda)$ $(k=1, 2, \ldots, 2n)$ is the adjoint to the fundamental system $u_k(t; \lambda)$ $(k=1, 2, \ldots, 2n)$ of solutions of equation (4). The constants $c_k (k=1, 2, \ldots, 2n)$ in formula (5) are uniquely determined if we require that $\varphi(t) = \tilde{R}^0_\lambda g$. We now determine the values of the constants for which this condition is satisfied under the assumption that $g(t)$ vanish outside of a finite interval.

First, we must make sure that $\varphi(t)$ belongs to $L^2(0, \infty)$. For $t \geqq a$, $g(t) = 0$ and, hence,

$$\varphi(t) = \sum_{k=1}^{2n} u_k(t; \lambda) \int_0^a v_k(s; \lambda) g(s)\, ds + \sum_{k=1}^{2n} c_k u_k(t; \lambda).$$

Then $\varphi(t)$ belongs to $L^2(0, \infty)$ if and only if

$$c_k = -\int_0^a v_k(s; \lambda) g(s)\, ds = -\int_0^\infty v_k(s; \lambda) g(s)\, ds \quad (k=m+1, m+2, \ldots, 2n).$$

For this choice of these constants $\tilde{R}^0_\lambda$ has the representation

$$\begin{aligned} \tilde{R}^0_\lambda g &= \sum_{k=1}^{2n} u_k(t; \lambda) \int_0^t v_k(s; \lambda) g(s)\, ds + \\ &\quad + \sum_{k=1}^{m} c_k u_k(t; \lambda) - \sum_{k=m+1}^{2n} u_k(t; \lambda) \int_0^\infty v_k(s; \lambda) g(s)\, ds = \\ &= \sum_{k=1}^{m} u_k(t; \lambda) \int_0^t v_k(s; \lambda) g(s)\, ds - \\ &\quad - \sum_{k=m+1}^{2n} u_k(t; \lambda) \int_t^\infty v_k(s; \lambda) g(s)\, ds + \sum_{k=1}^{m} c_k u_k(t; \lambda), \end{aligned}$$

or

$$\tilde{R}^0_\lambda g = \int_0^\infty K(t, s; \lambda) g(s)\, ds + \sum_{k=1}^{m} c_k u_k(t; \lambda), \tag{6}$$

where

$$K(t,s;\lambda) = \begin{cases} \sum_{k=1}^{m} u_k(t;\lambda)\, v_k(s;\lambda) & (s \leqq t) \\ -\sum_{k=m+1}^{2n} u_k(t;\lambda)\, v_k(s;\lambda) & (s > t). \end{cases}$$

In representation (6) $g(t)$ is any function of $L^2(0, \infty)$ which vanishes outside of some finite interval and the $c_k(k = 1, 2, \ldots, m)$ are the constants which correspond to it. From this representation it follows immediately that the constants $c_k(k = 1, 2, \ldots, m)$ are homogeneous and additive functionals defined on all such functions in $L^2(0, \infty)$. We show now that these functionals $c_k = c_k(g)$ $(k = 1, 2, \ldots, m)$ satisfy the conditions of lemma 2, i.e., that they are bounded in $L^2(0, a)$ for each $a < \infty$. We apply to both members of equation (6) the operator P_a of projection on the subspace $L^2(0, a)$ and then form the scalar products with $u_i(t; \lambda)$ $(i=1, 2, \ldots, m)$ to obtain

$$(7) \qquad (P_a\tilde{R}^0_\lambda g, u_i) = (K_a g, u_i) + \sum_{k=1}^{m} c_k(P_a u_k, u_i) \quad (i = 1, 2, \ldots, m),$$

where

$$K_a g = P_a \int_0^\infty K(t, s; \lambda)\, g(s)\, ds.$$

In the system (7) of m equations with m unknowns c_k $(k=1, 2, \ldots, m)$, the norms of the operators $P_a\tilde{R}^0_\lambda$ are bounded for all positive $a<\infty$ by a number which does not depend on a, the norms of the integral operators K_a in $L^2(0, \infty)$ with Hilbert-Schmidt kernel are bounded for all $a<\infty$ by a number which depends on a and, finally, the Gram determinant

$$\operatorname{Det}\left((P_a u_k, u_i) \right)$$

is different from zero by the linear independence of the solutions $u_k(t; \lambda)$ $(k=1, 2, \ldots, m)$. It follows that the $c_k(k=1, 2, \ldots, m)$ are functionals which are bounded in $L^2(0, a)$ for each $a<\infty$ so that, by Lemma 2, they can be represented in the form

$$c_k(g) = \int_0^\infty g(s)\, \psi_k(s; \lambda)\, ds \qquad (k = 1, 2, \ldots, m),$$

where the functions $\psi_k(s; \lambda)$ $(k=1, 2, \ldots, m)$ belong to $L^2(0, a)$ for each $a<\infty$. Now the representation (6) assumes the form

$$\tilde{R}^0_\lambda g = \int_0^\infty K(t, s; \lambda)\, g(s)\, ds + \sum_{k=1}^{m} u_k(t; \lambda) \int_0^\infty g(s)\, \psi_k(s; \lambda)\, ds,$$

or

$$\tilde{R}^0_\lambda g = \int_0^\infty K_0(t, s; \lambda)\, g(s)\, ds, \tag{8}$$

where

$$K_0(t, s; \lambda) = \begin{cases} \sum_{k=1}^{m} u_k(t;\lambda)\,[v_k(s;\lambda) + \psi_k(s;\lambda)] & (s \leqq t), \\ \sum_{k=1}^{m} u_k(t;\lambda)\,\psi_k(s;\lambda) - \sum_{k=m+1}^{2n} u_k(t;\lambda) v_k(s;\lambda) & (s > t). \end{cases} \tag{9}$$

Formula (8) shows that the resolvent $\tilde{R}^0_\lambda$ is an integral operator on all functions $g(t)$ in $L^2(0, \infty)$ which vanish outside of a finite interval.

From these considerations it still does not follow that $\tilde{R}^0_\lambda$ is an integral operator. By the boundedness of the operator $\tilde{R}^0_\lambda$ it follows from formula (8) only that, for each function in $L^2(0, \infty)$,

$$\tilde{R}^0_\lambda g = \underset{n\to\infty}{\text{l.i.m.}} \int_0^n K_0(t, s; \lambda)\, g(s)\, ds. \tag{10}$$

In order to derive from this the desired representation of the resolvent in the form

$$R^0_\lambda g = \int_0^\infty K_0(t, s; \lambda)\, g(s)\, ds \qquad (g(t) \in L^2(0, \infty)), \tag{11}$$

it is sufficient to prove that the integral

$$\int_0^\infty K_0(t, s; \lambda)\, g(s)\, ds$$

exists for each function $g(t) \in L^2(0, \infty)$. For this, in turn, it is sufficient that

$$\int_0^\infty |K_0(t, s;\lambda)|^2 ds < \infty \qquad (0 \leqq t < \infty), \tag{12}$$

or, equivalently, that the kernel $K_0(t, s; \lambda)$ belongs to $L^2(0, \infty)$ as a function of s for each t.

From formula (9) it follows that the kernel $K_0(t, s; \lambda)$ belongs to $L^2(0, \infty)$ as a function of t for each s since $u_k(t; \lambda) \in L^2(0, \infty)$ for $k=1, 2, \ldots, m$. By Lemma 1, the function $K_0(t, s; \lambda)$, which corresponds to the resolvent of a real self-adjoint extension, is symmetric with respect to the arguments s and t. Therefore, the integral (12) exists and the representation (11) is established.[5]

To complete the proof of the theorem it remains for us to generalize representation (11) to the case of the resolvent $\tilde{R}_\lambda$ of an arbitrary (not necessarily real) extension $\tilde{L}$ of the operator L.

Since the element $\tilde{R}_\lambda g$ $(g(t) \in L^2(0, \infty))$ belongs to the solution manifold of equation (3), we can represent it in the form

$$\tilde{R}_\lambda g = \tilde{R}^0_\lambda g + \sum_{k=1}^{2n} c_k u_k, \tag{13}$$

where $\tilde{R}^0_\lambda$ is the resolvent of a real self-adjoint extension. From this representation, in correspondence with the choice of the fundamental system $u_k(t; \lambda)$ $(k=1, 2, \ldots, 2n)$, it follows that for $k \leqq m$ the c_k vanish and for $k > m$ the c_k are linear (i.e., homogeneous, additive, and bounded) functionals defined on $L^2(0, \infty)$ and, hence, have the representation

$$c_k = c_k(g) = \int_0^\infty g(s)\, \chi_k(s; \lambda)\, ds, \tag{14}$$

where the functions $\chi_k(s; \lambda)$ $(k=1, 2, \ldots, m)$ belong to $L^2(0, \infty)$.

Substituting (14) into (13) and defining the kernel $\tilde{K}(t, s; \lambda)$ by the equation

$$\tilde{K}(t, s; \lambda) = K_0(t, s; \lambda) + \sum_{k=1}^{m} u_k(t; \lambda) \chi_k(s; \lambda), \tag{15}$$

we finally obtain the integral representation of the resolvent in the form

$$\tilde{R}_\lambda g = \int_0^\infty \tilde{K}(t, s; \lambda)\, g(s)\, ds \qquad (g(t) \in L^2(0, \infty)).$$

Thus, Theorem 1 is completely proved.

THEOREM 2: *The kernel $\tilde{K}(s, t; \lambda)$ of the resolvent $\tilde{R}_\lambda$ of an arbitrary self-adjoint extension $\tilde{L}$ of the operator L with deficiency indices (m, m) satisfies the conditions*

$$\int_0^\infty |\tilde{K}(t, s; \lambda)|^2 ds < \infty, \quad \int_0^\infty |\tilde{K}(t, s; \lambda)|^2 dt < \infty, \tag{16}$$

---

[5]In the above considerations the symmetry of $K_0(s, t; \lambda)$ is essential in the transition from (10) to (11) in order to avoid the question of the behavior of the functions $v_k(s; \lambda)$ $(k=m+1, m+2, \ldots, 2n)$ and $\psi_k(s; \lambda)$ $(k=1, 2, \ldots, m)$ as $s \to \infty$. Furthermore, the symmetry of the kernel $K_0(t, s; \lambda)$ implies certain properties of these functions at infinity (cf. Theorem 3 page 184).

and if the operator is quasi-regular ($m=2n$) *also satisfies the condition*

$$\int_0^\infty \int_0^\infty |K(t, s; \lambda)|^2 \, ds dt < \infty. \tag{17}$$

Proof: Property (16) follows immediately from formula (15), since the function $K_0(t, s; \lambda)$ (in each of the variables s and t) and the functions $\chi_k(s; \lambda)$ $(k = 1, 2, \ldots, m)$ belong to $L^2(0, \infty)$.

We come now to property (17). We establish first that for any deficiency indices (m, m) $(n \leqq m \leqq 2n)$ of the operator L, the functions $\psi_k(t; \lambda)$ $(k= 1, 2, \ldots, m)$ and $v_k(t; \lambda)$ $(k=m+1, m+2, \ldots, 2n)$ which appears in formula (9) for $K_0(t, s; \lambda)$ belong to $L^2(0, \infty)$. With this aim we again use the symmetry of the function $K_0(t, s; \lambda)$, which implies that

$$\sum_{k=1}^{m} u_k(s; \lambda) \psi_k(t; \lambda) - \sum_{k=m+1}^{2n} u_k(s; \lambda) v_k(t; \lambda) = \tag{18}$$

$$= \sum_{k=1}^{m} [v_k(s; \lambda) + \psi_k(s; \lambda)] u_k(t; \lambda).$$

We choose numbers $s_1, s_2, \ldots, s_{2n}$ such that the determinant

$$\operatorname{Det}(u_k(s_i; \lambda)) \neq 0$$

(this is possible since the functions $u_k(s; \lambda)$ are linearly independent) and we let $s=s_i$ $(i=1, 2, \ldots, 2n)$ in equation (18). Solving the resulting system of $2n$ equations in terms of the $2n$ functions $\psi_k(t; \lambda)$ $(k=1, 2, \ldots, m)$ and $v_k(t; \lambda)$ $(k=m+1, m+2, \ldots, 2n)$ we find that each of these functions is some linear combination of the functions $u_1(t; \lambda), \ldots, u_m(t; \lambda)$ and, hence, belongs to $L^2(0, \infty)$.

We now establish property (17). In the case of a quasi-regular operator $(m=2n)$ all summands in the first part of formula (9) belong to the space L^2 of functions of two variables in the quadrant $t > 0$, $s > 0$ and, hence,

$$\int_0^\infty \int_0^\infty |K_0(t, s; \lambda)|^2 ds dt < \infty. \tag{19}$$

Property (17) follows from (19) by means of (15).

From the proof of the theorem it follows that, in the case of a quasi-regular operator L, the resolvent of an arbitrary self-adjoint extension $\tilde{L}$ of L is defined in terms of a Hilbert-Schmidt kernel and therefore is a completely continuous operator.

From Theorem 2 it follows that the real and imaginary parts of the kernel of the resolvent of any self-adjoint extension of the operator are Carleman kernels.[6]

We note that the arguments in the proof of Theorem 2 yield the following theorem.

THEOREM 3: *Let m denote the maximal number of linearly independent solutions of equation* (4) *which belong to* $L^2(0, \infty)$. *If the first m functions in the fundamental system of solutions of equation* (4) *belong to* $L^2(0, \infty)$, *then the last* $2n-m$ *functions in the adjoint fundamental system of this equation also belong to* $L^2(0, \infty)$.

The following theorem is concerned with a case when it is possible to deduce the deficiency indices of the operator L from the number of functions in $L^2(0, \infty)$ which satisfy equation (4) with λ real.

THEOREM 4: *In order that an operator L of order 2n have deficiency indices* $(2n, 2n)$ *it is necessary that the equation*

$$l[u] - \lambda u = 0$$

have $2n$ *solutions in* $L^2(0, \infty)$ *for each real* λ *and sufficient that this equation have* $2n$ *solutions in* $L^2(0, \infty)$ *for at least one real* λ.

Proof: For the proof of the necessity, we recall that in the case of a quasi-regular operator, the resolvent of its self-adjoint extension $\tilde{L}$ is a completely continuous operator and, hence, the spectrum of $\tilde{L}$ consists only of eigenvalues $\lambda_r (r=1, 2, 3, \ldots)$ with the unique limit point $\lambda=\infty$. It follows that all points of the plane are of regular type with respect to L except perhaps for the points $\lambda_r (r = 1, 2, 3, \ldots)$. If λ_r is not of regular type, then λ_r is an eigenvalue of L. This is impossible, since the equation

$$L\varphi - \lambda_r \varphi = 0$$

is equivalent to

$$l\,[\varphi] - \lambda_r \varphi = 0$$

under the conditions

$$\varphi\,(0) = \varphi^{[1]}(0) = \ldots = \varphi^{[2n-1]}\,(0) = 0$$

and, by the existence theorem of Section 1 (Appendix II), has only the solution $\varphi\,(t) \equiv 0$.

[6]A Carleman kernel is a measurable (in general, complex valued) function $K(s, t)\ (-\infty < {s \atop t} < \infty)$ for which:

$$1^\circ \qquad \overline{K(s, t)} = K(t, s)$$

almost everywhere in the (s, t)–plane;

$$2^\circ \qquad \int_{-\infty}^{\infty} |K(s, t)|^2\, dt < \infty$$

almost everywhere on the s-axis.

Thus, in the case of deficiency indices $(2n, 2n)$, the whole λ-plane is the field of regularity of the operator L. By the theorem on the invariance of the deficiency indices (Section 78) every solution of equation (4) with arbitrary λ (real or not real) belongs to $L^2(0, \infty)$.

The sufficiency of the theorem follows immediately from Theorem 4 of Section 83.

It is not difficult to transfer the results obtained in the present section to the case of an interval with two singular end points. We sketch the proof of the fundamental Theorem 1 for this case. Here, as in the case of one singular end point, it is sufficient to consider the resolvent of an arbitrary real self-adjoint extension. Let (cf. the proof of Theorem 3, Section 4, Appendix II) the deficiency numbers of the operators L, $L^{(-)}$ and $L^{(+)}$ be m, $m^{(-)}$ and $m^{(+)}$, respectively. By Theorem 3 of Section 4,

$$\left.\begin{aligned} m^{(-)} &= n + p, \\ m^{(+)} &= n + m - p. \end{aligned}\right. \qquad (0 \leqq p \leqq n)$$

We choose a fundamental system of solutions of equation (4) such that the first m solutions $u_k(t; \lambda)$ belong to $L^2(-\infty, \infty)$. In the $(2n-m)$ dimensional linear envelope of the remaining solutions of this system there exist $n-p-m$ linearly independent solutions in $L^2(-\infty, \infty)$, which we denote by $u_{m+1}(t; \lambda), \ldots, u_{n+p}(t; \lambda)$, and $n-p$ linearly independent solutions in $L^2(0, \infty)$, which we denote by $u_{n+p+1}(t; \lambda), \ldots, u_{2n}(t; \lambda)$. We remark that the functions

$$u_{m+1}(t; \lambda), \ldots, u_{n+p}(t; \lambda), u_{n+p+1}(t; \lambda), \ldots, u_{2n}(t; \lambda)$$

are linearly independent. Otherwise a linear combination of them would belong to $L^2(-\infty, \infty)$, which would imply, contrary to assumption, that the deficiency numbers of the operator L exceed m. With the aid of the chosen fundamental system $u_k(t; \lambda)$ $(k=1, 2, \ldots, 2n)$ each solution of the equation

$$l[y] - \lambda y = g,$$

(where $g(t)$ is a function which vanishes outside of a finite interval) can be represented in the form

$$\varphi(t) = \sum_{k=1}^{2n} u_k(t; \lambda) \int_{-\infty}^{t} v_k(s; \lambda) g(s)\, ds + \sum_{k=1}^{2n} c_k u_k(t; \lambda).$$

For suitable choice of the constants c_k, this function coincides with $\tilde{R}_\lambda^0 g$.

An argument similar to that for the case of one singular end point yields a complete proof of Theorem 1 and Theorem 2 for the case of two singular end points.

7. Inversion Formulas Related to Differential Operators of the Second Order

In Section 69 it was proved that every self-adjoint operator A with a simple spectrum determines an isometric mapping V of the space H onto some space L^2_σ. Under this mapping the operator A goes into the operator of the multiplication by the independent variable. If E_λ is the spectral function and g is any generating element of the operator A, then $\sigma(\lambda) = (E_\lambda g, g)$ and the indicated isometric correspondence is defined by the formulas

$$\Phi(\lambda) = Vf, \tag{1}$$

$$f = V^{-1}\Phi(\lambda) = \int_{-\infty}^{\infty} \Phi(\lambda)\, dE_\lambda g, \tag{2}$$

where the element f and the function $\Phi(\lambda)$ are in H and L^2_σ, respectively. If $f \in \mathrm{D}_A$, then instead of formula (2) the following equation holds:

$$Af = \int_{-\infty}^{\infty} \lambda\Phi(\lambda)\, dE_\lambda g.$$

We call formulas (1) and (2) the *inversion formulas* related to the self-adjoint operator A.

In Example C of Section 77 we obtained as the inversion formulas related to the operator of differentiation on the entire real axis, the mutually inverse Fourier-Plancherel transformations of the space $L^2(-\infty, \infty)$ onto itself

$$\Phi(\lambda) = \underset{N\to\infty}{\mathrm{l.i.m.}} \frac{1}{\sqrt{2\pi}} \int_{-N}^{N} f(t)\, e^{i\lambda t} dt,$$

$$f(t) = \mathrm{l.i.m.} \frac{1}{\sqrt{2\pi}} \int_{-N}^{N} \Phi(\lambda)\, e^{-i\lambda t} dt.$$

Here any improper element may be taken as a generating element.

Operators with multiple spectrum also generate inversion formulas which are analogous to formulas (1), (2).

The problem of the present and the following section is to obtain inversion formulas connected with singular *quasi-differential operators* on the semi-axis. This problem was solved first for the operator of second order by Weyl.[7] We begin also by considering operators of the second

[7] Cf. H. Weyl [2], Vol. 1. The results of Weyl were obtained recently by B. M. Levitan [2], Vol. I, by another method.

order but we use a very general and powerful method recently discovered by Krein.[8] This method was called by its author the method of the *direction functionals*. It permits us to obtain inversion formulas for quasi-differential operators of any order (cf. the following section).

Thus, let L be a quasi-differential operator with minimal domain, generated by the operation

$$l = -\frac{d}{dt} p \frac{d}{dt} + q$$

on the interval $(0, \infty)$ with one singular end point. We assume that the deficiency indices of L are $(1, 1)$.[9] As we shall see, in this case the spectrum of each of the self-adjoint extensions of L is simple.

Let $\tilde{L}$ be a self-adjoint extension of the operator L which is determined by the boundary conditions[10]

$$p(t)\,\varphi'(t)\,|_{t=0} = \theta\varphi(0) \qquad (\mathfrak{J}\theta = 0),$$

and let $u_1(t; \lambda)$ and $u_2(t; \lambda)$ be the solutions of the equation

$$l[u] - \lambda u = 0 \qquad (\mathfrak{J}\lambda = 0),$$

which satisfy the initial conditions

$$u_1(0; \lambda) = 1, \quad p(t)\,u_1'(t; \lambda)\,|_{t=0} = 0,$$

$$u_2(0; \lambda) = 0, \quad p(t)\,u_2'(t; \lambda)\,|_{t=0} = 1.$$

We denote the set of all functions $f(t)$ in $L^2(0, \infty)$ which vanish outside of a finite interval by $\mathfrak{D}$ and we define on $\mathfrak{D}$ a homogeneous and additive functional

$$\Phi(f; \lambda) = \int_0^\infty f(t)\,u(t; \lambda)\,dt,$$

where

$$u(t; \lambda) = u_1(t; \lambda) + \theta u_2(t; \lambda).$$

Following M. G. Krein, we call $\Phi(f; \lambda)$ a *direction functional.*

---

[8] Cf. M. G. Krein [2], [5], Vol. I; M. S. Lifschitz [1], Vol. I, and A. J. Povzner [1], Vol. I.

[9] If the deficiency indices of the operator L are $(2, 2)$, then each self-adjoint extension of this operator has a discrete spectrum. In this case the question of the structure of the inversion formulas does not arise since the role of formula (2) is played by the expansion of the function into orthogonal eigenfunctions, and the role of formula (1) is played by the expressions for the coefficients in the Fourier expansion of the function.

[10] Thus, there remains to consider only one extension, that which is determined by the condition $\varphi(0)=0$; it corresponds to $\theta=\infty$. The inversion formula established below is extended to this case by means of a passage to a limit.

The direction functional $\Phi(f; \lambda)$ has the following three properties which are essential for what follows.

1° $\Phi(f; \lambda)$ is an analytic function of λ $(-\infty < \lambda < \infty)$ for each fixed function $f \in \mathfrak{D}$.

2° If for some function $f \in \mathfrak{D}$ and some real λ

$$\Phi(f; \lambda) = 0,$$

then the equation

$$(\tilde{L} - \lambda E)\varphi = f$$

has a solution in the class of functions which vanish outside a finite interval.

3° If f is a function in $D_{\tilde{L}}$ which vanishes outside a finite interval, then for each real λ

$$\Phi(\tilde{L}f; \lambda) = \lambda \Phi(f; \lambda).$$

We prove only properties 2° and 3°. We establish 2° first. Let $f(t) = 0$ for $t > a$ and assume that

$$\Phi(f; \lambda_0) = 0. \tag{3}$$

We denote by $\varphi(t)$ $(0 \leqq t < \infty)$ the solution of the equation

$$l[\varphi] - \lambda_0 \varphi = f,$$

which satisfies the condition

$$\varphi(a) = p(a)\varphi'(a) = 0,$$

(by the uniqueness theorem $\varphi(t) = 0$ for $t \geqq a$). Using the Lagrange identity, we get

$$\Phi(f; \lambda_0) = \int_0^\infty \{l[\varphi] - \lambda_0 \varphi\} u(t; \lambda_0)\, dt = [\varphi, u]_0^\infty = [\varphi, u]_0 =$$

$$= \left[p\left(u \frac{d\varphi}{dt} - \varphi \frac{du}{dt} \right) \right]_{t=0} = p(t)\varphi'(t) - \theta\varphi(t)\,|_{t=0},$$

which in combination with (3) yields

$$p(t)\varphi'(t)\,|_{t=0} = \theta\varphi(0),$$

so that $\varphi(t) \in D_{\tilde{L}}$. Thus, property 2° is established.

The validity of 3° follows from the equations

$$\Phi(\tilde{L}f; \lambda) = \int_0^\infty l[f]\, u(t; \lambda)\, dt = [f, u]_0^\infty + \lambda \int_0^\infty f(t)\, u(t; \lambda) dt = \lambda \Phi(f; \lambda).$$

The direction functional $\Phi(f; \lambda)$ generates a mapping of the linear manifold $\mathfrak{D}$ of functions $f(t)$ onto some linear manifold $\mathfrak{D}'$ of functions

$\Phi(\lambda) = \Phi(f; \lambda)$. If a function $f(t)$ belongs to $D_{\tilde{L}}$ then by 3° this mapping takes $\tilde{L}f$ into the function $\lambda\Phi(\lambda)$. This fact permits us to assume that the spectrum of the operator $\tilde{L}$ is simple and that for a proper choice of the generating function the first of the inversion formulas connected with the operator $\tilde{L}$ will have the form

$$\Phi(\lambda) = \int_0^\infty f(t)\, u(t; \lambda)\, dt$$

for $f(t) \in \mathfrak{D}$. This formula is analogous to the Fourier-Plancherel transform.

We now leave these considerations and proceed to the solution of the problem which was stated above. We shall need two propositions due to Krein.[11]

LEMMA: *For each finite interval Δ_0 of the λ axis there exists a function $\psi(t) \in \mathfrak{D}$ such that the function $\Phi(\lambda) = \Phi(\psi; \lambda)$ does not vanish in the interval Δ_0.*

Proof: Let λ_0 be a fixed point of the interval Δ_0 and let $\chi_0(t)$ be a function in $\mathfrak{D}$ such that

$$\Phi(\chi_0; \lambda_0) \neq 0.$$

If $\Phi(\chi_0; \lambda) \neq 0$ in the whole interval, then for the proof of the lemma it is sufficient to choose $\psi = \chi_0$. If at some point $\lambda_1 \in \Delta_0$

$$\Phi(\chi_0; \lambda_1) = 0,$$

then by property 2° there exists a function $\chi_1 \in D_{\tilde{L}}$ such that

$$\tilde{L}\chi_1 - \lambda_1\chi_1 = \chi_0.$$

Applying to both sides of this equation the direction functional and using property 3° we obtain

$$\lambda\Phi(\chi_1; \lambda) - \lambda_1\Phi(\chi_1; \lambda) = \Phi(\chi_0; \lambda),$$

which yields

$$\Phi(\chi_1; \lambda) = \frac{\Phi(\chi_0; \lambda)}{\lambda - \lambda_1}.$$

Thus, replacing χ_0 by χ_1 we see that the function $\Phi(\chi_1; \lambda)$ does not vanish at $\lambda = \lambda_0$ and at $\lambda = \lambda_1$ it has a zero of multiplicity one less than that of the function $\Phi(\chi_0; \lambda)$. Since the analytic function $\Phi(\chi_0; \lambda)$ has only a finite number of zeros in the finite interval Δ_0, each of finite multiplicity, a finite number of repetitions of the above procedure yields a function $\psi \in \mathfrak{D}$ for which the functional $\Phi(\psi; \lambda)$ vanishes for no $\lambda \in \Delta_0$.

[11]Cf. M. G. Krein [5], Vol. I.

THEOREM 1: *Let Δ_0 be a finite interval and let ψ be a function in $\mathfrak{D}$ such that*

$$\Phi(\psi;\lambda) \neq 0$$

if $\lambda \in \Delta_0$. Furthermore, let E_λ be the spectral function of $\tilde{L}$ and let

$$g(t) = \int_{\Delta_0} \frac{1}{\Phi(\psi;\lambda)}\, dE_\lambda \psi.$$

Then for each function $f \in \mathfrak{D}$ and each interval $\Delta \subset \Delta_0$,

$$E(\Delta) f = \int_{\Delta} \Phi(f;\lambda)\, dE_\lambda g. \tag{4}$$

Proof: If we define the functional $F(f;\lambda)$ in terms of the direction functional $\Phi(f;\lambda)$ by the relation

$$F(f;\lambda) = \frac{\Phi(f;\lambda)}{\Phi(\psi;\lambda)},$$

then the formula to be proved is equivalent to the equation

$$E(\Delta) f = \int_{\Delta} F(f;\lambda)\, dE_\lambda \psi. \tag{4'}$$

We proceed to the proof of this equation. Assume that the left end points of the intervals Δ_0 and Δ coincide and let the point α be the common left end point of these intervals.

Consider the element $w_\mu = w_\mu(t)$ of the space $L^2(0,\infty)$ which is defined by the equation

$$w_\mu = \int_\alpha^\mu dE_\lambda f - \int_\alpha^\mu F(f;\lambda)\, dE_\lambda \psi$$

as a function of the parameter μ in the interval Δ_0. To establish equation (4′) it is necessary to prove that $w_\mu = 0$ for $\mu \in \Delta_0$ and, for this, it is obviously sufficient to show that

$$\frac{d}{d\mu} w_\mu = 0.$$

This equation must be understood in the sense of strong convergence in $L^2(0,\infty)$, i.e.,

$$\lim_{\delta\to 0} \frac{1}{\delta^2} \| w_{\mu+\delta} - w_\mu \|^2 = \lim_{\delta\to 0} \frac{1}{\delta^2} \int_0^\infty | w_{\mu+\delta}(t) - w_\mu(t) |^2 dt = 0.$$

We introduce the projection operator

$$E(\Delta_{\mu,\delta}) = E_{\mu+\delta} - E_{\mu},$$

and estimate $\| w_{\mu+\delta} - w_{\mu} \|$. We have

$$\begin{aligned} \| w_{\mu+\delta} - w_{\mu} \| &= \| E(\Delta_{\mu,\delta}) f - \int_{\mu}^{\mu+\delta} F(f;\lambda)\, dE_{\lambda}\psi \| \leqq \\ &\leqq \| E(\Delta_{\mu,\delta}) [f - F(f;\mu)\psi] \| + \\ &+ \| E(\Delta_{\mu,\delta}) F(f;\mu)\psi - \int_{\mu}^{\mu+\delta} F(f;\lambda)\, dE_{\lambda}\psi \|. \end{aligned} \tag{5}$$

We estimate the first and second summands on the right side of this inequality. Since

$$\Phi(f - F(f;\mu)\psi;\mu) = \Phi(f;\mu) - F(f;\mu)\,\Phi(\psi;\mu) = 0,$$

it follows by 2° that the difference $f - F(f;\mu)\psi$ can be represented in the form

$$f - F(f;\mu)\psi = \tilde{L}\varphi - \mu\varphi, \tag{6}$$

and, hence,

$$\begin{aligned} \| E(\Delta_{\mu,\delta}) [f - F(f;\mu)\psi] \|^2 &= \| E(\Delta_{\mu,\delta}) [\tilde{L}\varphi - \mu\varphi] \|^2 = \\ &= \int_{\mu}^{\mu+\delta} |\lambda - \mu|^2 d(E_{\lambda}\varphi, \varphi) \leqq \delta^2 \| E(\Delta_{\mu,\delta}\varphi \|^2. \end{aligned}$$

As $\delta \to 0$, the element $E(\Delta_{\mu,\delta})\varphi$ tends to the element $E(\Delta_{\mu,0})\varphi$ which either is zero or is an eigenfunction of the operator $\tilde{L}$ corresponding to the eigenvalue μ. In the latter case, formula (6) yields the equation

$$E(\Delta_{\mu,0}) f - F(f;\mu)\, E(\Delta_{\mu,0})\psi = 0.$$

This shows that formula (4′) is valid for $\Delta = \Delta_{\mu,0}$. Therefore, it was possible to assume from the beginning that $E(\Delta_{\mu,0})\varphi = 0$ which means that

$$\lim_{\delta \to 0} \frac{1}{\delta} \| E(\Delta_{\mu,\delta}) [f - F(f;\mu)\psi] \| = 0. \tag{7}$$

For the second summand on the right side of the inequality (5), we obtain

$$\| E(\Delta_{\mu,\delta}) F(f;\mu)\psi - \int_{\mu}^{\mu+\delta} F(f;\lambda)\, dE_\lambda\psi \|^2 =$$

$$= \left\| \int_{\mu}^{\mu+\delta} [F(f;\mu) - F(f;\lambda)]\, dE_\lambda \psi \right\|^2 \leqq$$

(8)

$$\leqq \int_{\mu}^{\mu+\delta} | F(f;\mu) - F(f;\lambda)|^2 d(E_\lambda\psi, \psi) =$$

$$= \int_{\mu}^{\mu+\delta} \left| \frac{\partial F(f;\lambda)}{\partial\lambda} \right|^2_{\lambda=\xi} (\lambda-\mu)^2 d(E_\lambda\psi,\psi) = M^2\delta^2(E(\Delta_{\mu,\delta})\psi,\psi),$$

where

$$\mu < \xi < \mu + \delta.$$

In (8), M denotes the maximum of the absolute value of the derivative $\frac{\partial F(f;\lambda)}{\partial\lambda}$ in the interval Δ_0.

We remark that if μ is an eigenvalue of the operator $\tilde{L}$, then ψ is not an eigenfunction of this operator corresponding to μ. Otherwise,

$$\Phi(\psi;\lambda) = \frac{1}{\mu}\Phi(\tilde{L}\psi;\lambda) = \frac{\lambda}{\mu}\Phi(\psi;\lambda),$$

which implies that $\Phi(\psi;\lambda) = 0$ for $\lambda \neq \mu$, and this contradicts the choice of the function ψ.

Thus, from inequality (8) it follows that

$$\lim_{\delta\to 0} \frac{1}{\delta} \left\| E(\Delta_{\mu,\delta}) F(f;\mu)\psi - \int_{\mu}^{\mu+\delta} F(f;\lambda)\, dE_\lambda\psi \right\| = 0. \tag{9}$$

From (5), (7) and (9) we conclude that $w_\mu = 0$. This yields equation (4′), which was to be proved.

Now we come to the fundamental theorem of the present section.

THEOREM 2: *If $L, \tilde{L}$ and $u(t;\lambda)$ have the meanings given in the beginning of the present section, then the inversion formulas which are related to the operator $\tilde{L}$ have the forms*

$$\Phi(\lambda) = \int_0^\infty f(t)\, u(t;\lambda)\, dt, \tag{10}$$

$$f(t) = \int_{-\infty}^{\infty} \Phi(\lambda)\, u(t;\lambda)\, d\sigma(\lambda). \tag{11}$$

Here $\sigma(\lambda)$ is some nondecreasing function determined by the operator $\tilde{L}$; $\Phi(\lambda)$ and $f(t)$ run through the spaces $L_\sigma^2(-\infty, \infty)$ and $L^2(0, \infty)$ respectively. If one of the functions $\Phi(\lambda)$ and $f(t)$ vanishes outside of a finite interval, then the integral on the right side of the corresponding formula can be understood in the ordinary sense. In the case of arbitrary functions $\Phi(\lambda)$ and $f(t)$ the integrals must be understood as limits in the norms of L_σ^2 and $L^2(0, \infty)$, respectively, of the proper integrals

$$\int_0^C f(t)\, u(t; \lambda)\, dt,$$

$$\int_A^B \Phi(\lambda)\, u(t; \lambda)\, d\sigma(\lambda).$$

Proof: We note first that in formula (4) it is possible to take the element $g_0 = E(\Delta_0)g$ instead of the element g. We subdivide the λ-axis into finite intervals $\Delta_k(\pm k = 0, 1, 2, \ldots)$ and we choose for each interval Δ_k an element g_k in the same way as we choose the element g_0 for the interval Δ_0. We let

$$g = \sum_{k=-\infty}^{\infty} g_k,$$

and in case the orthogonal series on the right side diverges, we understand g to be an improper element[12] (cf. Section 77) such that, for each finite interval Δ,

$$E(\Delta)\, g = \sum_k E(\Delta)\, g_k,$$

where the series on the right side contains only a finite number of summands.

If f is an arbitrary function in $\mathfrak{D}$ and Δ is an arbitrary finite interval of the λ-axis then

$$\Delta = \Delta_m^* + \sum_{k=m+1}^{n-1} \Delta_k + \Delta_n^* \qquad (\Delta_m^* \subset \Delta_m;\ \Delta_n^* \subset \Delta_n),$$

and

$$E(\Delta) f = E(\Delta_m^*) f + \sum_{k=m+1}^{n-1} E(\Delta_k) f + E(\Delta_n^* f).$$

Therefore, for each finite interval Δ of the λ-axis,

$$E(\Delta) f = \int_\Delta \Phi(f; \lambda)\, dE_\lambda g.$$

[12]A. J. Povzner and W. A. Martschenko communicated to us a proof of the fact that for $p(t) = 1$, this element is always improper.

Let $\Delta \to [-\infty, \infty]$ to obtain the equation

$$f(t) = \int_{-\infty}^{\infty} \Phi(f; \lambda)\, dE_\lambda g. \tag{12}$$

Since the manifold $\mathfrak{D}$ is dense in $L^2(0, \infty)$, this formula implies that the operator $\tilde{L}$ has a simple spectrum and the function $g = g(t)$ is its generating element.

From the theorem of Section 69 on the canonical representation of a self-adjoint operator with simple spectrum it follows that there exists an isometric mapping

$$f(t) = \int_{-\infty}^{\infty} \Phi(\lambda)\, dE_\lambda g$$

of the space L^2_σ where $\sigma(\lambda) = (E_\lambda g, g)$ onto the space $L^2(0, \infty)$. Under this mapping, an arbitrary function $f(t)$ in $\mathfrak{D}$ corresponds to the direction functional $\Phi(f; \lambda) = \Phi(\lambda)$. Thus, the operator $\tilde{L}$, is isomorphic to the operator of multiplication by λ in the space L^2_σ, so that

$$\tilde{L} f = \int_{-\infty}^{\infty} \lambda\, \Phi(\lambda)\, dE_\lambda g$$

for $f(t) \in D_{\tilde{L}}$.

We turn now to the derivation of formulas (10) and (11). If $f(t)$ is an arbitrary function in $L^2(0, \infty)$ and

$$f_n(t) = \begin{cases} f(t) & (t < n), \\ 0 & (t \geqq n), \end{cases}$$

then, by (12)

$$f_n(t) = \int_{-\infty}^{\infty} \Phi(f_n; \lambda)\, dE_\lambda g. \tag{13}$$

Let

$$\Phi_n(\lambda) = \Phi(f_n; \lambda) = \int_0^n f(t)\, u(t; \lambda)\, dt. \tag{14}$$

Since the sequence of functions $\{f_n(t)\}$ converges in the norm of $L^2(0, \infty)$ to the function $f(t)$, the isometric correspondence between the spaces $L^2(0, \infty)$ and L^2_σ implies that the sequence $\{\Phi(\lambda)\}$ converges in the norm of L^2_σ to some function $\Phi(\lambda)$. Passing to the limit in the sense of the indicated norm in equations (13) and (14), we obtain the formulas

$$f(t) = \int_{-\infty}^{\infty} \Phi(\lambda)\, dE_\lambda g,$$

(15)

$$\Phi(\lambda) = \int_{0}^{\infty} f(t)\, u(t; \lambda)\, dt.$$

The second of these formulas coincides with (10) and, therefore, it remains only to prove formula (11).

Obviously, it is sufficient to prove that formula (11) is the inversion of formula (10) if $\Phi(\lambda)$ runs through the manifold $\mathfrak{M}$ of functions in L_σ^2 which vanish outside of a finite interval. Thus, let the function $\Phi(\lambda)$ belong to $\mathfrak{M}$ and let $f(t)$ be the function of $L^2(0, \infty)$ corresponding to it, which is defined by formula (15). We introduce the function

$$\eta_t(s) = \begin{cases} 1 & (s \leqq t), \\ 0 & (s > t), \end{cases}$$

and use the fact that the correspondence between $L^2(0, \infty)$ and L_σ^2 is isometric to obtain the equation

$$\int_0^\infty f(s)\, \eta_t(s)\, ds = \int_{-\infty}^{\infty} \Phi(\lambda)\, \Phi(\eta_t; \lambda)\, d\sigma(\lambda),$$

or

$$\int_0^t f(s)\, ds = \int_{-\infty}^{\infty} \Phi(\lambda) \left\{ \int_0^t u(s; \lambda)\, ds \right\} d\sigma(t).$$

The left member has the derivative $f(t)$ almost everywhere and the right integral in λ may be expressed with finite limits since $\Phi(\lambda) \in \mathfrak{M}$. Therefore, the derivative with respect to t of the right member is

$$\int_{-\infty}^{\infty} \Phi(\lambda)\, u(t; \lambda)\, d\sigma(\lambda),$$

and the proof is complete.

Our result concerning the inversion formula would be incomplete if we did not show how to actually find the nondecreasing function $\sigma(\lambda)$. We now concern ourselves with this question; with this aim we denote by $u(t; \lambda)$ that solution of the equation

$$l[y] - \lambda y = 0,$$

which for $t = 0$ satisfies the conditions which characterize the given extension $\tilde{L}$ of the operator L. As was indicated above, this condition has the form

$$p(t)\,\varphi'(t)\,|_{t=0} = \theta\varphi(0).$$

The real parameter θ we take as finite so that $u(0, \lambda) \neq 0$ and we can let

$$u(0; \lambda) = 1.$$

Furthermore, we denote by $v(t; \lambda)$ a second solution of the homogeneous equation which is normalized by means of the equations

$$v(0; \lambda) = 0,\ p(t)\,v'(t; \lambda)\,|_{t=0} = -1.$$

The kernel $K(t, s; \lambda)$ of the integral operator which is the resolvent of the operator $\tilde{L}$, i.e., the Green's function, has the form

$$(16)\quad K(t,s;z) = \begin{cases} u(s;z)\,[v(t;z) + m(z)u(t;z)\,] & (s \leqq t), \\ u(t;z)\,[v(s;z) + m(z)\,u(s;z)\,] & (s > t), \end{cases} \quad (\Im z > 0).$$

The function $m(z)$ is uniquely defined by the condition

$$v(t; z) + m(z)\,u(t; z) \in L^2(0, \infty).$$

We now let $\varphi(t)$ run through the set of all functions in $D_{\tilde{L}}$ which vanish outside of a finite interval. Let

$$(\tilde{L} - zE)\,\varphi = f,$$

where z $(\Im z > 0)$ is fixed. The set $\mathfrak{N}$ through which the function $f(t)$ runs obviously is dense in $L^2(0, \infty)$. If $g(t)$ is an arbitrary function in $\mathfrak{D}$, then

$$\int_0^\infty \varphi(t)\overline{g(t)}\,dt = \int_{-\infty}^\infty \Phi(\varphi; \lambda)\,\overline{\Phi(g; \lambda)}\,d\sigma(\lambda).$$

But since

$$\varphi(t) = \int_0^\infty K(t, s; z)\,f(s)\,ds,$$

and, by property 3° of the direction functional,

$$\Phi(\varphi; \lambda) = \frac{\Phi(f; \lambda)}{\lambda - z},$$

we have

$$\int_0^\infty\int_0^\infty K(t, s; z) f(s)\overline{g(t)}\,ds\,dt = \int_{-\infty}^\infty \frac{\Phi(f; \lambda)\,\overline{\Phi(g; \lambda)}}{\lambda - z}\,d\sigma(\lambda).$$

This equation is valid for all $f, g \in \mathfrak{D}$, since $\mathfrak{N}$ is dense in $L^2(0, \infty)$. Let

$$f(t) = g(t) = \begin{cases} \dfrac{1}{\delta} \text{ for } 0 \leqq t \leqq \delta, \\ 0 \text{ for } t > \delta. \end{cases}$$

Then our equation takes the form

$$\frac{1}{\delta^2}\int_0^\delta \int_0^\delta K(t, s; z)\, dt\, ds = \int_{-\infty}^{\infty} \omega_\delta(\lambda)\, \frac{d\sigma(\lambda)}{\lambda - z}, \tag{17}$$

where

$$\omega_\delta(\lambda) = \left[\frac{1}{\delta}\int_0^\delta u(t; \lambda)\, dt\right]^2.$$

As $\delta \to 0$, the left member of (17) tends to the limit

$$K(0, 0; z) = m(z).$$

Therefore,

$$\lim_{\delta \to 0} \int_{-\infty}^{\infty} \omega_\delta(\lambda)\, \frac{d\sigma(\lambda)}{\lambda - z} = m(z),$$

which yields

$$\lim_{\delta \to 0} \int_{-\infty}^{\infty} \omega_\delta(\lambda)\, \frac{d\sigma(\lambda)}{\lambda^2 - 1} = \Im m(i).$$

It is easy to conclude that the integral

$$\int_{-\infty}^{\infty} \frac{d\sigma(\lambda)}{\lambda^2 + 1}$$

exists and equals $\Im m(i)$. Therefore,

$$\int_{-\infty}^{\infty} \left(\frac{1}{\lambda - z} - \frac{1}{\lambda - i}\right) d\sigma(\lambda) = m(z) - m(i).$$

It follows that

$$m(z) = \Re m(i) + \int_{-\infty}^{\infty} \frac{1 + \lambda z}{\lambda + z}\, \frac{d\sigma(\lambda)}{\lambda^2 + 1}. \tag{18}$$

Applying the Stieltjes inversion formula (cf. Section 59), we obtain

$$\frac{\sigma(\lambda - 0) + \sigma(\lambda + 0)}{2} = \text{const.} + \lim_{y \to 0} \frac{1}{\pi} \int_0^\lambda \Im m(x + iy)\, dx.$$

It is possible to show that this formula is valid in the case when the boundary condition has the form $\varphi(0) = 0$.

Completing the present section, we make several remarks concerning the case when the deficiency indices of the operator are (2, 2). A class of self-adjoint extensions of such an operator which was first studied by Weyl in the paper quoted on page 186 will be discussed.

Let L have the deficiency indices (1, 1) or (2, 2). Let L_θ denote an extension of L determined by the boundary condition

$$p(t)\,\varphi'(t)\,|_{t=0} = \theta\varphi(0), \tag{19}$$

so that D_{L_θ} consists of all the functions $\varphi(t) \in D_{L^*}$ which satisfy condition (19). Obviously L_θ is a symmetric extension of L. If the deficiency indices of the operator L are (1, 1), then the operator L_θ is self-adjoint. Letting θ vary from $-\infty$ to ∞, we obtain all self-adjoint extensions of L. If the deficiency indices of the operator L are (2, 2), then the deficiency indices of the operator L_θ will be (1, 1). Each self-adjoint extension $\tilde{L}_\theta$ of the operator L_θ is an extension of L. Varying θ in the interval $(-\infty, \infty)$ and taking for each value θ all possible extensions $\tilde{L}_\theta$, we obtain a class of self-adjoint extensions of L. This class is characterized by the fact that one of the two boundary conditions (2) of Section 5 which define the self-adjoint extension has the form (19); obviously, the second of these conditions involves only the singular end point $t = \infty$. Thus, the class of self-adjoint extensions obtained are characterized by *reduced conditions.*

The assumption made in the beginning of this section that the deficiency indices of the operator L are (1, 1) was used only once, namely, in the proof of property 2° of the direction functional. However, it is easy to see that this property 2° is maintained in the case of the deficiency indices (2, 2), if only those self-adjoint extensions $\tilde{L}$ are considered which are characterized by the reduced conditions. We denote such extensions by $\tilde{L}_\theta$.

Thus, all of the above propositions remain valid for extensions $\tilde{L}_\theta$ of the operator L with the deficiency indices (2, 2). In particular, from Theorem 2 it follows that the self-adjoint extensions $\tilde{L}_\theta$ have simple spectra.

It is not difficult to see that as in the case of the deficiency indices (1, 1), the kernel of the resolvent $\tilde{R}_z$ of the operator $\tilde{L}_\theta$ (for $\theta \neq \infty$) is defined by the formula (16) where the function $m(z)$ is uniquely determined by the choice of the self-adjoint extension $\tilde{L}_\theta$ of L_θ. In particular, $K(0, 0; z) = m(z)$. Reasoning further as in the case of the deficiency indices (1, 1) we obtain formula (18). Since, by Theorem 2 of Section 6, the operator $\tilde{L}_\theta$ has a pure point spectrum with the unique limit point at infinity, the function $\sigma(\lambda)$ is piecewise constant and formula (18) shows that $m(z)$ is a meromorphic function of z.

In correspondence with the results of Section 4 of Appendix I, the scalar product $(\tilde{R}_z f, f)$ for fixed $f(t) \in L^2(0, \infty)$ and z ($\Im z > 0$) runs through a circle $C(f; z)$ as $\tilde{R}_z$ runs through the set of all orthogonal resolvents of the operator L_θ. But from formula (16) it follows that the scalar product $(\tilde{R}_z f, f)$ has the form $\alpha + \beta m(z)$, where the constants α and β do not depend on the choice of the resolvents. Therefore, as the point $(\tilde{R}_z f, f)$ runs through the circle $C(f; z)$, the point $\omega = m(z)$ also runs through a certain circle which is called the *Weyl limit circle*. In the case of the deficiency indices (1, 1) (and only in that case) the Weyl limit circle degenerates to a point. Now we clarify the meaning of the terms "*limit circle*" and "*limit point*" mentioned in the end of Section 5 of Appendix II. We do not present here the method by which Weyl arrived at the limit circle.[13]

We remark that although the circle $C(f; z)$ depends on the choice of f and z, the corresponding limit circle C depends only on z but does not depend on f, because the function $m(z)$ does not depend on f. We shall find the equation of the limit circle. With this aim, following Krein, we use the Hilbert functional equation for the resolvent of the operator L_θ which yields, without difficulty,

$$K(t, s; z) - K(t, s; \bar{z}) = (z - \bar{z}) \int_0^\infty K(t, \xi; z) K(\xi, s; \bar{z}) \, d\xi.$$

Let $t = s = 0$ and use formula (16) to obtain

$$m(z) - \overline{m(z)} = (z - \bar{z}) \int_0^\infty | v(\xi; z) + m(z) u(\xi; z) |^2 d\xi.$$

Hence, the equation of the limit circle has the form

$$\int_0^\infty | v(\xi; z) + \omega u(\xi; z) |^2 d\xi = \frac{\omega - \bar{\omega}}{z - \bar{z}} .$$

The case $\theta = \infty$ involves a modification of the above arguments and leads to a similar result. We shall not go into this further.

---

[13] Cf. B. M. Levitan [2], Vol. I.

8. Generalization to Differential Operators of Arbitrary Order

In Section 72 we saw that every self-adjoint operator A with a spectrum of multiplicity $r<\infty$ determines an isometric mapping of H onto the space of vector functions L^2_S. Under this mapping the operator A goes into the operator of multiplication by the independent variable. We recall that the matrix distribution function $S(\lambda)$ is determined by the spectral function E_λ of the operator A if one chooses some generating basis $g_1, g_2, \ldots, g_p$ $(r \leqq p < \infty)$ of this operator and lets

$$S(\lambda) = ((E_\lambda g_i, g_k))_{i,k=1}^p.$$

The indicated isometric mapping is defined by the formulas

$$\overrightarrow{\Phi(\lambda)} = Vf, \tag{1}$$

$$f = V^{-1}\overrightarrow{\Phi(\lambda)} = \text{l.i.m.} \int_{-N}^{N} \sum_{k=1}^{p} \Phi_k(\lambda)\, dE_\lambda g_k, \tag{2}$$

where the element f and the vector function $\overrightarrow{\Phi(\lambda)} = \{\Phi_1(\lambda), \ldots, \Phi_p(\lambda)\}$ run through the spaces H and L^2_S respectively.

Formulas (1) and (2) we call the *inversion formulas* related to the self-adjoint operator A. We remark that this definition is more general than the definition given in the preceding section for the case of operators with a simple spectrum since it does not require that the basis which generates the distribution function $S(\lambda)$ be minimal. In particular, by the definition of the present section, an operator with a simple spectrum has related inversion formulas which define an isometric correspondence between H and L^2_S, where the order of the matrix $S(\lambda)$ is larger than one.

We now derive the inversion formulas for quasi-differential operators of arbitrary order. Let L be a quasi-differential operator with minimal domain, generated by the operation l of order $2n$ on the interval $(0, \infty)$ with a singular end point and let $\tilde{L}$ be an arbitrary self-adjoint extension of the operator L. We make no assumptions now about the deficiency indices of the operator L.

We introduce on the manifold $\mathfrak{D}$ of functions of the space $L^2(0, \infty)$ which vanish outside a finite interval, the $2n$ direction functionals

$$\Phi_j(f; \lambda) = \int_0^\infty f(t)\, u_j(t; \lambda)\, dt \qquad (j = 1, 2, \ldots, 2n), \tag{3}$$

where $\{u_j(t; \lambda)\}_{j=1}^{2n}$ is the fundamental system of solutions of the equation

$$l[u] - \lambda u = 0,$$

which satisfy the initial conditions

$$u_j^{[k-1]}(0; \lambda) = \begin{cases} 1 & (j=k) \\ 0 & (j \neq k) \end{cases} \qquad (j, k = 1, 2, \ldots, 2n).$$

It is easy to verify that each of the functionals $\Phi_j(f; \lambda)$ has properties 1° and 3° of Section 7. As far as property 2° is concerned, it is now formulated in the following form.

2° If for some function $f \in \mathfrak{D}$ and some real value λ the $2n$ equations

$$\Phi_j(f; \lambda) = 0 \qquad (j = 1, 2, \ldots, 2n)$$

hold, then the equation

$$(\tilde{L} - \lambda E)\varphi = f$$

has a solution in the class of functions which vanish outside a finite interval.

LEMMA 1:[14] *For each finite interval Δ_0 of the λ-axis there exists a system of functions $\psi_1(t), \psi_2(t), \ldots, \psi_{2n}(t)$ in $\mathfrak{D}$ such that the determinant*

$$\mathrm{D}(\lambda) = \mathrm{Det}\,(\Phi_j(\psi_k; \lambda))_{j,k=1}^{2n}$$

does not vanish in the interval Δ_0.

The proof of the lemma is easy to outline by the pattern of the proof of the corresponding lemma of Section 7 which is based on the analyticity of the determinant $\mathrm{D}(\lambda)$.

THEOREM 1: *Let Δ_0 be a finite interval and let $\psi_1, \psi_2, \ldots, \psi_{2n}$ be a system of functions in $\mathfrak{D}$ such that*

$$\mathrm{Det}\,(\Phi_j(\psi_k; \lambda)) \neq 0 \qquad (\lambda \in \Delta_0).$$

Furthermore, let E_λ be the spectral function of the operator $\tilde{L}$ and let $g_1, g_2, \ldots, g_{2n}$ be a system of functions defined by the equations

$$g_i(t) = \int_{\Delta_0} \sum_{k=1}^{2n} \Omega_{ik}(\lambda)\, dE_\lambda \psi_k,$$

where $\Omega_{ik}(\lambda)$ are the elements of the matrix which is the inverse of the matrix

$$(\Phi_j(\psi_k; \lambda))_{j,k=1}^{2n}.$$

Then for each function $f \in \mathfrak{D}$ and each interval $\Delta \subset \Delta_0$,

$$E(\Delta)f = \int_{\Delta} \sum_{k=1}^{2n} \Phi_j(f; \lambda)\, dE_\lambda g_j. \tag{4}$$

[14] This and the following proposition are due to M. G. Krein (cf. the work cited in the beginning of Section 7).

The proof of this theorem is completely analogous to the proof of Theorem 1 of Section 7. In the same way as there, we replace the condition (4) to be proved by the equation

$$E(\Delta)f = \int_{\Delta} \sum_{j=1}^{2n} F_j(f;\lambda)\, dE_\lambda \psi_j,$$

where

$$F_j(f;\lambda) = \sum_{k=1}^{2n} \Omega_{jk}(\lambda)\, \Phi_k(f;\lambda),$$

and we introduce the element

$$w_\mu = \int_a^\mu dE_\lambda f - \int_a^\mu \sum_{j=1}^{2n} F_j(f;\lambda)\, dE_\lambda\, \psi_j.$$

Inequality (5) of Section 7 now has the form

$$\| w_{\mu+\delta} - w_\mu \| \leqq \left\| E(\Delta_{\mu,\delta}) \left[f - \sum_{j=1}^{2n} F_j(f;\mu)\, \psi_j \right] \right\| +$$

$$+ \left\| E(\Delta_{\mu,\delta}) \sum_{j=1}^{2n} F_j(f;\mu)\, \psi_j - \int_\mu^{\mu+\delta} \sum_{j=1}^{2n} F_j(f,\lambda)\, dE_\lambda\, \psi_j \right\|.$$

The first summand on the right side of the inequality is estimated in the same way as in Section 7. As far as the second summand is concerned,

$$\text{(5)} \qquad \sum_{j=1}^{2n} \left\| E(\Delta_{\mu,\delta})\, F_j(f;\mu)\, \psi_j - \int_\mu^{\mu+\delta} F_j(f;\lambda)\, dE_\lambda \psi_j \right\|^2 \leqq$$

$$\leqq \sum_{j=1}^{2n} M_j^2 \delta^2 (E(\Delta_{\mu,\delta})\, \psi_j, \psi_j),$$

where M_j denotes the maximum of the absolute value of the derivative $\frac{\partial F_j}{\partial \lambda}$ in the interval Δ_0. If μ is an eigenvalue of the operator $\tilde{L}$, then the function ψ_k is not an eigenfunction belonging to this operator since, otherwise, by property (3),

$$\Phi_j(\psi_k;\lambda) = \frac{1}{\mu}\, \Phi_j(\tilde{L}\psi_k;\lambda) = \frac{\lambda}{\mu}\, \Phi_j(\psi_k;\lambda) \quad (j = 1, 2, \ldots, 2n),$$

which would imply that $\Phi_j(\psi_k;\lambda)=0$ for $\lambda \neq \mu$ and, hence, D $(\lambda)=0$ for $\lambda \neq \mu$. This contradicts the choice of the system $\psi_1, \psi_2, \ldots, \psi_{2n}$.

Thus, the right side of inequality (5) has magnitude greater than δ^2, which yields the validity of the theorem.

THEOREM 2: *If L, $\tilde{L}$ and $u_j(t;\lambda)$ $(j=1, 2, \ldots, 2n)$ have the meanings given in the beginning of the present section, then there exists a matrix distribution function $S(\lambda) = (\sigma_{ik}(\lambda))_{i,k=1}^{n}$ for which the formulas*

$$\Phi_j(\lambda) = \text{l.i.m.} \int_{-N}^{N} f(t)\, u_j(t;\lambda)\, dt \qquad (j = 1, 2, \ldots, 2n),$$

$$f(t) = \text{l.i.m.} \int_{-N}^{N} \sum_{i,k=1}^{2n} \Phi_i(\lambda) u_k(t;\lambda)\, d\sigma_{ik}(\lambda)$$

establish an isometric mapping of the space $L^2(0, \infty)$ onto the space of vector functions L_S^2.

Proof: We substitute $\Delta = \Delta_0$ in formula (4) and we replace the system of elements $\{g_j\}_{j=1}^{2n}$ by the system $g_j^{(0)} = E(\Delta_0)\, g_j$ $(j=1, 2, \ldots, 2n)$. Then we subdivide the λ-axis into finite intervals $\Delta_k (\pm k=0, 1, 2, \ldots)$ and we choose for each of these a system of elements $\{g_j^{(k)}\}_{j=1}^{2n}$ in the same way as the system $\{g_j^{(0)}\}_{j=1}^{2n}$ was chosen for the interval Δ_0.

Let

$$g_i = \sum_{k=-\infty}^{\infty} g_i^{(k)} \qquad (i = 1, 2, \ldots, 2n).$$

Then, as in Section 7, we obtain a system of elements (proper or improper) $g_1, g_2, \ldots, g_{2n}$ which satisfy the condition

$$E(\Delta)\, g_i = \sum_k E(\Delta)\, g_i^{(k)}$$

for each finite interval Δ. (The series on the right contains only a finite number of summands.) Furthermore, for each finite interval Δ of the λ-axis we obtain the formula

$$E(\Delta) f = \int_{\Delta} \sum_{j=1}^{2n} \Phi_j(f;\lambda)\, dE_\lambda g_j,$$

and for $\Delta = [-\infty, \infty]$ we obtain

$$f = \text{l.i.m.} \int_{-N}^{N} \sum_{j=1}^{2n} \Phi_j(f;\lambda)\, dE_\lambda g_j.$$

Since the manifold $\mathfrak{D}$ is dense in $L^2(0, \infty)$, the last formula implies that the multiplicity of the spectrum of the operator $\tilde{L}$ does not exceed $2n$ and that $g_1, g_2, \ldots, g_{2n}$ is a generating basis.

The closure relation for $f \in \mathfrak{D}$ has the form

$$\int_0^\infty |f(t)|^2 dt = \int_{-\infty}^{\infty} \sum_{i,k=1}^{2n} \Phi_j(f;\lambda)\,\Phi_k(f;\lambda)\,d\sigma_{ik}(\lambda).$$

All further calculations mentioned in the end of the preceding section are carried over without difficulty to the case now under consideration. Hence, Theorem 2 is established.

9. Examples

I. The Trigonometric Functions. The differential operator:

$$l = -\frac{d^2}{dt^2} \qquad (0 \leqq t < \infty).$$

The deficiency indices of the operator L are (1, 1). The boundary condition which characterizes a self-adjoint extension of the operator are

$$y'(0) = \theta y(0).$$

The fundamental system of solutions:

$$u(t;\lambda) = \cos(\sqrt{\lambda} t) + \frac{\theta}{\sqrt{\lambda}} \sin(\sqrt{\lambda}\, t),$$

$$v(t;\lambda) = -\frac{1}{\sqrt{\lambda}} \sin(\sqrt{\lambda}\, t).$$

The determination of the function $m(z)$ ($\Im z > 0$):

$$v(t;z) + m(z)u(t;z) =$$

$$= m(z)\left\{\cos(\sqrt{\lambda}\, t) + \frac{\theta}{\sqrt{z}} \sin(\sqrt{\lambda}\, t)\right\} - \frac{1}{\sqrt{z}} \sin(\sqrt{z}\, t) =$$

$$= \left\{m(z)\left[\frac{1}{2} - \frac{\theta}{2i\sqrt{z}}\right] + \frac{1}{2i\sqrt{z}}\right\} e^{-i\sqrt{z}t} +$$

$$+ \left\{m(z)\left[\frac{1}{2} + \frac{\theta}{2i\sqrt{z}}\right] - \frac{1}{2i\sqrt{z}}\right\} e^{i\sqrt{z}t};$$

the condition that this belongs to $L^2(0, \infty)$ yields

$$m(z)\left[\frac{1}{2} - \frac{\theta}{2i\sqrt{z}}\right] + \frac{1}{2i\sqrt{z}} = 0,$$

so that

$$m(z) = \frac{1}{\theta - i\sqrt{z}}.$$

The determination of $\sigma(\lambda)$:

$$\sigma(\lambda) = \text{const.} + \lim_{y\to 0} \frac{1}{\pi} \int_0^\lambda \Im \frac{1}{\theta - i\sqrt{x+iy}}\, dx\, ;$$

if $\theta \geqq 0$, then

$$\sigma'(\lambda) = \frac{\sqrt{\lambda}}{\pi(\lambda+\theta^2)} \qquad (\lambda \geqq 0),$$

$$\sigma(\lambda) = 0 \qquad (\lambda < 0);$$

if $\theta < 0$, then

$$\sigma'(\lambda) = \frac{\sqrt{\lambda}}{\pi(\lambda+\theta^2)} \qquad (\lambda \geqq 0),$$

and at the point

$$\lambda_0 = -\theta^2$$

there is a jump

$$\sigma(\lambda_0 + 0) - \sigma(\lambda_0) = -2\theta.$$

The inversion formulas:

(a) $(\theta \geqq 0)$

$$\Phi(\lambda) = \int_0^\infty f(t) \left\{ \cos(\sqrt{\lambda}\, t) + \frac{\theta}{\sqrt{\lambda}} \sin(\sqrt{\lambda}\, t) \right\} dt \quad (\lambda \geqq 0),$$

$$f(t) = \frac{1}{\pi} \int_0^\infty \Phi(t) \left\{ \cos(\sqrt{\lambda}\, t) + \frac{\theta}{\sqrt{\lambda}} \sin(\sqrt{\lambda}\, t) \right\} \frac{\sqrt{\lambda}}{\lambda+\theta^2}\, d\lambda;$$

(b) $(\theta < 0)$

$$\Phi(\lambda) = \begin{cases} \displaystyle\int_0^\infty f(t) \left\{ \cos(\sqrt{\lambda}\, t) + \frac{\theta}{\sqrt{\lambda}} \sin(\sqrt{\lambda}\, t) \right\} dt & (\lambda \geqq 0), \\ \displaystyle\int_0^\infty f(t) e^{\theta t}\, dt & (\lambda = -\theta^2), \\ 0 & (\lambda \neq -\theta^2, \lambda < 0), \end{cases}$$

$$f(t) = -2\theta\Phi(-\theta^2)e^{\theta t} + \frac{1}{\pi} \int_0^\infty \Phi(\lambda) \left\{ \cos(\sqrt{\lambda}\, t) + \frac{\theta}{\sqrt{\lambda}} \sin(\sqrt{\lambda}\, t) \right\} \frac{\sqrt{\lambda}}{\lambda+\theta^2}\, d\lambda.$$

If the boundary condition is of the form

$$y(0) = 0,$$

then the inversion formulas are obtained by letting $\theta \to \infty$:

$$\Phi(\lambda) = \int_0^\infty f(t) \frac{\sin(\sqrt{\lambda}\, t)}{\sqrt{\lambda}}\, dt,$$

$$f(t) = \frac{1}{\pi} \int_0^\infty \Phi(\lambda) \sin(\sqrt{\lambda}\, t)\, d\lambda.$$

In this case as for $\theta = 0$, we obtain the ordinary Fourier-Plancherel transformation for the semi-axis.

II. The Legendre functions. The differential operation:

$$l = -\frac{d}{dt}(1 - t^2)\frac{d}{dt} \qquad (-1 < t < 1).$$

Both endpoints of the interval are singular.

An orthonormal system of solutions of the equation

$$(L^* - \lambda E)\, u = 0$$

for nonreal λ has the form

$$u_1(t;\lambda) = A_\mu\{P_\mu(t) + P_\mu(-t)\},$$

$$(A_\mu, B_\mu > 0),$$

$$u_2(t;\lambda) = B_\mu\{P_\mu(t) - P_\mu(-t)\},$$

where μ is any root of the equation

$$\mu(\mu + 1) = \lambda,$$

and $P_\mu(t)$ is a Legendre function of the first kind which can be defined by means of the following series[15]

$$P_\mu(t) =$$

$$= 1 + \sum_{k=1}^{\infty} \frac{(\mu+1)(\mu+2)\ldots(\mu+k)(-\mu)(-\mu+1)\ldots(-\mu+k-1)}{k!^2}\left(\frac{1}{2} - \frac{t}{2}\right)^k.$$

It converges for t in the circle $|t - 1| < 2$. This series is the hypergeometric function

$$F\left(\mu + 1, -\mu, 1; \frac{1-t}{2}\right).$$

[15] Cf. E. T. Whittaker and N. G. Watson [1], Vol. I.

Furthermore,

$$P'_\mu(t) = \frac{\mu(\mu+1)}{2} F\left(\mu + 2, -\mu + 1, 2; \frac{1-t}{2}\right).$$

Therefore, as $t \to -1 + 0$,

$$\frac{P_\mu(t)}{\ln \dfrac{1-t}{1+t}} \to -\frac{\sin \pi\mu}{\pi},$$

$$(1 - t^2) P'_\mu(t) \to \frac{2 \sin \pi\mu}{\pi}.$$

It follows that both solutions belong to $L^2(-1, 1)$, i.e., the deficiency indices of the operator L are (2, 2).

We let

$$\varphi^*(t) = \varphi(t) - \frac{1}{2} \varphi^{[1]}(t) \ln \frac{1+t}{1-t},$$

where the quasi-derivative $\varphi^{[1]}(t)$ equals

$$(1 - t^2)\, \varphi'(t),$$

and then we can express the Lagrange bilinear form by

$$[\varphi, \psi]_t = \varphi^{[1]}(t) \overline{\psi^*(t)} - \varphi^*(t) \overline{\psi^{[1]}(t)}.$$

For every function $\varphi(t) \in D_{L^*}$ there exist the finite limits

$$\lim_{t \to \pm 1} \varphi^{[1]}(t), \quad \lim_{t \to \pm 1} \varphi^*(t).$$

The manifold D_L is the set of all those functions $\varphi(t) \in D_{L^*}$ for which these four limits equal zero. These two results follow from the fact that for each function $f(t) \in L^2(-1, 1)$, the general integral of the equation

$$l[y] = f(t)$$

has the form

$$y(t) = \left\{ C_1 + \frac{1}{2} \int_{-1}^{t} f(s) \ln \frac{1+s}{1-s}\, ds \right\} + \left\{ C_2 - \frac{1}{2} \int_{-1}^{t} f(s)\, ds \right\} \ln \frac{1+t}{1-t}.$$

In order to obtain a self-adjoint extension of the operator L it is necessary, by the general theory, to choose some λ in the upper half-plane (we take $\lambda = \mu(\mu+1)$, where μ is purely imaginary) and, choosing a unitary matrix $(\theta_{ik})^2_{i,\,k=1}$, to let

$$w_1(t; \lambda) = u_1(t; \lambda) + \theta_{11} \overline{u_1(t; \lambda)} + \theta_{12} \overline{u_2(t; \lambda)},$$

$$w_2(t; \lambda) = u_2(t; \lambda) + \theta_{21} \overline{u_1(t; \lambda)} + \theta_{22} \overline{u_2(t; \lambda)}.$$

Furthermore, it is necessary to find the values

$$w_k^{[1]}(\pm 1;\lambda),\ w_k^*(\pm 1;\lambda) \qquad (k=1,2),$$

and, in terms of them, to express the boundary conditions

$$[\varphi, w_k]_{-1}^{1}=0 \qquad (k=1,2)$$

which distinguish the domain D_{L^*} from the domain D_{L_θ}.

Using the formulas mentioned it is not difficult to establish the following table:

$$w_1^{[1]}(\pm 1)=\frac{2\sin \pi\mu}{\pi}\{\mp A_\mu(1-\theta_{11})-B_\mu\theta_{12}\},$$

$$w_2^{[1]}(\pm 1)=\frac{2\sin \pi\mu}{\pi}\{\pm A_\mu\theta_{21}+B_\mu(1-\theta_{22})\},$$

$$w_1^*(\pm 1)=A_\mu\{[1+\gamma(\mu)]+\theta_{11}[1+\overline{\gamma(\mu)}]\}\pm\theta_{12}B_\mu[1-\overline{\gamma(\mu)}],$$

$$w_2^*(\pm 1)=\theta_{21}A_\mu[1+\overline{\gamma(\mu)}]\pm B_\mu\{[1-\gamma(\mu)]+\theta_{22}[1-\overline{\gamma(\mu)}]\}.$$

We do not calculate the value of $\gamma(\mu)$, but from what follows it will be clear that $\gamma(\mu)\neq\pm 1$.

We obtain the simplest and most important self-adjoint extension by letting

$$\theta_{11}=\theta_{22}=1,\ \ \theta_{12}=\theta_{21}=0.$$

Then

$$w_1^{[1]}(\pm 1)=0,\ \ w_2^{[1]}(\pm 1)=0,$$

and

$$w_1^*(\pm 1)=A_\mu[2+\gamma(\mu)+\overline{\gamma(\mu)}],$$

$$w_2^*(\pm 1)=\pm B_\mu[2-\gamma(\mu)-\overline{\gamma(\mu)}].$$

Since neither $w_1(t)$ nor $w_2(t)$ belong to D_L, the last expressions are distinct from zero, which implies that $\gamma(\mu)\neq\pm 1$.

The boundary conditions which characterize the extension considered have the form

$$\varphi^{[1]}(1)=\varphi^{[1]}(-1)=0. \tag{1}$$

It is possible to express them differently. First, they are equivalent to the condition

$$\frac{1}{\sqrt{1-t^2}}\varphi^{[1]}(t)\in L^2(-1,1).$$

Second, they are equivalent to the following requirement:

$$\varphi(t) \text{ tends to finite limits as } t \to \pm 1. \tag{3}$$

We now show that these conditions are equivalent. It is sufficient to consider the case of real functions. Then

$$\begin{aligned}
&-\int_\alpha^\beta \varphi(t)\,[(1-t^2)\,\varphi'(t)]'\,dt = \\
&= -(1-t^2)\,\varphi'(t)\,\varphi(t)\,\big|_\alpha^\beta + \int_\alpha^\beta (1-t^2)\,\varphi'^2(t)\,dt
\end{aligned} \qquad (-1<\alpha<\beta<1), \tag{4}$$

the left side of which has finite limits for $\alpha \to -1$ and $\beta \to 1$, for each function $\varphi(t) \in D_{L^*}$. On the other hand,

$$\varphi^{[1]}(t) = \varphi^{[1]}(-1) - \int_{-1}^{t} l\,[\varphi(s)\,]\,ds = \varphi^{[1]}(1) + \int_{1}^{t} l\,[\varphi(s)]\,ds.$$

Therefore, condition (1) is satisfied, so that

$$|\,\varphi^{[1]}(t)\,| \leqq \sqrt{1 \pm t}\,\|\,L^*\varphi\,\|,$$

and, consequently,

$$\lim_{t \to \pm 1} \varphi^{[1]}(t) \ln \frac{1+t}{1-t} = 0.$$

It follows that, as $t \to \pm 1$, $\varphi(t)$ tends to finite limits, i.e., condition (3) is satisfied. From the identity (4) follows also the existence of the integral[16]

$$I(\varphi) = \int_{-1}^{1} (1-t^2)\,\varphi'^2(t)\,dt \tag{5}$$

so that condition (2) is satisfied. Thus, (1) → (3) and (1) → (2). It remains to prove that (2) → (1). But if the integral (5) exists, then $\varphi^{[1]}(1)$ cannot be different from zero, since otherwise the finiteness of $\varphi^*(1)$ would imply that $\varphi(t)$ has an infinite limit as $t \to 1$. This would contradict identity (4).

The self-adjoint extension characterized by any one of the conditions (1), (2), and (3) reduces to the classical expansion in terms of Legendre polynomials. Since the boundary conditions (1) are reduced in the sense defined earlier, the spectrum of the extension is simple. The spectrum is

---

[16] We remark that the equation $l\,[\varphi]=0$ coincides with the Euler equation for the integral $I(\varphi)$. In this connection cf. Friedrichs [2], Vol. I.

discrete and the inverse operator is completely continuous. The Legendre polynomials

$$P_n(t) = \frac{1}{2^n n!}\frac{d^n(t^2-1)^n}{dt^n} \qquad (n = 0, 1, 2, \ldots)$$

satisfy the equations

$$-\frac{d}{dt}(1-t^2)\frac{du}{dt} - n(n+1)u = 0$$

and, in addition, they satisfy condition (3). But, since the set of all polynomials $P_n(t)$ form a complete orthogonal system in $L^2(-1, 1)$, the spectrum of the extension considered consists of the points

$$n(n+1) \qquad (n = 0, 1, 2, \ldots),$$

and the inversion formulas corresponding to this case follow from the expansion in the series of Legendre polynomials.

III. The Tchebysheff-Hermite functions are connected with the differential operation

$$l = -\frac{d^2}{dt^2} + t^2 \qquad (-\infty < t < \infty), \tag{6}$$

which, in particular, is very important in the theory of so-called linear oscillators.

For the deficiency indices of the operator L we use Theorem 4 of Section 6, applying this theorem to an operation of the more general form

$$-\frac{d^2}{dt^2} + q(t) \tag{7}$$

where $q(t) \geqq 0$.

We apply the method of successive approximations to the equation

$$u'' = q(t)u \qquad (0 \leqq t < \infty),$$

and this shows that the solution $u(t)$ which equals one for $t = 0$ satisfies the inequality

$$u(t) \geqq 1$$

for all $t \geqq 0$, and hence does not belong to $L^2(0, \infty)$.

Thus, the equation

$$-u'' + q(t)u - \lambda u = 0$$

has for $\lambda = 0$ a solution which does not belong to $L^2(0, \infty)$. By Theorem

4 of Section 6, the operator with minimal domain which is generated by the operator (7) on the semi-axis $(0, \infty)$ has deficiency indices (1,1). Obviously, the operator with minimal domain generated by the operator (7) on the negative semi-axis also has deficiency indices (1, 1). Applying Theorem 3 of Section 4, we find that the operator with minimal domain which is generated by operation (7) on the whole axis has deficiency indices (0, 0).

Obviously we would obtain the same result if we had replaced the requirement of the non-negativity of the function $q(t)$ by the condition of its semi-boundedness from below.

In particular, the operator L which is generated by the operation (6) is a self-adjoint operator. It is easy to verify that the equation

$$-u'' + (t^2 - \lambda) u = 0$$

with

$$\lambda = 2n + 1 \qquad (n = 0, 1, 2, \ldots)$$

is satisfied by the sequence of Tchebysheff-Hermite functions (cf. Section 11) which form a complete system in $L^2(-\infty, \infty)$. Therefore, the operator L has a pure point spectrum with the unique limit point at infinity. Thus, the inversion formulas connected with the operator L are indicated by the expansion in Tchebysheff-Hermite functions.

The example considered is particularly instructive. It shows that the property that the resolvent is a completely continuous operator, which, by Theorem 2 of Section 6 of Appendix II, always holds in the quasi-regular case, also can hold in other cases (even in case of minimal deficiency indices, as in the above example).

IV. The Bessel functions. Among the various differential operations which lead to Bessel functions the most important one has the form

$$l = -\frac{d^2}{dt^2} + \frac{\nu^2 - \frac{1}{4}}{t^2}. \tag{8}$$

We consider this operation with the parameter $\nu \geqq 0$. As far as the interval is concerned it is natural to study the following three cases:

(α)	$0 < t \leqq 1$	one singular end point,
(β)	$1 \leqq t < \infty$	one singular end point,
(γ)	$0 < t < \infty$	two singular end points.

The general integral of the equation

$$l[u] - \lambda u = 0 \tag{9}$$

with $\lambda \neq 0$ has the form

$$u(t;\lambda) = At^{\frac{1}{2}}J_\nu(t\sqrt{\lambda}) + Bt^{\frac{1}{2}}Y_\nu(t\sqrt{\lambda}). \tag{10}$$

Here A and B are arbitrary constants and $J_\nu(z)$ and $Y_\nu(z)$ are the Bessel functions of the first and second kind, respectively. They are defined by the formulas

$$J_\nu(z) = \sum_{k=0}^{\infty} \frac{(-1)^k \left(\frac{z}{2}\right)^{\nu+2k}}{k!\,\Gamma(\nu+k+1)},$$

$$Y_\nu(z) = \frac{J_\nu(z)\cos \pi\nu - J_{-\nu}(z)}{\sin \pi\nu}.$$

We assume that $\Im\lambda > 0$ and $t > 0$. Then the value $z = t\sqrt{\lambda}$ satisfies the inequality

$$0 < \arg z < \frac{\pi}{2}.$$

For $|\arg z| < \pi$ and $|z| \to \infty$ the following asymptotic formulas hold:

$$J_\nu(z) + iY_\nu(z) = H_\nu^{(1)}(z) \sim \sqrt{\frac{2}{\pi z}}\, e^{i\left(z-\frac{\pi\nu}{2}-\frac{\pi}{4}\right)},$$

$$J_\nu(z) - iY_\nu(z) = H_\nu^{(2)}(z) \sim \sqrt{\frac{2}{\pi z}}\, e^{-i\left(z-\frac{\pi\nu}{2}-\frac{\pi}{4}\right)}.$$

We substitute these formulas in (10) and find that equation (9) has not more than one solution which belongs to $L^2(1, \infty)$. But one such solution must exist. Hence, this solution is the function

$$u_1(t;\lambda) = AH_\nu^{(1)}(t\sqrt{\lambda}).$$

The deficiency indices of the operator L in case (β) are (1, 1).

We shall find the deficiency indices of the operator L in case α. Since, as $z \to 0$,

$$z^{\frac{1}{2}}J_\nu(z) \sim \frac{z^{\nu+\frac{1}{2}}}{2^\nu\,\Gamma(\nu+1)},$$

and, on the other hand, $z^{\frac{1}{2}}Y_\nu(z)$ does not belong to $L^2(0, 1)$ for $\nu \geqq 1$ and belongs to $L^2(0, 1)$ for $0 \leqq \nu < 1$; in the case α the deficiency indices of L are

$$(2, 2) \text{ for } 0 \leqq \nu < 1,$$

$$(1, 1) \text{ for } \nu \geqq 1.$$

Therefore, in case γ the deficiency indices of L are

$$\begin{array}{ll} (1, 1) & \text{for } 0 \leqq \nu < 1, \\ (0, 0) & \text{for } \nu \geqq 1. \end{array}$$

In each of the enumerated cases the differential operation (8) generates certain inversion formulas.

We obtain a unique pair of formulas for the interval $(0, \infty)$ if $\nu \geqq 1$. These formulas have the form

$$\left\{ \begin{aligned} g(\lambda) &= \int_0^\infty \sqrt{\lambda t}\, J_\nu(\lambda t) f(t)\, dt, \\ f(t) &= \int_0^\infty \sqrt{\lambda t}\, J_\nu(\lambda t) g(\lambda)\, d\lambda. \end{aligned} \right. \tag{11}$$

We have here not only a unitary but a self-adjoint operator in $L^2(0, \infty)$. It is called the *Hankel transform*. To derive the inversion formulas (11) rather simply, we can apply one of the methods indicated for obtaining the Fourier-Plancherel inversion formulas. It turns out that formulas (11) hold also for $\nu \geqq 0$ but, for $0 \leqq \nu < 1$, are not the pair of unique inversion formulas on the semi-axis[17] generated by formula (8).

We consider the case of the interval $(0, 1]$ for $\nu \geqq 1$ and we assume that the boundary condition at the regular end point has the form

$$\varphi(1) = 0.$$

Hence, we can let

$$u(t;\lambda) = \frac{\pi}{2}\sqrt{t}\{J_\nu(t\sqrt{\lambda})\, Y_\nu(\sqrt{\lambda}) - Y_\nu(t\sqrt{\lambda}) J_\nu(\sqrt{\lambda})\}.$$

Furthermore, let

$$v(t;\lambda) = \frac{\pi}{2}\sqrt{t\lambda}\,\{J_\nu(t\sqrt{\lambda})\, Y_\nu'(\sqrt{\lambda}) - Y_\nu(t\sqrt{\lambda}) J_\nu'(\sqrt{\lambda})\}.$$

Then

$$K(t,s;z) = \begin{cases} u(t;z)\,[v(s;z) + m(z)\, u(s;z)] & (t \leqq s) \\ u(s;z)\,[v(t;z) + m(z)\, u(t;z)] & (t > s). \end{cases}$$

The function $m(z)$ is defined with the aid of the condition that the solution

$$v(t;z) + m(z)\; u(t;z)$$

[17] The formulas (11) are valid also for $\nu \geqq -\frac{1}{2}$ and also in a slightly different form for complex ν such that $\nu > -\frac{1}{2}$.

belongs to the space $L^2(0, 1)$. This means that the function

$$\frac{\pi}{2}\sqrt{t}\{J_\nu(t\sqrt{z})\,[\sqrt{z}\,Y'_\nu(\sqrt{z}) + m(z)\,Y_\nu(\sqrt{z})]\} -$$

$$-\frac{\pi}{2}\sqrt{t}\{Y_\nu(t\sqrt{z})\,[\sqrt{z}J'_\nu(\sqrt{z}) + m(z)J_\nu(\sqrt{z})\,]\}$$

should reduce to the first term. Thus,

$$m(z) = -\sqrt{z}\,\frac{J'_\nu(\sqrt{z})}{J_\nu(\sqrt{z})}.$$

This function can be represented in the form

$$m(z) = -2\sum_{n=1}^{\infty}\frac{\lambda_n}{z-\lambda_n}$$

where the λ_n are the nonzero roots of the function $J_\nu(\sqrt{z})$. It follows that $\sigma(\lambda)$ is a piecewise constant function and that the sequence $\lambda_1, \lambda_2, \lambda_3, \ldots$ forms the spectrum. The eigenfunctions are

$$u_n(t) = u(t;\lambda_n) = \frac{\pi}{2}\sqrt{t}\,Y_\nu(\sqrt{\lambda_n})J_\nu(t\sqrt{\lambda_n}) = -\frac{\sqrt{t}J_\nu(t\sqrt{\lambda_n})}{\sqrt{\lambda_n}J'_\nu(\sqrt{\lambda_n})}.$$

The inversion formulas reduce to the expansion formula

$$f(t) = \sum_{n=1}^{\infty}\frac{2\sqrt{t}J_\nu(t\sqrt{\lambda_n})}{J'^2_\nu(\sqrt{\lambda_n})}\int_0^1\sqrt{s}\,J_\nu(s\sqrt{\lambda_n})f(s)\,ds,$$

which is known as the *Fourier-Bessel series*.

If at the regular end point the general condition

$$\varphi'(1) = -\theta\,\varphi(1)$$

was assumed, then we would arrive at the so-called *Fourier-Dienes series*.

We do not consider the case of the interval $(0, 1)$ for $0 \leqq \nu < 1$. We remark only that in this case the method that we used in connection with the Legendre operation can be adapted.

In addition, we leave to the reader the considerations of the case of the interval $(1, \infty)$ and the deficiency indices $(1, 1)$. If we assume at the regular end point the condition

$$\varphi(1) = 0,$$

then we find, by the method used repeatedly,

$$m(z) = -\sqrt{z}\,\frac{J'_\nu(\sqrt{z}) + iY'_\nu(\sqrt{z})}{J_\nu(\sqrt{z}) + iY_\nu(\sqrt{z})},$$

and we obtain the *Weber inversion formulas*

$$\Phi(\lambda) = \int_1^\infty \sqrt{t}\,\{J_\nu(t\sqrt{\lambda})\,Y_\nu(\sqrt{\lambda}) - Y_\nu(t\sqrt{\lambda})\,J_\nu(\sqrt{\lambda})\}\,f(t)\,dt,$$

$$f(t) = \int_0^\infty \sqrt{t}\,\frac{J_\nu(t\sqrt{\lambda})\,Y_\nu(\sqrt{\lambda}) - Y_\nu(t\sqrt{\lambda})\,J_\nu(\sqrt{\lambda})}{2\{J_\nu^2(\sqrt{\lambda}) + Y_\nu^2(\sqrt{\lambda})\}}\,\Phi(\lambda)\,d\lambda.$$

INDEX

Translator's Note: No bibliography is included in this volume. All references to bibliography in this book are to the complete bibliography of the German edition, which is reproduced in Volume I of this translation.

A CATALOG OF SELECTED

DOVER BOOKS

IN SCIENCE AND MATHEMATICS

Astronomy

BURNHAM'S CELESTIAL HANDBOOK, Robert Burnham, Jr. Thorough guide to the stars beyond our solar system. Exhaustive treatment. Alphabetical by constellation: Andromeda to Cetus in Vol. 1; Chamaeleon to Orion in Vol. 2; and Pavo to Vulpecula in Vol. 3. Hundreds of illustrations. Index in Vol. 3. 2,000pp. 6⅛ x 9¼.
Vol. I: 0-486-23567-X
Vol. II: 0-486-23568-8
Vol. III: 0-486-23673-0

EXPLORING THE MOON THROUGH BINOCULARS AND SMALL TELESCOPES, Ernest H. Cherrington, Jr. Informative, profusely illustrated guide to locating and identifying craters, rills, seas, mountains, other lunar features. Newly revised and updated with special section of new photos. Over 100 photos and diagrams. 240pp. 8¼ x 11. 0-486-24491-1

THE EXTRATERRESTRIAL LIFE DEBATE, 1750–1900, Michael J. Crowe. First detailed, scholarly study in English of the many ideas that developed from 1750 to 1900 regarding the existence of intelligent extraterrestrial life. Examines ideas of Kant, Herschel, Voltaire, Percival Lowell, many other scientists and thinkers. 16 illustrations. 704pp. 5⅜ x 8½. 0-486-40675-X

THEORIES OF THE WORLD FROM ANTIQUITY TO THE COPERNICAN REVOLUTION, Michael J. Crowe. Newly revised edition of an accessible, enlightening book recreates the change from an earth-centered to a sun-centered conception of the solar system. 242pp. 5⅜ x 8½. 0-486-41444-2

A HISTORY OF ASTRONOMY, A. Pannekoek. Well-balanced, carefully reasoned study covers such topics as Ptolemaic theory, work of Copernicus, Kepler, Newton, Eddington's work on stars, much more. Illustrated. References. 521pp. 5⅜ x 8½.
0-486-65994-1

A COMPLETE MANUAL OF AMATEUR ASTRONOMY: TOOLS AND TECHNIQUES FOR ASTRONOMICAL OBSERVATIONS, P. Clay Sherrod with Thomas L. Koed. Concise, highly readable book discusses: selecting, setting up and maintaining a telescope; amateur studies of the sun; lunar topography and occultations; observations of Mars, Jupiter, Saturn, the minor planets and the stars; an introduction to photoelectric photometry; more. 1981 ed. 124 figures. 25 halftones. 37 tables. 335pp. 6½ x 9¼. 0-486-40675-X

AMATEUR ASTRONOMER'S HANDBOOK, J. B. Sidgwick. Timeless, comprehensive coverage of telescopes, mirrors, lenses, mountings, telescope drives, micrometers, spectroscopes, more. 189 illustrations. 576pp. 5⅝ x 8¼. (Available in U.S. only.)
0-486-24034-7

STARS AND RELATIVITY, Ya. B. Zel'dovich and I. D. Novikov. Vol. 1 of *Relativistic Astrophysics* by famed Russian scientists. General relativity, properties of matter under astrophysical conditions, stars, and stellar systems. Deep physical insights, clear presentation. 1971 edition. References. 544pp. 5⅝ x 8¼. 0-486-69424-0

Chemistry

THE SCEPTICAL CHYMIST: THE CLASSIC 1661 TEXT, Robert Boyle. Boyle defines the term "element," asserting that all natural phenomena can be explained by the motion and organization of primary particles. 1911 ed. viii+232pp. 5⅜ x 8½.
0-486-42825-7

RADIOACTIVE SUBSTANCES, Marie Curie. Here is the celebrated scientist's doctoral thesis, the prelude to her receipt of the 1903 Nobel Prize. Curie discusses establishing atomic character of radioactivity found in compounds of uranium and thorium; extraction from pitchblende of polonium and radium; isolation of pure radium chloride; determination of atomic weight of radium; plus electric, photographic, luminous, heat, color effects of radioactivity. ii+94pp. 5⅜ x 8½. 0-486-42550-9

CHEMICAL MAGIC, Leonard A. Ford. Second Edition, Revised by E. Winston Grundmeier. Over 100 unusual stunts demonstrating cold fire, dust explosions, much more. Text explains scientific principles and stresses safety precautions. 128pp. 5⅜ x 8½. 0-486-67628-5

THE DEVELOPMENT OF MODERN CHEMISTRY, Aaron J. Ihde. Authoritative history of chemistry from ancient Greek theory to 20th-century innovation. Covers major chemists and their discoveries. 209 illustrations. 14 tables. Bibliographies. Indices. Appendices. 851pp. 5⅜ x 8½. 0-486-64235-6

CATALYSIS IN CHEMISTRY AND ENZYMOLOGY, William P. Jencks. Exceptionally clear coverage of mechanisms for catalysis, forces in aqueous solution, carbonyl- and acyl-group reactions, practical kinetics, more. 864pp. 5⅜ x 8½.
0-486-65460-5

ELEMENTS OF CHEMISTRY, Antoine Lavoisier. Monumental classic by founder of modern chemistry in remarkable reprint of rare 1790 Kerr translation. A must for every student of chemistry or the history of science. 539pp. 5⅜ x 8½. 0-486-64624-6

THE HISTORICAL BACKGROUND OF CHEMISTRY, Henry M. Leicester. Evolution of ideas, not individual biography. Concentrates on formulation of a coherent set of chemical laws. 260pp. 5⅜ x 8½. 0-486-61053-5

A SHORT HISTORY OF CHEMISTRY, J. R. Partington. Classic exposition explores origins of chemistry, alchemy, early medical chemistry, nature of atmosphere, theory of valency, laws and structure of atomic theory, much more. 428pp. 5⅜ x 8½. (Available in U.S. only.) 0-486-65977-1

GENERAL CHEMISTRY, Linus Pauling. Revised 3rd edition of classic first-year text by Nobel laureate. Atomic and molecular structure, quantum mechanics, statistical mechanics, thermodynamics correlated with descriptive chemistry. Problems. 992pp. 5⅜ x 8½. 0-486-65622-5

FROM ALCHEMY TO CHEMISTRY, John Read. Broad, humanistic treatment focuses on great figures of chemistry and ideas that revolutionized the science. 50 illustrations. 240pp. 5⅜ x 8½. 0-486-28690-8

Engineering

DE RE METALLICA, Georgius Agricola. The famous Hoover translation of greatest treatise on technological chemistry, engineering, geology, mining of early modern times (1556). All 289 original woodcuts. 638pp. 6¾ x 11. 0-486-60006-8

FUNDAMENTALS OF ASTRODYNAMICS, Roger Bate et al. Modern approach developed by U.S. Air Force Academy. Designed as a first course. Problems, exercises. Numerous illustrations. 455pp. 5⅜ x 8½. 0-486-60061-0

DYNAMICS OF FLUIDS IN POROUS MEDIA, Jacob Bear. For advanced students of ground water hydrology, soil mechanics and physics, drainage and irrigation engineering and more. 335 illustrations. Exercises, with answers. 784pp. 6⅛ x 9¼. 0-486-65675-6

THEORY OF VISCOELASTICITY (Second Edition), Richard M. Christensen. Complete consistent description of the linear theory of the viscoelastic behavior of materials. Problem-solving techniques discussed. 1982 edition. 29 figures. xiv+364pp. 6⅛ x 9¼. 0-486-42880-X

MECHANICS, J. P. Den Hartog. A classic introductory text or refresher. Hundreds of applications and design problems illuminate fundamentals of trusses, loaded beams and cables, etc. 334 answered problems. 462pp. 5⅜ x 8½. 0-486-60754-2

MECHANICAL VIBRATIONS, J. P. Den Hartog. Classic textbook offers lucid explanations and illustrative models, applying theories of vibrations to a variety of practical industrial engineering problems. Numerous figures. 233 problems, solutions. Appendix. Index. Preface. 436pp. 5⅜ x 8½. 0-486-64785-4

STRENGTH OF MATERIALS, J. P. Den Hartog. Full, clear treatment of basic material (tension, torsion, bending, etc.) plus advanced material on engineering methods, applications. 350 answered problems. 323pp. 5⅜ x 8½. 0-486-60755-0

A HISTORY OF MECHANICS, René Dugas. Monumental study of mechanical principles from antiquity to quantum mechanics. Contributions of ancient Greeks, Galileo, Leonardo, Kepler, Lagrange, many others. 671pp. 5⅜ x 8½. 0-486-65632-2

STABILITY THEORY AND ITS APPLICATIONS TO STRUCTURAL MECHANICS, Clive L. Dym. Self-contained text focuses on Koiter postbuckling analyses, with mathematical notions of stability of motion. Basing minimum energy principles for static stability upon dynamic concepts of stability of motion, it develops asymptotic buckling and postbuckling analyses from potential energy considerations, with applications to columns, plates, and arches. 1974 ed. 208pp. 5⅜ x 8½. 0-486-42541-X

METAL FATIGUE, N. E. Frost, K. J. Marsh, and L. P. Pook. Definitive, clearly written, and well-illustrated volume addresses all aspects of the subject, from the historical development of understanding metal fatigue to vital concepts of the cyclic stress that causes a crack to grow. Includes 7 appendixes. 544pp. 5⅜ x 8½. 0-486-40927-9

Mathematics

FUNCTIONAL ANALYSIS (Second Corrected Edition), George Bachman and Lawrence Narici. Excellent treatment of subject geared toward students with background in linear algebra, advanced calculus, physics and engineering. Text covers introduction to inner-product spaces, normed, metric spaces, and topological spaces; complete orthonormal sets, the Hahn-Banach Theorem and its consequences, and many other related subjects. 1966 ed. 544pp. 6⅛ x 9¼. 0-486-40251-7

ASYMPTOTIC EXPANSIONS OF INTEGRALS, Norman Bleistein & Richard A. Handelsman. Best introduction to important field with applications in a variety of scientific disciplines. New preface. Problems. Diagrams. Tables. Bibliography. Index. 448pp. 5⅜ x 8½. 0-486-65082-0

VECTOR AND TENSOR ANALYSIS WITH APPLICATIONS, A. I. Borisenko and I. E. Tarapov. Concise introduction. Worked-out problems, solutions, exercises. 257pp. 5⅝ x 8¼. 0-486-63833-2

AN INTRODUCTION TO ORDINARY DIFFERENTIAL EQUATIONS, Earl A. Coddington. A thorough and systematic first course in elementary differential equations for undergraduates in mathematics and science, with many exercises and problems (with answers). Index. 304pp. 5⅜ x 8½. 0-486-65942-9

FOURIER SERIES AND ORTHOGONAL FUNCTIONS, Harry F. Davis. An incisive text combining theory and practical example to introduce Fourier series, orthogonal functions and applications of the Fourier method to boundary-value problems. 570 exercises. Answers and notes. 416pp. 5⅜ x 8½. 0-486-65973-9

COMPUTABILITY AND UNSOLVABILITY, Martin Davis. Classic graduate-level introduction to theory of computability, usually referred to as theory of recurrent functions. New preface and appendix. 288pp. 5⅜ x 8½. 0-486-61471-9

ASYMPTOTIC METHODS IN ANALYSIS, N. G. de Bruijn. An inexpensive, comprehensive guide to asymptotic methods–the pioneering work that teaches by explaining worked examples in detail. Index. 224pp. 5⅜ x 8½ 0-486-64221-6

APPLIED COMPLEX VARIABLES, John W. Dettman. Step-by-step coverage of fundamentals of analytic function theory–plus lucid exposition of five important applications: Potential Theory; Ordinary Differential Equations; Fourier Transforms; Laplace Transforms; Asymptotic Expansions. 66 figures. Exercises at chapter ends. 512pp. 5⅜ x 8½. 0-486-64670-X

INTRODUCTION TO LINEAR ALGEBRA AND DIFFERENTIAL EQUATIONS, John W. Dettman. Excellent text covers complex numbers, determinants, orthonormal bases, Laplace transforms, much more. Exercises with solutions. Undergraduate level. 416pp. 5⅜ x 8½. 0-486-65191-6

RIEMANN'S ZETA FUNCTION, H. M. Edwards. Superb, high-level study of landmark 1859 publication entitled "On the Number of Primes Less Than a Given Magnitude" traces developments in mathematical theory that it inspired. xiv+315pp. 5⅜ x 8½. 0-486-41740-9

CALCULUS OF VARIATIONS WITH APPLICATIONS, George M. Ewing. Applications-oriented introduction to variational theory develops insight and promotes understanding of specialized books, research papers. Suitable for advanced undergraduate/graduate students as primary, supplementary text. 352pp. 5⅜ x 8½. 0-486-64856-7

COMPLEX VARIABLES, Francis J. Flanigan. Unusual approach, delaying complex algebra till harmonic functions have been analyzed from real variable viewpoint. Includes problems with answers. 364pp. 5⅜ x 8½. 0-486-61388-7

AN INTRODUCTION TO THE CALCULUS OF VARIATIONS, Charles Fox. Graduate-level text covers variations of an integral, isoperimetrical problems, least action, special relativity, approximations, more. References. 279pp. 5⅜ x 8½. 0-486-65499-0

COUNTEREXAMPLES IN ANALYSIS, Bernard R. Gelbaum and John M. H. Olmsted. These counterexamples deal mostly with the part of analysis known as "real variables." The first half covers the real number system, and the second half encompasses higher dimensions. 1962 edition. xxiv+198pp. 5⅜ x 8½. 0-486-42875-3

CATASTROPHE THEORY FOR SCIENTISTS AND ENGINEERS, Robert Gilmore. Advanced-level treatment describes mathematics of theory grounded in the work of Poincaré, R. Thom, other mathematicians. Also important applications to problems in mathematics, physics, chemistry and engineering. 1981 edition. References. 28 tables. 397 black-and-white illustrations. xvii + 666pp. 6⅛ x 9¼. 0-486-67539-4

INTRODUCTION TO DIFFERENCE EQUATIONS, Samuel Goldberg. Exceptionally clear exposition of important discipline with applications to sociology, psychology, economics. Many illustrative examples; over 250 problems. 260pp. 5⅜ x 8½. 0-486-65084-7

NUMERICAL METHODS FOR SCIENTISTS AND ENGINEERS, Richard Hamming. Classic text stresses frequency approach in coverage of algorithms, polynomial approximation, Fourier approximation, exponential approximation, other topics. Revised and enlarged 2nd edition. 721pp. 5⅜ x 8½. 0-486-65241-6

INTRODUCTION TO NUMERICAL ANALYSIS (2nd Edition), F. B. Hildebrand. Classic, fundamental treatment covers computation, approximation, interpolation, numerical differentiation and integration, other topics. 150 new problems. 669pp. 5⅜ x 8½. 0-486-65363-3

THREE PEARLS OF NUMBER THEORY, A. Y. Khinchin. Three compelling puzzles require proof of a basic law governing the world of numbers. Challenges concern van der Waerden's theorem, the Landau-Schnirelmann hypothesis and Mann's theorem, and a solution to Waring's problem. Solutions included. 64pp. 5⅜ x 8½. 0-486-40026-3

THE PHILOSOPHY OF MATHEMATICS: AN INTRODUCTORY ESSAY, Stephan Körner. Surveys the views of Plato, Aristotle, Leibniz & Kant concerning propositions and theories of applied and pure mathematics. Introduction. Two appendices. Index. 198pp. 5⅜ x 8½. 0-486-25048-2

TENSOR CALCULUS, J.L. Synge and A. Schild. Widely used introductory text covers spaces and tensors, basic operations in Riemannian space, non-Riemannian spaces, etc. 324pp. 5⅜ x 8¼. 0-486-63612-7

ORDINARY DIFFERENTIAL EQUATIONS, Morris Tenenbaum and Harry Pollard. Exhaustive survey of ordinary differential equations for undergraduates in mathematics, engineering, science. Thorough analysis of theorems. Diagrams. Bibliography. Index. 818pp. 5⅜ x 8½. 0-486-64940-7

INTEGRAL EQUATIONS, F. G. Tricomi. Authoritative, well-written treatment of extremely useful mathematical tool with wide applications. Volterra Equations, Fredholm Equations, much more. Advanced undergraduate to graduate level. Exercises. Bibliography. 238pp. 5⅜ x 8½. 0-486-64828-1

FOURIER SERIES, Georgi P. Tolstov. Translated by Richard A. Silverman. A valuable addition to the literature on the subject, moving clearly from subject to subject and theorem to theorem. 107 problems, answers. 336pp. 5⅜ x 8½. 0-486-63317-9

INTRODUCTION TO MATHEMATICAL THINKING, Friedrich Waismann. Examinations of arithmetic, geometry, and theory of integers; rational and natural numbers; complete induction; limit and point of accumulation; remarkable curves; complex and hypercomplex numbers, more. 1959 ed. 27 figures. xii+260pp. 5⅜ x 8½. 0-486-63317-9

POPULAR LECTURES ON MATHEMATICAL LOGIC, Hao Wang. Noted logician's lucid treatment of historical developments, set theory, model theory, recursion theory and constructivism, proof theory, more. 3 appendixes. Bibliography. 1981 edition. ix + 283pp. 5⅜ x 8½. 0-486-67632-3

CALCULUS OF VARIATIONS, Robert Weinstock. Basic introduction covering isoperimetric problems, theory of elasticity, quantum mechanics, electrostatics, etc. Exercises throughout. 326pp. 5⅜ x 8½. 0-486-63069-2

THE CONTINUUM: A CRITICAL EXAMINATION OF THE FOUNDATION OF ANALYSIS, Hermann Weyl. Classic of 20th-century foundational research deals with the conceptual problem posed by the continuum. 156pp. 5⅜ x 8½. 0-486-67982-9

CHALLENGING MATHEMATICAL PROBLEMS WITH ELEMENTARY SOLUTIONS, A. M. Yaglom and I. M. Yaglom. Over 170 challenging problems on probability theory, combinatorial analysis, points and lines, topology, convex polygons, many other topics. Solutions. Total of 445pp. 5⅜ x 8½. Two-vol. set.
Vol. I: 0-486-65536-9 Vol. II: 0-486-65537-7

INTRODUCTION TO PARTIAL DIFFERENTIAL EQUATIONS WITH APPLICATIONS, E. C. Zachmanoglou and Dale W. Thoe. Essentials of partial differential equations applied to common problems in engineering and the physical sciences. Problems and answers. 416pp. 5⅜ x 8½. 0-486-65251-3

THE THEORY OF GROUPS, Hans J. Zassenhaus. Well-written graduate-level text acquaints reader with group-theoretic methods and demonstrates their usefulness in mathematics. Axioms, the calculus of complexes, homomorphic mapping, p-group theory, more. 276pp. 5⅜ x 8½. 0-486-40922-8

Math–Decision Theory, Statistics, Probability

ELEMENTARY DECISION THEORY, Herman Chernoff and Lincoln E. Moses. Clear introduction to statistics and statistical theory covers data processing, probability and random variables, testing hypotheses, much more. Exercises. 364pp. 5⅜ x 8½. 0-486-65218-1

STATISTICS MANUAL, Edwin L. Crow et al. Comprehensive, practical collection of classical and modern methods prepared by U.S. Naval Ordnance Test Station. Stress on use. Basics of statistics assumed. 288pp. 5⅜ x 8½. 0-486-60599-X

SOME THEORY OF SAMPLING, William Edwards Deming. Analysis of the problems, theory and design of sampling techniques for social scientists, industrial managers and others who find statistics important at work. 61 tables. 90 figures. xvii +602pp. 5⅜ x 8½. 0-486-64684-X

LINEAR PROGRAMMING AND ECONOMIC ANALYSIS, Robert Dorfman, Paul A. Samuelson and Robert M. Solow. First comprehensive treatment of linear programming in standard economic analysis. Game theory, modern welfare economics, Leontief input-output, more. 525pp. 5⅜ x 8½. 0-486-65491-5

PROBABILITY: AN INTRODUCTION, Samuel Goldberg. Excellent basic text covers set theory, probability theory for finite sample spaces, binomial theorem, much more. 360 problems. Bibliographies. 322pp. 5⅜ x 8½. 0-486-65252-1

GAMES AND DECISIONS: INTRODUCTION AND CRITICAL SURVEY, R. Duncan Luce and Howard Raiffa. Superb nontechnical introduction to game theory, primarily applied to social sciences. Utility theory, zero-sum games, n-person games, decision-making, much more. Bibliography. 509pp. 5⅜ x 8½. 0-486-65943-7

INTRODUCTION TO THE THEORY OF GAMES, J. C. C. McKinsey. This comprehensive overview of the mathematical theory of games illustrates applications to situations involving conflicts of interest, including economic, social, political, and military contexts. Appropriate for advanced undergraduate and graduate courses; advanced calculus a prerequisite. 1952 ed. x+372pp. 5⅜ x 8½. 0-486-42811-7

FIFTY CHALLENGING PROBLEMS IN PROBABILITY WITH SOLUTIONS, Frederick Mosteller. Remarkable puzzlers, graded in difficulty, illustrate elementary and advanced aspects of probability. Detailed solutions. 88pp. 5⅜ x 8½. 65355-2

PROBABILITY THEORY: A CONCISE COURSE, Y. A. Rozanov. Highly readable, self-contained introduction covers combination of events, dependent events, Bernoulli trials, etc. 148pp. 5⅝ x 8¼. 0-486-63544-9

STATISTICAL METHOD FROM THE VIEWPOINT OF QUALITY CONTROL, Walter A. Shewhart. Important text explains regulation of variables, uses of statistical control to achieve quality control in industry, agriculture, other areas. 192pp. 5⅜ x 8½. 0-486-65232-7

Math–Geometry and Topology

ELEMENTARY CONCEPTS OF TOPOLOGY, Paul Alexandroff. Elegant, intuitive approach to topology from set-theoretic topology to Betti groups; how concepts of topology are useful in math and physics. 25 figures. 57pp. 5⅜ x 8½. 0-486-60747-X

COMBINATORIAL TOPOLOGY, P. S. Alexandrov. Clearly written, well-organized, three-part text begins by dealing with certain classic problems without using the formal techniques of homology theory and advances to the central concept, the Betti groups. Numerous detailed examples. 654pp. 5⅜ x 8½. 0-486-40179-0

EXPERIMENTS IN TOPOLOGY, Stephen Barr. Classic, lively explanation of one of the byways of mathematics. Klein bottles, Moebius strips, projective planes, map coloring, problem of the Koenigsberg bridges, much more, described with clarity and wit. 43 figures. 210pp. 5⅜ x 8½. 0-486-25933-1

THE GEOMETRY OF RENÉ DESCARTES, René Descartes. The great work founded analytical geometry. Original French text, Descartes's own diagrams, together with definitive Smith-Latham translation. 244pp. 5⅜ x 8½. 0-486-60068-8

EUCLIDEAN GEOMETRY AND TRANSFORMATIONS, Clayton W. Dodge. This introduction to Euclidean geometry emphasizes transformations, particularly isometries and similarities. Suitable for undergraduate courses, it includes numerous examples, many with detailed answers. 1972 ed. viii+296pp. 6⅛ x 9¼. 0-486-43476-1

PRACTICAL CONIC SECTIONS: THE GEOMETRIC PROPERTIES OF ELLIPSES, PARABOLAS AND HYPERBOLAS, J. W. Downs. This text shows how to create ellipses, parabolas, and hyperbolas. It also presents historical background on their ancient origins and describes the reflective properties and roles of curves in design applications. 1993 ed. 98 figures. xii+100pp. 6½ x 9¼. 0-486-42876-1

THE THIRTEEN BOOKS OF EUCLID'S ELEMENTS, translated with introduction and commentary by Sir Thomas L. Heath. Definitive edition. Textual and linguistic notes, mathematical analysis. 2,500 years of critical commentary. Unabridged. 1,414pp. 5⅜ x 8½. Three-vol. set.
Vol. I: 0-486-60088-2 Vol. II: 0-486-60089-0 Vol. III: 0-486-60090-4

SPACE AND GEOMETRY: IN THE LIGHT OF PHYSIOLOGICAL, PSYCHOLOGICAL AND PHYSICAL INQUIRY, Ernst Mach. Three essays by an eminent philosopher and scientist explore the nature, origin, and development of our concepts of space, with a distinctness and precision suitable for undergraduate students and other readers. 1906 ed. vi+148pp. 5⅜ x 8½. 0-486-43909-7

GEOMETRY OF COMPLEX NUMBERS, Hans Schwerdtfeger. Illuminating, widely praised book on analytic geometry of circles, the Moebius transformation, and two-dimensional non-Euclidean geometries. 200pp. 5⅝ x 8¼. 0-486-63830-8

DIFFERENTIAL GEOMETRY, Heinrich W. Guggenheimer. Local differential geometry as an application of advanced calculus and linear algebra. Curvature, transformation groups, surfaces, more. Exercises. 62 figures. 378pp. 5⅜ x 8½. 0-486-63433-7

History of Math

THE WORKS OF ARCHIMEDES, Archimedes (T. L. Heath, ed.). Topics include the famous problems of the ratio of the areas of a cylinder and an inscribed sphere; the measurement of a circle; the properties of conoids, spheroids, and spirals; and the quadrature of the parabola. Informative introduction. clxxxvi+326pp. 5⅜ x 8½.
0-486-42084-1

A SHORT ACCOUNT OF THE HISTORY OF MATHEMATICS, W. W. Rouse Ball. One of clearest, most authoritative surveys from the Egyptians and Phoenicians through 19th-century figures such as Grassman, Galois, Riemann. Fourth edition. 522pp. 5⅜ x 8½. 0-486-20630-0

THE HISTORY OF THE CALCULUS AND ITS CONCEPTUAL DEVELOPMENT, Carl B. Boyer. Origins in antiquity, medieval contributions, work of Newton, Leibniz, rigorous formulation. Treatment is verbal. 346pp. 5⅜ x 8½. 0-486-60509-4

THE HISTORICAL ROOTS OF ELEMENTARY MATHEMATICS, Lucas N. H. Bunt, Phillip S. Jones, and Jack D. Bedient. Fundamental underpinnings of modern arithmetic, algebra, geometry and number systems derived from ancient civilizations. 320pp. 5⅜ x 8½. 0-486-25563-8

A HISTORY OF MATHEMATICAL NOTATIONS, Florian Cajori. This classic study notes the first appearance of a mathematical symbol and its origin, the competition it encountered, its spread among writers in different countries, its rise to popularity, its eventual decline or ultimate survival. Original 1929 two-volume edition presented here in one volume. xxviii+820pp. 5⅜ x 8½. 0-486-67766-4

GAMES, GODS & GAMBLING: A HISTORY OF PROBABILITY AND STATISTICAL IDEAS, F. N. David. Episodes from the lives of Galileo, Fermat, Pascal, and others illustrate this fascinating account of the roots of mathematics. Features thought-provoking references to classics, archaeology, biography, poetry. 1962 edition. 304pp. 5⅜ x 8½. (Available in U.S. only.) 0-486-40023-9

OF MEN AND NUMBERS: THE STORY OF THE GREAT MATHEMATICIANS, Jane Muir. Fascinating accounts of the lives and accomplishments of history's greatest mathematical minds–Pythagoras, Descartes, Euler, Pascal, Cantor, many more. Anecdotal, illuminating. 30 diagrams. Bibliography. 256pp. 5⅜ x 8½. 0-486-28973-7

HISTORY OF MATHEMATICS, David E. Smith. Nontechnical survey from ancient Greece and Orient to late 19th century; evolution of arithmetic, geometry, trigonometry, calculating devices, algebra, the calculus. 362 illustrations. 1,355pp. 5⅜ x 8½. Two-vol. set. Vol. I: 0-486-20429-4 Vol. II: 0-486-20430-8

A CONCISE HISTORY OF MATHEMATICS, Dirk J. Struik. The best brief history of mathematics. Stresses origins and covers every major figure from ancient Near East to 19th century. 41 illustrations. 195pp. 5⅜ x 8½. 0-486-60255-9

Physics

OPTICAL RESONANCE AND TWO-LEVEL ATOMS, L. Allen and J. H. Eberly. Clear, comprehensive introduction to basic principles behind all quantum optical resonance phenomena. 53 illustrations. Preface. Index. 256pp. 5⅜ x 8½. 0-486-65533-4

QUANTUM THEORY, David Bohm. This advanced undergraduate-level text presents the quantum theory in terms of qualitative and imaginative concepts, followed by specific applications worked out in mathematical detail. Preface. Index. 655pp. 5⅜ x 8½. 0-486-65969-0

ATOMIC PHYSICS (8th EDITION), Max Born. Nobel laureate's lucid treatment of kinetic theory of gases, elementary particles, nuclear atom, wave-corpuscles, atomic structure and spectral lines, much more. Over 40 appendices, bibliography. 495pp. 5⅜ x 8½. 0-486-65984-4

A SOPHISTICATE'S PRIMER OF RELATIVITY, P. W. Bridgman. Geared toward readers already acquainted with special relativity, this book transcends the view of theory as a working tool to answer natural questions: What is a frame of reference? What is a "law of nature"? What is the role of the "observer"? Extensive treatment, written in terms accessible to those without a scientific background. 1983 ed. xlviii+172pp. 5⅜ x 8½. 0-486-42549-5

AN INTRODUCTION TO HAMILTONIAN OPTICS, H. A. Buchdahl. Detailed account of the Hamiltonian treatment of aberration theory in geometrical optics. Many classes of optical systems defined in terms of the symmetries they possess. Problems with detailed solutions. 1970 edition. xv + 360pp. 5⅜ x 8½. 0-486-67597-1

PRIMER OF QUANTUM MECHANICS, Marvin Chester. Introductory text examines the classical quantum bead on a track: its state and representations; operator eigenvalues; harmonic oscillator and bound bead in a symmetric force field; and bead in a spherical shell. Other topics include spin, matrices, and the structure of quantum mechanics; the simplest atom; indistinguishable particles; and stationary-state perturbation theory. 1992 ed. xiv+314pp. 6⅛ x 9¼. 0-486-42878-8

LECTURES ON QUANTUM MECHANICS, Paul A. M. Dirac. Four concise, brilliant lectures on mathematical methods in quantum mechanics from Nobel Prize-winning quantum pioneer build on idea of visualizing quantum theory through the use of classical mechanics. 96pp. 5⅜ x 8½. 0-486-41713-1

THIRTY YEARS THAT SHOOK PHYSICS: THE STORY OF QUANTUM THEORY, George Gamow. Lucid, accessible introduction to influential theory of energy and matter. Careful explanations of Dirac's anti-particles, Bohr's model of the atom, much more. 12 plates. Numerous drawings. 240pp. 5⅜ x 8½. 0-486-24895-X

ELECTRONIC STRUCTURE AND THE PROPERTIES OF SOLIDS: THE PHYSICS OF THE CHEMICAL BOND, Walter A. Harrison. Innovative text offers basic understanding of the electronic structure of covalent and ionic solids, simple metals, transition metals and their compounds. Problems. 1980 edition. 582pp. 6⅛ x 9¼. 0-486-66021-4

HYDRODYNAMIC AND HYDROMAGNETIC STABILITY, S. Chandrasekhar. Lucid examination of the Rayleigh-Benard problem; clear coverage of the theory of instabilities causing convection. 704pp. 5⅜ x 8¼. 0-486-64071-X

INVESTIGATIONS ON THE THEORY OF THE BROWNIAN MOVEMENT, Albert Einstein. Five papers (1905–8) investigating dynamics of Brownian motion and evolving elementary theory. Notes by R. Fürth. 122pp. 5⅜ x 8½. 0-486-60304-0

THE PHYSICS OF WAVES, William C. Elmore and Mark A. Heald. Unique overview of classical wave theory. Acoustics, optics, electromagnetic radiation, more. Ideal as classroom text or for self-study. Problems. 477pp. 5⅜ x 8½. 0-486-64926-1

GRAVITY, George Gamow. Distinguished physicist and teacher takes reader-friendly look at three scientists whose work unlocked many of the mysteries behind the laws of physics: Galileo, Newton, and Einstein. Most of the book focuses on Newton's ideas, with a concluding chapter on post-Einsteinian speculations concerning the relationship between gravity and other physical phenomena. 160pp. 5⅜ x 8½. 0-486-42563-0

PHYSICAL PRINCIPLES OF THE QUANTUM THEORY, Werner Heisenberg. Nobel Laureate discusses quantum theory, uncertainty, wave mechanics, work of Dirac, Schroedinger, Compton, Wilson, Einstein, etc. 184pp. 5⅜ x 8½. 0-486-60113-7

ATOMIC SPECTRA AND ATOMIC STRUCTURE, Gerhard Herzberg. One of best introductions; especially for specialist in other fields. Treatment is physical rather than mathematical. 80 illustrations. 257pp. 5⅜ x 8½. 0-486-60115-3

AN INTRODUCTION TO STATISTICAL THERMODYNAMICS, Terrell L. Hill. Excellent basic text offers wide-ranging coverage of quantum statistical mechanics, systems of interacting molecules, quantum statistics, more. 523pp. 5⅜ x 8½. 0-486-65242-4

THEORETICAL PHYSICS, Georg Joos, with Ira M. Freeman. Classic overview covers essential math, mechanics, electromagnetic theory, thermodynamics, quantum mechanics, nuclear physics, other topics. First paperback edition. xxiii + 885pp. 5⅜ x 8½. 0-486-65227-0

PROBLEMS AND SOLUTIONS IN QUANTUM CHEMISTRY AND PHYSICS, Charles S. Johnson, Jr. and Lee G. Pedersen. Unusually varied problems, detailed solutions in coverage of quantum mechanics, wave mechanics, angular momentum, molecular spectroscopy, more. 280 problems plus 139 supplementary exercises. 430pp. 6½ x 9¼. 0-486-65236-X

THEORETICAL SOLID STATE PHYSICS, Vol. 1: Perfect Lattices in Equilibrium; Vol. II: Non-Equilibrium and Disorder, William Jones and Norman H. March. Monumental reference work covers fundamental theory of equilibrium properties of perfect crystalline solids, non-equilibrium properties, defects and disordered systems. Appendices. Problems. Preface. Diagrams. Index. Bibliography. Total of 1,301pp. 5⅜ x 8½. Two volumes. Vol. I: 0-486-65015-4 Vol. II: 0-486-65016-2

WHAT IS RELATIVITY? L. D. Landau and G. B. Rumer. Written by a Nobel Prize physicist and his distinguished colleague, this compelling book explains the special theory of relativity to readers with no scientific background, using such familiar objects as trains, rulers, and clocks. 1960 ed. vi+72pp. 5⅜ x 8½. 0-486-42806-0

A TREATISE ON ELECTRICITY AND MAGNETISM, James Clerk Maxwell. Important foundation work of modern physics. Brings to final form Maxwell's theory of electromagnetism and rigorously derives his general equations of field theory. 1,084pp. 5⅜ x 8½. Two-vol. set. Vol. I: 0-486-60636-8 Vol. II: 0-486-60637-6

QUANTUM MECHANICS: PRINCIPLES AND FORMALISM, Roy McWeeny. Graduate student-oriented volume develops subject as fundamental discipline, opening with review of origins of Schrödinger's equations and vector spaces. Focusing on main principles of quantum mechanics and their immediate consequences, it concludes with final generalizations covering alternative "languages" or representations. 1972 ed. 15 figures. xi+155pp. 5⅜ x 8½. 0-486-42829-X

INTRODUCTION TO QUANTUM MECHANICS With Applications to Chemistry, Linus Pauling & E. Bright Wilson, Jr. Classic undergraduate text by Nobel Prize winner applies quantum mechanics to chemical and physical problems. Numerous tables and figures enhance the text. Chapter bibliographies. Appendices. Index. 468pp. 5⅜ x 8½. 0-486-64871-0

METHODS OF THERMODYNAMICS, Howard Reiss. Outstanding text focuses on physical technique of thermodynamics, typical problem areas of understanding, and significance and use of thermodynamic potential. 1965 edition. 238pp. 5⅜ x 8½. 0-486-69445-3

THE ELECTROMAGNETIC FIELD, Albert Shadowitz. Comprehensive undergraduate text covers basics of electric and magnetic fields, builds up to electromagnetic theory. Also related topics, including relativity. Over 900 problems. 768pp. 5⅝ x 8¼. 0-486-65660-8

GREAT EXPERIMENTS IN PHYSICS: FIRSTHAND ACCOUNTS FROM GALILEO TO EINSTEIN, Morris H. Shamos (ed.). 25 crucial discoveries: Newton's laws of motion, Chadwick's study of the neutron, Hertz on electromagnetic waves, more. Original accounts clearly annotated. 370pp. 5⅜ x 8½. 0-486-25346-5

EINSTEIN'S LEGACY, Julian Schwinger. A Nobel Laureate relates fascinating story of Einstein and development of relativity theory in well-illustrated, nontechnical volume. Subjects include meaning of time, paradoxes of space travel, gravity and its effect on light, non-Euclidean geometry and curving of space-time, impact of radio astronomy and space-age discoveries, and more. 189 b/w illustrations. xiv+250pp. 8⅜ x 9¼. 0-486-41974-6

STATISTICAL PHYSICS, Gregory H. Wannier. Classic text combines thermodynamics, statistical mechanics and kinetic theory in one unified presentation of thermal physics. Problems with solutions. Bibliography. 532pp. 5⅜ x 8½. 0-486-65401-X

Paperbound unless otherwise indicated. Available at your book dealer, online at **www.doverpublications.com**, or by writing to Dept. GI, Dover Publications, Inc., 31 East 2nd Street, Mineola, NY 11501. For current price information or for free catalogues (please indicate field of interest), write to Dover Publications or log on to **www.doverpublications.com** and see every Dover book in print. Dover publishes more than 500 books each year on science, elementary and advanced mathematics, biology, music, art, literary history, social sciences, and other areas.